C & DATA STRUCTURES

LIMITED WARRANTY AND DISCLAIMER OF LIABILITY

THE CD-ROM WHICH ACCOMPANIES THE BOOK MAY BE USED ON A SINGLE PC ONLY. THE LICENSE DOES NOT PERMIT THE USE ON A NETWORK (OF ANY KIND). YOU FURTHER AGREE THAT THIS LICENSE GRANTS PERMISSION TO USE THE PRODUCTS CONTAINED HEREIN, BUT DOES NOT GIVE YOU RIGHT OF OWNERSHIP TO ANY OF THE CONTENT OR PRODUCT CONTAINED ON THIS CD-ROM. USE OF THIRD PARTY SOFTWARE CONTAINED ON THIS CD-ROM IS LIMITED TO AND SUBJECT TO LICENSING TERMS FOR THE RESPECTIVE PRODUCTS.

CHARLES RIVER MEDIA, INC. ("CRM") AND/OR ANYONE WHO HAS BEEN INVOLVED IN THE WRITING, CREATION, OR PRODUCTION OF THE ACCOMPANYING CODE ("THE SOFTWARE") OR THE THIRD PARTY PRODUCTS CONTAINED ON THE CD-ROM OR TEXTUAL MATERIAL IN THE BOOK, CANNOT AND DO NOT WARRANT THE PERFORMANCE OR RESULTS THAT MAY BE OBTAINED BY USING THE SOFTWARE OR CONTENTS OF THE BOOK. THE AUTHOR AND PUBLISHER HAVE USED THEIR BEST EFFORTS TO ENSURE THE ACCURACY AND FUNCTIONALITY OF THE TEXTUAL MATERIAL AND PROGRAMS CONTAINED HEREIN. WE HOWEVER, MAKE NO WARRANTY OF ANY KIND, EXPRESS OR IMPLIED, REGARDING THE PERFORMANCE OF THESE PROGRAMS OR CONTENTS. THE SOFTWARE IS SOLD "AS IS " WITHOUT WARRANTY (EXCEPT FOR DEFECTIVE MATERIALS USED IN MANUFACTURING THE DISK OR DUE TO FAULTY WORKMANSHIP).

THE AUTHOR, THE PUBLISHER, DEVELOPERS OF THIRD PARTY SOFTWARE, AND ANYONE INVOLVED IN THE PRODUCTION AND MANUFACTURING OF THIS WORK SHALL NOT BE LIABLE FOR DAMAGES OF ANY KIND ARISING OUT OF THE USE OF (OR THE INABILITY TO USE) THE PROGRAMS, SOURCE CODE, OR TEXTUAL MATERIAL CONTAINED IN THIS PUBLICATION. THIS INCLUDES, BUT IS NOT LIMITED TO, LOSS OF REVENUE OR PROFIT, OR OTHER INCIDENTAL OR CONSEQUENTIAL DAMAGES ARISING OUT OF THE USE OF THE PRODUCT.

THE SOLE REMEDY IN THE EVENT OF A CLAIM OF ANY KIND IS EXPRESSLY LIMITED TO REPLACEMENT OF THE BOOK AND/OR CD-ROM, AND ONLY AT THE DISCRETION OF CRM.

THE USE OF "IMPLIED WARRANTY" AND CERTAIN "EXCLUSIONS" VARY FROM STATE TO STATE, AND MAY NOT APPLY TO THE PURCHASER OF THIS PRODUCT.

C & DATA STRUCTURES

P. S. DESHPANDE
O. G. KAKDE

CHARLES RIVER MEDIA, INC.
Hingham, Massachusetts

Copyright © 2003 by Dreamtech Press
Reprint Copyright © 2004 by CHARLES RIVER MEDIA, INC.
All rights reserved.

No part of this publication may be reproduced in any way, stored in a retrieval system of any type, or transmitted by any means or media, electronic or mechanical, including, but not limited to, photocopy, recording, or scanning, without *prior permission in writing* from the publisher.

Acquisitions Editor: James Walsh
Production: Dreamtech Press
Cover Design: Sherry Stinson

CHARLES RIVER MEDIA, INC.
10 Downer Avenue
Hingham, Massachusetts 02043
781-740-0400
781-740-8816 (FAX)
info@charlesriver.com
www.charlesriver.com

This book is printed on acid-free paper.

P.S. Deshpande and O.G. Kakde. *C & Data Structures*
ISBN: 1-58450-338-6

All brand names and product names mentioned in this book are trademarks or service marks of their respective companies. Any omission or misuse (of any kind) of service marks or trademarks should not be regarded as intent to infringe on the property of others. The publisher recognizes and respects all marks used by companies, manufacturers, and developers as a means to distinguish their products.

Library of Congress Cataloging-in-Publication Data

Deshpande, P. S.
C & data structures / P.S. Deshpande, O.G. Kakde.
p. cm.
ISBN 1-58450-338-6 (Paperback with CD-ROM : alk. paper)
1. C (Computer program language) 2. Data structures (Computer science) I. Kakde, O. G. II. Title.
QA76.73.C15D48 2003
005.7'3—dc22
2003021572

QA
76.73
.C15
D48
2004

Printed in the United States of America
04 7 6 5 4 3 2 First Edition

CHARLES RIVER MEDIA titles are available for site license or bulk purchase by institutions, user groups, corporations, etc. For additional information, please contact the Special Sales Department at 781-740-0400.

Requests for replacement of a defective CD-ROM must be accompanied by the original disc, your mailing address, telephone number, date of purchase and purchase price. Please state the nature of the problem, and send the information to CHARLES RIVER MEDIA, INC., 10 Downer Avenue, Hingham, Massachusetts 02043. CRM's sole obligation to the purchaser is to replace the disc, based on defective materials or faulty workmanship, but not on the operation or functionality of the product.

Contents

Acknowledgments

Writing any book is not an easy task. We spent about one year designing the contents, implementing the programs and testing the programs. Our 12 years of teaching experience has helped us to explain the issues in the language and complex problems in data structures.

We are thankful to our student Rupesh Nasre who, after receiving an M.Tech. degree in computer science from IIT Mumbai, helped us in implementing advanced problems in data structures. We are also thankful to our students Ms. Usha Agrawal and Mr. Ramchandra Vibhute for helping us in writing programs for Parts I and II.

University of Ottawa
Université d'Ottawa
Check Out / Prêt
29003007605778 00:03 2017/12/10
1. Data structures and abstraction using C / Geoff Whale.
39003020742200 Due / Dû: 18-01-08
2. C & data structures / P S Deshpande, O G Kakde
39003029217709 Due / Dû: 18-01-08
Total 2 article(s)
Canada's University
L'Université Canadienne
www.biblio.uOttawa.ca

Preface

The book is written for both undergraduate and graduate students of core computer science areas. The book would be very useful for other students to learn programming for the purpose of making a career in computer science. It covers all those topics that generally appear in aptitude tests and interviews. It not only gives the language syntax but also discusses its behavior by showing the internal implementation. We have covered almost the entire range of data structures and programming such as non-recursive implementation of tree traversals, A* algorithm in Artificial Intelligence, 8-queens problems, etc. We also have supplied a CD-ROM which contains all the source material that appears in the book. We welcome comments from our readers.

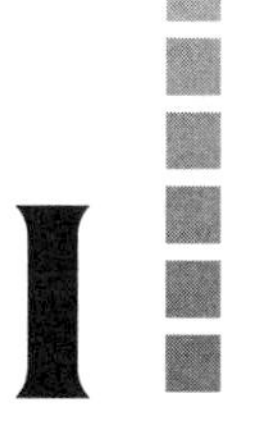

C Language

1 Introduction to the C Language

THE FIRST PROGRAM IN C

Introduction

The C program is a set of functions. The program execution begins by executing the function `main ()`. You can compile the program and execute using Turbo C compiler or using the following commands in Unix/Linux:

```
$ cc  -o a  a.c
```

where a.c is the file in which you have written the program. This will create an executable file named a.exe.

`$./a.` This will execute the program.

Program

```
#include <stdio.h>
main()
{
printf("Hello \n"); /* prints Hello on standard output */
}
```

```
Output : Hello
```

Explanation

1. The program execution begins with the function `main()`.
2. The executable statements are enclosed within a block that is marked by '{' and '}'.

3. The `printf()` function redirects the output to a standard output, which in most cases is the output on screen.
4. Each executable statement is terminated by ';'
5. The comments are enclosed in '/*...*/'

Variables

Introduction

When you want to process some information, you can save the values temporarily in variables. In the following program you can define two variables, save the values, and put the addition in the third variable.

Program

```
#include <stdio.h>
main()
{
int i,j,k; // Defining variables  Statement A
i = 6;    // Statement  B
j = 8;
k = i + j;
printf("sum of two numbers is %d \n",k); // Printing results
}
output : sum of two numbers is 14
```

Explanation

1. Statement A defines variables of the type *integer*. For each variable you have to attach some data type. The *data type* defines the amount of storage allocated to variables, the values that they can accept, and the operations that can be performed on variables.
2. The '//' is used as single line comment.
3. The '%d' is used as format specifier for the integer. Each data type has a format specifier that defines how the data of that data type will be printed.
4. The assignment operator is '=' and the statement is in the format:

   Var = expression;

Points to Remember

1. The variables are defined at the begining of the block.
2. The data type is defined at the begining of declaration and followed by a list of variables.
3. It is the data type that assigns a property to a variable.

INPUTTING THE DATA

Introduction

In C, the input from a standard input device, such as a keyboard, is taken by using the function scanf. In scanf, you have to specify both the variables in which you can take input, and the format specifier, such as %d for the integer.

Program

```
#include <stdio.h>
main()
{
int i,j,k;
scanf("%d%d",&i,&j);  // statement A
k = i + j;
printf("sum of two numbers is %d \n",k);
}
Input 3 4
Output:  sum of two numbers is 7
```

Explanation

1. Statement A indicates the scanf statement that is used for taking input. In scanf you have to specify a list of addresses of variables (&i,&j) which can take input. Before specifying the list of variables you have to include a list of format specifiers that indicate the format of the input. For example, for the integer data type the format specifier is %d.
2. In scanf, you have to specify the address of the variable, such as &i. The address is the memory location where a variable is stored. The reason you must specify the address will be discussed later.

3. The number of format specifiers must be the same as the number of variables.

Point to Remember

In scanf, you have to specify the address of a variable, such as `&i`, `&j`, and a list of format specifiers.

THE CONTROL STATEMENT (if STATEMENT)

Introduction

You can conditionally execute statements using the `if` or the `if...else` statement. The control of the program is dependent on the outcome of the Boolean condition that is specified in the `if` statement.

Program

```
#include <stdio.h>
main()
{
int i,j,big;  //variable declaration
scanf("%d%d",&i,&j); big = i;
if(big < j)   // statement A
{      // C
big = j;      // Part Z , then part
}     // D
printf("biggest of two numbers is %d \n",big);
if(i < j)  // statement B
    {
    big = j;    // Part X
    }
  else
    {
    big = i;    // Part Y
    }
printf("biggest of two numbers(using else) is %d \n",big);
}
```

Explanation

1. Statement A indicates the `if` statement. The general form of the `if` statement is

```
if(expr)
{
s1 ;
s2 ;
....
}
```

2. expr is a Boolean expression that returns true (nonzero) or false (zero).
3. In C, the value nonzero is true while zero is taken as false.
4. If you want to execute only one statement, opening and closing braces are not required, which is indicated by C and D in the current program.
5. The `else` part is optional. If the `if` condition is true then the part that is enclosed after the `if` is executed (Part X). If the `if` condition is false then the `else` part is executed (Part Y).
6. Without the `else` statement (in the first `if` statement), if the condition is true then Part Z is executed.

Points to Remember

1. `if` and `if...else` are used for conditional execution. After the `if` statement the control is moved to the next statement.
2. If the `if` condition is satisfied, then the "then" part is executed; otherwise the `else` part is executed.
3. You can include any operators such as <, >, <=, >=, = = (for equality). Note that when you want to test two expressions for equality, use = = instead of =.

THE ITERATION LOOP (`for` LOOP)

Introduction

When you want to execute certain statements repeatedly you can use iteration statements. C has provided three types of iteration statements: the `for` loop, `while` loop, and `do...while` loop. Generally, the statements in the loop are executed until the specified condition is true or false.

Program

```
#include <stdio.h>
main()
{
int i,n;  //the
scanf("%d",&n);
for(i = 0; i<n; i= i+1)   // statement A
{
printf("the numbers are %d \n",i); // statement B
}
}
/* input and output
5
the numbers are 0
the numbers are 1
the numbers are 2
the numbers are 3
the numbers are 4
*/
```

Explanation

1. Statement A indicates the for loop. The statements in the enclosing braces, such as statement B, indicate the statements that are executed repeatedly because of the for loop.
2. The format of the for loop is

```
for (expr1;expr2;expr3)
{
s1;
s2 ;  // repeat section
}
```

3. expr2 is a Boolean expression. If it is not given, it is assumed to be true.
4. The expressions expr1, expr2 and expr3 are optional.
5. expr1 is executed only once, the first time the for loop is invoked.
6. expr2 is executed each time before the execution of the repeat section.
7. When expr2 is evaluated false, the loop is terminated and the repeat section is not executed.
8. After execution of the repeat section, expr3 is executed. Generally, this is the expression that is used to ensure that the loop will be terminated after certain iterations.

Points to Remember

1. The for loop is used for repeating the execution of certain statements.
2. The statements that you want to repeat should be written in the repeat section.
3. Generally, you have to specify any three expressions in the for loop.
4. While writing expressions, ensure that expr2 is evaluated to be false after certain iterations; otherwise your loop will never be terminated, resulting in infinite iterations.

THE do...while LOOP

Introduction

The do...while loop is similar to the while loop, but it checks the conditional expression only after the repetition part is executed. When the expression is evaluated to be false, the repetition part is not executed. Thus it is guaranteed that the repetition part is executed at least once.

```
Program
#include <stdio.h>
main()
{
int i,n;  //the
scanf("%d",&n);
i = 0;
do                    // statement A
{
printf("the numbers are %d \n",i);
i = i +1;
}while( i<n)  ;

}
/*
5
the numbers are 0
the numbers are 1
the numbers are 2
```

```
the numbers are 3
the numbers are 4
*/
```

Explanation

1. Statement A indicates the `do...while` loop.
2. The general form of the `do...while` loop is

   ```
   do
   {
   repetition part;
   } while (expr);
   ```

   When the `do...while` loop is executed, first the repetition part is executed, and then the conditional expression is evaluated. If the conditional expression is evaluated as true, the repetition part is executed again. Thus the condition is evaluated after each iteration. In the case of a normal `while` loop, the condition is evaluated before making the iteration.
3. The loop is terminated when the condition is evaluated to be false.

Point to Remember

The `do...while` loop is used when you want to make at least one iteration. The condition should be checked after each iteration.

THE `switch` STATEMENT

Introduction

When you want to take one of a number of possible actions and the outcome depends on the value of the expression, you can use the `switch` statement. `switch` is preferred over multiple `if...else` statements because it makes the program more easily read.

Program

```
#include <stdio.h>
main()
```

```
{
int i,n;  //the
scanf("%d",&n); for(i = 1; i<n; i= i+1)
{
switch(i%2) // statement A
{
case 0 : printf("the number %d is even \n",i); // statement B
            break;   // statement C
case 1 : printf("the number %d is odd \n",i);
            break;
}
}
}
/*
5
the number 1 is odd
the number 2 is even
the number 3 is odd
the number 4 is even

*/
```

Explanation

The program demonstrates the use of the switch statement.

1. The general form of a switch statement is

```
Switch(switch_expr)
 {
    case  constant expr1 :  S1;
                            S2;
                            break;
    case  constant expr1 :  S3;
                            S4;
                            break;
     .....
      default            : S5;
                            S6;
                            break;
}
```

2. When control transfers to the `switch` statement then `switch_expr` is evaluated and the value of the expression is compared with `constant_expr1` using the equality operator.
3. If the value is equal, the corresponding statements (S1 and S2) are executed. If `break` is not written, then S3 and S4 are executed. If `break` is used, only S1 and S2 are executed and control moves out of the `switch` statement.
4. If the value of `switch_expr` does not match that of `constant_expr1`, then it is compared with the next `constant_expr`. If no values match, the statements in default are executed.
5. In the program, statement A is the `switch` expression. The expression `i%2` calculates the remainder of the division. For 2, 4, 6 etc., the remainder is 0 while for 1, 3, 5 the remainder is 1. Thus `i%2` produces either 0 or 1.
6. When the expression evaluates to 0, it matches that of `constant_expr1`, that is, 0 as specified by statement B, and the corresponding `printf` statement for an even number is printed. `break` moves control out of the `switch` statement.
7. The clause '`default`' is optional.

Point to Remember

The `switch` statement is more easily read than multiple `if...else` statements and it is used when you want to selectively execute one action among multiple actions.

2 Data Types

THE BUILT-IN DATA TYPES IN C

Introduction

Data types are provided to store various types of data that is processed in real life. A student's record might contain the following data types: name, roll number, and grade percentage. For example, a student named Anil might be assigned roll number 5 and have a grade percentage of 78.67. The roll number is an integer without a decimal point, the name consists of all alpha characters, and the grade percentage is numerical with a decimal point. C supports representation of this data and gives instructions or statements for processing such data. In general, data is stored in the program in variables, and the kind of data the variable can have is specified by the data type. Using this example, grade percentage has a *float* data type, and roll number has an *integer* data type. The data type is attached to the variable at the time of declaration, and it remains attached to the variable for the lifetime of the program. Data type indicates what information is stored in the variable, the amount of memory that can be allocated for storing the data in the variable, and the available operations that can be performed on the variable. For example, the operation S1 * S2, where S1 and S2 are character strings, is not valid for character strings because character strings cannot be multipled.

Program

```
// the program gives maximum and minimum values of data type
#include <stdio.h>
main()
```

```
{
int i,j ;// A
i = 1;
while (i > 0)
{
j = i;
i++;
}
printf ("the maximum value of integer is %d\n",j);
printf ("the value of integer after overflow is %d\n",i);
}
```

Explanation

1. In this program there are two variables, i and j, of the type integer, which is declared in statement A.
2. The variables should be declared in the declaration section at the beginning of the block.
3. If you use variables without declaring them, the compiler returns an error.

Points to Remember

1. C supports various data types such as float, int, char, etc., for storing data.
2. The variables should be declared by specifying the data type.
3. The data type determines the number of bytes to be allocated to the variable and the valid operations that can be performed on the variable.

VARIOUS DATA TYPES IN C

Introduction

C supports various data types for processing information. There is a family of integer data types and floating-point data types. Characters are stored internally as integers, and they are interpreted according to the character set. The most commonly used character set is ASCII. In the ASCII character set, A is represented by the number 65.

Program/Examples

The data type families are as follows:

Integer family

`char` data type

`int` data type

`short int` data type

`long int` data type

These data types differ in the amount of storage space allocated to their respective variables. Additionally, each type has two variants, signed and unsigned, which will be discussed later.

Float family (real numbers with decimal points)

`Float` data type

`Double` data type

(ANSI has also specified `long double`, which occupies the same storage space as `double`)

Explanation

1. Data type determines how much storage space is allocated to variables.
2. Data type determines the permissible operations on variables.

Points to Remember

1. C has two main data type families: *integer* for representing whole numbers and characters of text data, and *float* for representing the real-life numbers.
2. Each family has sub-data types that differ in the amount of storage space allocated to them.
3. In general, the data types that are allocated more storage space can store larger values.

THE INTEGER DATA TYPE FAMILY

Introduction

Integer data types are used for storing whole numbers and characters. The integers are internally stored in binary form.

Program/Example

Here is an example that shows how integers are stored in the binary form.

Number =13

- Decimal representation = $1*10^1 + 3*10^0$
- Binary representation = 1101 = $1*2^3 + 1*2^2 + 0*2^1 + 1*1$

Each 1 or 0 is called a bit, thus the number 13 requires 4 bits.

In the same way, the number 130 is 1000 0010 in binary.

If the general data type is `char`, 8 bits are allocated. Using 8 bits, you can normally represent decimal numbers from 0 to 255 (0000 0000 to 1111 1111). This is the case when the data type is unsigned `char`. However, with signed `char`, the leftmost bit is used to represent the sign of the number. If the sign bit is 0, the number is positive, but if it is 1, the number is negative.

Binary representation of the following numbers in signed `char` is as follows:

Number = 127 Binary representation = 0111 1111 (leftmost bit is 0, indicating positive.)

Number = –128 Binary representation = 1000 0000 (leftmost bit is 1, indicating negative.)

The negative numbers are stored in a special form called "2's complement." It can be explained as follows:

Suppose you want to represent –127:

1. Convert 127 to binary form, i.e. 0111 1111.
2. Complement each bit: put a 0 wherever there is 1 and for 0 put 1. So you will get 1000 0000.

3. Add 1 to the above number

```
   1000 0000
+          1
 ———————————
  1000 0001 (–127)
```

Thus in the signed char you can have the range –128 to +127, i.e. (-2^8 to 2^8-1).

The binary representation also indicates the values in the case of overflow. Suppose you start with value 1 in char and keep adding 1. You will get the following values in binary representation:

```
0000 0001 (1)
0111 1111 (127)
1000 0000 (-128)
1000 0001 (-127)
```

In the case of unsigned char you will get

```
0000 0001 (1)
0111 1111 (127)
1000 0000 (128)
1000 0001 (129)
1111 1111 (255)
0000 0000 (0)
```

This concept is useful in finding out the behavior of the integer family data types.

The bytes allocated to the integer family data types are (1 byte = 8 bits) shown in Table 2.1.

TABLE 2.1 Integer data type storage allocations

Data Type	Allocation	Range
signed char	1 byte	-2^7 to 2^7-1 (–128 to 127)
Unsigned char	1 byte	0 to 2^8-1 (0 to 255)
short	2 bytes	-2^{15} to $2^{15}-1$ (–32768 to 32767)
Unsigned short	2 bytes	0 to $2^{16}-1$ (0 to 65535)
long int	4 bytes	2^{31} to $2^{31}-1$ (2,147,483,648 to 2,147,483,647)
int	2 or 4 bytes depending on implementation	Range for 2 or 4 bytes as given above

Explanation

1. In C, the range of the number depends on the number of bytes allocated and whether the number is signed.
2. If the data type is unsigned the lower value is 0 and the upper depends on the number of bytes allocated.
3. If the data type is signed then the leftmost bit is used as a sign bit.
4. The negative number is stored in 2's complement form.
5. The overflow behavior is determined by the binary presentation and its interpretation, that is, whether or not the number is signed.

Points to Remember

1. The behavior of a data type can be analyzed according to its binary representation.
2. In the case of binary representation, you have to determine whether the number is positive or negative.

OVERFLOW IN char AND UNSIGNED char DATA TYPES

Introduction

Overflow means you are carrying out an operation such that the value either exceeds the maximum value or is less than the minimum value of the data type.

Program

```
// the program gives maximum and minimum values of data type
#include <stdio.h>
main()
{
char i,j ;
i = 1;
while (i > 0) // A
{
j = i; //  B
i++; //  C
```

```
}
printf ("the maximum value of char is %d\n",j);
printf ("the value of char after overflow is %d\n",i);
}
```

Explanation

1. This program is used to calculate the maximum positive value of char data type and the result of an operation that tries to exceed the maximum positive value.
2. The while loop is terminated when the value of i is negative, as given in statement A. This is because if you try to add 1 to the maximum value you get a negative value, as explained previously (127 + 1 gives –128).
3. The variable j stores the previous value of i as given in statement B.
4. The program determines the maximum value as 127. The value after overflow is -128.
5. The initial value of i is 1 and it is incremented by 1 in the while loop. After i reaches 127, the next value is -128 and the loop is terminated.

Points to Remember

1. In the case of signed char, if you continue adding 1 then you will get the maximum value, and if you add 1 to the maximum value then you will get the most negative value.
2. You can try this program for short and int, but be careful when you are using int. If the implementation is 4 bytes it will take too much time to terminate the while loop.
3. You can try this program for unsigned char. Here you will get the maximum value, 255. The value after overflow is 0.

THE char TYPE

Introduction

Alpha characters are stored internally as integers. Since each character can have 8 bits, you can have 256 different character values (0–255). Each integer is associated with a character using a character set. The most commonly used character set is ASCII. In ASCII, "A" is represented as decimal value 65, octal value 101, or hexadecimal value 41.

Explanation

If you declared C as a character as

```
char c;
```

then you can assign A as follows:

```
c = 'A';
c = 65;
c = '\x41';  // Hexadecimal representation
c = '\101';  // Octal representation
```

You cannot write c = 'A' because 'A' is interpreted as a string.

Escape Sequence

Certain characters are not printable but can be used to give directive to functions such as printf. For example, to move printing to the next line you can use the character "\n". These characters are called escape sequences. Though the escape sequences look like two characters, each represents only a single character.

The complete selection of escape sequences is shown here.

\a	alert (bell) character	\\	backslash
\b	backspace	\?	question mark
\f	form feed	\'	single quote
\n	new line	\"	double quote
\r	carriage return	\ooo	octal number
\t	horizontal tab	\xhh	hexadecimal number
\v	vertical tab		

Points to Remember

1. Characters are stored as a set of 255 integers and the integer value is interpreted according to the character set.
2. The most common character set is ASCII.

3. You can give directive to functions such as `printf` by using escape sequence characters.

OCTAL NUMBERS

Introduction

You can represent a number by using the octal number system; that is, base 8. For example, if the number is 10, it can be represented in the octal as 12, that is, $1*8^1 + 2*8^0$.

Explanation

When octal numbers are printed they are preceeded by "%o".

HEXADECIMAL NUMBERS

Introduction

Hexadecimal numbers use base 16. The characters used in hexadecimal numbers are 0, 1, 2, 3, 4, 5, 6, 7, 8, 9, A, B, C, D, E, and F. For example, if the decimal number is 22, it is represented as 16 in the hexadecimal representation: $1*16^1 + 6*16^0$.

Explanation

You can print numbers in hexadecimal form by using the format "0x."

REPRESENTATION OF FLOATING-POINT NUMBERS

Introduction

Floating-point numbers represent two components: one is an exponent and the other is fraction. For example, the number 200.07 can be represented as

$0.20007*10^3$, where 0.2007 is the fraction and 3 is the exponent. In a binary form, they are represented similarly. There are two types of representation: `short` or single-precision floating-point number and `long` or double-precision floating-point number. `short` occupies 4 bytes or 32 bits while `long` occupies 8 bytes or 64 bits.

Program/Example

In C, `short` or single-precision floating point is represented by the data type `float` and appears as:

```
float f ;
```

A single-precision floating-point number is represented as follows:

31	30 — 23	22 — 0
s	exponent	fraction

Here the fractional part occupies 23 bits from 0 to 22. The exponent part occupies 8 bits from 23 to 30 (bias exponent, that is, exponent + 01111111). The sign bit occupies the 31st bit.

Suppose the decimal number is 100.25. It can be converted as follows:

1. Convert 100.25 into its equivalent binary representation: 1100100.01.
2. Then represent this number so that there is only 1 bit on the left side of the decimal point: $1.0010001*2^6$
3. In a binary representation, exponent 6 means the number 110. Now add the bias, 0111 1111, to get the exponent: 1000 0101

Since the number is positive, the sign bit is 0. The significant, or fractional, part is:

1001 0001 0000 0000 0000 000

Note that up until the fractional part, only those bits that are on the right side of the decimal point are present. The 0s are added to the right side to make the fractional part take up 23 bits.

Special rules are applied for some numbers:

1. The number 0 is stored as all 0s, but the sign bit is 1.
2. Positive infinity is represented as all 1s in the exponent and all 0s in the fractional part with the sign bit 0.
3. Negative infinity is represented as all 1s in the exponent and all 0s in fractional part with the sign bit 1.

4. A NAN (not a number) is an invalid floating number in which all the exponent bits are 1, and in the fractional part you may have 1s or 0s.

The range of the float data type is 10^{-38} to 10^{38} for positive values and -10^{38} to -10^{-38} for negative values.

The values are accurate to 6 or 7 significant digits depending on the actual implementation.

Conversion of a number in the floating-point form to a decimal number

Suppose the number has the following components:

a. Sign bit: 1
b. Exponent: 1000 0011
c. Significant or fractional part: 1001 0010 0000 0000 0000 000

   Since the exponent is bias, find out the unbiased exponent.
d. 100 = 1000 0011 – 0111 1111 (number 4)

   Represent the number as $1.1001001*2^4$

   Represent the number without the exponent as 11001.001

   Convert the binary number to decimal: –25.125

For double precision, you can declare the variable as double d; it is represented as

63	62 ... 52	51 ... 0
s	exponent	fraction

Here the fractional part occupies 52 bits from 0 to 51. The exponent part occupies 11 bits from 52 to 62 (the bias exponent is the exponent plus 011 1111 1111). The sign bit occupies bit 63. The range of double representation is $+10^{-308}$ to $+10^{308}$ and -10^{308} to -10^{-308}. The precision is to 10 or more digits.

Formats for representing floating points

Following are the valid representions of floating points:

0.23456

2.3456E–1

.23456

.23456e–2

2.3456E–4

–.232456E–4

2345.6

23.456E2

–23456

23456e3

Following are the invalid formats:

e1

2.5e–.5

25.2–e5

2.5.3

You can determine whether a format is valid or invalid based on the following rules:

1. The value can include a sign, it must include a numerical part, and it may or may not have exponent part.
2. The numerical part can be of following form:

   d.d, d., .d, d, where d is a set of digits.
3. If the exponent part is present, it should be represented by 'e' or 'E', which is followed by a positive or negative integer. It should not have a decimal point and there should be at least 1 digit after 'E'.
4. All floating numbers have decimal points or 'e' (or both).
5. When 'e' or 'E' is used, it is called scientific notation.
6. When you write a constant, such as 50, it is interpreted as an integer. To interpret it as floating point you have to write it as 50.0 or 50, or 50e0.

   You can use the format `%f` for printing floating numbers. For example,

   ```
   printf("%f\n", f);
   ```

`%f` prints output with 6 decimal places. If you want to print output with 8 columns and 3 decimal places, you can use the format `%8.3f`. For printing double you can use `%lf`.

Floating-point computation may give incorrect results in the following situations:

1. If the calculated value has a precision that exceeds the precision limit of the type;

2. If the calculated value exceeds the range allowable for the type;
3. If the two calculated values involve approximation then their operation may involve approximation.

Points to Remember

1. C provides two main floating-point representations: float (single precision) and double (double precision).
2. A floating-point number has a fractional part and a biased exponent.
3. Float occupies 4 bytes and double occupies 8 bytes.

TYPE CONVERSION

Introduction

Type conversion occurs when the expression has data of mixed data types, for example, converting an integer value into a float value, or assigning the value of the expression to a variable with different data types.

Program/Example

In type conversion, the data type is promoted from lower to higher because converting higher to lower involves loss of precision and value.

For type conversion, C maintains a hierarchy of data types using the following rules:

1. Integer types are lower than floating-point types.
2. Signed types are lower than unsigned types.
3. Short whole-number types are lower than longer types.
4. The hierarchy of data types is as follows: double, float, long, int, short, char.

These general rules are accompanied by specific rules, as follows:

1. If the mixed expression is of the double data type, the other operand is also converted to double and the result will be double.
2. If the mixed expression is of the unsigned long data type, then the other operand is also converted to double and the result will be double.

3. Float is promoted to double.
4. If the expression includes long and unsigned integer data types, the unsigned integer is converted to unsigned long and the result will be unsigned long.
5. If the expression contains long and any other data type, that data type is converted to long and the result will be long.
6. If the expression includes unsigned integer and any other data type, the other data type is converted to an unsigned integer and the result will be unsigned integer.
7. Character and short data are promoted to integer.
8. Unsigned char and unsigned short are converted to unsigned integer.

FORCED CONVERSION

Introduction

Forced conversion occurs when you are converting the value of the larger data type to the value of the smaller data type, for example, if the declaration is

```
char c;
```

and you use the expression `c = 300;` Since the maximum possible value for `c` is 127, the value 300 cannot be accommodated in `c`. In such a case, the integer 300 is converted to `char` using forced conversion.

Program/Example

In general, forced conversion occurs in the following cases:

1. When an expression gives a larger data type but the variable has a smaller data type.
2. When a function is written using a smaller data type but you call the function by using larger data type. For example, in `printf` you specify `%d`, but you provide floating-point value.

Forced conversion is performed according to following rules:

1. Normally, when floating points are converted to integers, truncation occurs. For example, 10.76 is converted to 10.

2. When double is converted to float, the values are rounded or truncated, depending on implementation.
3. When longer integers are converted to shorter ones, only the lower bits are preserved and high-order bits are skipped. For example, the bit representation of 300 is 1 0010 1100. If it is assigned to character, the lower bits are preserved since a character can have 8 bits. So you will get the number 0010 1100 (44 in decimal).

In the case of type conversion, lower data types are converted to higher data types, so it is better to a write a function using higher data types such as `int` or `double` even if you call the function with `char` or `float`. C provides built-in mathematical functions such as `sqrt` (square root) which take the argument as double data type. Suppose you want to call the function by using the integer variable 'k'. You can call the function

```
sqrt((double) n)
```

This is called *type casting*, that is, converting the data type explicitly. Here the value 'k' is properly converted to the double data type value.

Points to Remember

1. C makes forced conversion when it converts from higher data type to lower data type.
2. Forced conversion may decrease the precision or convert the value to one that doesn't have a relation with the original value.
3. Type casting is the preferred method of forced conversion.

TYPE CASTING

Introduction

Type casting is used when you want to convert the value of a variable from one type to another. Suppose you want to print the value of a double data type in integer form. You can use type casting to do this. Type casting is done to cast an operator which is the name of the target data type in parentheses.

Program

```
#include <stdio.h>
main()
{
double d1 = 123.56;   \\ A
int i1=456;                 \\ B

printf("the value of d1  as int without cast operator %d\n",d1); \\ C
printf("the value of d1 as int with cast operator %d\n",(int)d1);
\\ D
printf("the value of  i1 as double without cast operator %f\n",i1); \\
E
printf("the value of i1 as double with cast operator %f\n",(double)i1);
\\ F
i1 = 10;
printf("effect of multiple unary operator %f\n",(double)++i1);  \\ G
i1 = 10;       \\ H
//printf("effect of multiple unary operator %f\n",(double) ++ -i1);
error  \\ I i1 = 10;
printf("effect of multiple unary operator %f\n",(double)- ++i1);\\ J
i1 = 10;       \\ K
printf("effect of multiple unary operator %f\n",(double)- –i1); \\ L
i1 = 10;       \\ M
printf("effect of multiple unary operator %f\n",(double)-i1++); \\ N

}
```

Explanation

1. Statement A defines variable d1 as double.
2. Statement B defines variable i1 as int.
3. Statement C tries to print the integer value of d1 using the placeholder %d. You will see that some random value is printed.
4. Statement D prints the value of d1 using a cast operator. You will see that it will print that value correctly.
5. Statements E and F print the values of i1 using a cast operator. These will print correctly as well.
6. Statements from G onwards give you the effects of multiple unary operators. A cast operator is also a unary operator.

7. Unary operators are associated from right to left, that is, the left unary operator is applied to the right value.
8. Statement G gives the effect of the cast operator double. The increment operator, in this case i1, is first incremented and then type casting is done.
9. If you do not comment out statement I you will get errors. This is because if unary +, - is included with the increment and decrement operator, it may introduce ambiguity. For example, +++i may be taken as unary + and increment operator ++, or it may be taken as increment operator ++ and unary +. Any such ambiguous expressions are not allowed in the language.
10. Statement J will not introduce any error because you put the space in this operator, which is used to resolve any ambiguity.

Points to Remember

1. Type casting is used when you want to convert the value of one data type to another.
2. Type casting does not change the actual value of the variable, but the resultant value may be put in temporary storage.
3. Type casting is done using a cast operator that is also a unary operator.
4. The unary operators are associated from right to left.

3 C Operators

ASSIGNMENT OPERATOR

Introduction

The assignment operator is used for assigning the value of an expression to a variable. The general format for an assignment operator is var = expression.

You can use other formats such as var += expression, which means var = var + expression.

Program

```
#include<stdio.h>

main( )
{
  int a,b,c,d;
  printf("ENTER VALUES OF a,b, c, d");
  scanf("%d%d%d",&a,&b,&c);
   a  += b*c+d;
  printf("\n a = %d",a);
  }
```

```
Input
a = 5, b= 5, c = 7, d = 8.
Output
ENTER VALUES OF a,b, c, d
5
5
7
8
a =  48
```

Explanation

The assignment operators have the lowest priority and they are evaluated from right to left. The assignment operators are as follows:

`=, +=, -=, *=, /=, %=.`

Suppose the expression is

```
a = 5;
a += 5*7+8;
```

You will get the value 48. It is evaluated by the following steps:

1. 5*7 = 35.
2. 35+8 = 43.
3. `a += 43` means `a = a + 43` which gives the value 48.

You can assign a value to multiple variables in one statement as:
`i = j = k = 10` which gives value 10 to `i, j, k`.

ARITHMETIC OPERATOR

Introduction

You can process data using arithmetic operators such as `+, -, *, \` and the modulus operator `%`. `%` indicates the remainder after integer division; `%` cannot be used for float data type or double data type. If both operands `i1` and `i2` are integers, the expression `i1/i2` provides integer division, even if the target is a floating point variable. The operators have normal precedence rules, as follows:

1. Unary operators such as –, + are evaluated.
2. The multiplication (*) and division (/,%) operators are evaluated.
3. The addition (+) and subtraction (–) operators are evaluated.
4. The assignment operator is evaluated.
5. The expressions are evaluated from left to right for unary operators. The assignment is from right to left.

Program

```
#include<stdio.h>

main( )
{
  int a,b,c,d;
  int sum,sub,mul,rem;
  float div;
    printf("ENTER VALUES OF b, c, d");
     scanf("%d%d%d",&b&c,&d);
  sum = b+c;
  sub = b-c;
  mul = b*c;
  div = b/c;
  rem = b%d;
  a = b/c * d;
  printf("\n sum = %d, sub = %d, mul = %d, div = %f",sum,sub,mul,div);
  printf("\n remainder of division of b & d is %d",rem);
  printf("\n a = %d",a);
  }
```

Input

```
b = 10, c = 5, d= 3.
```

Output

ENTER VALUES OF b, c, d

```
10
5
3

sum = 15, sub = 5, mul = 50, div = 2.0
remainder of division of b & d is 1
a = 6
```

Explanation

1. Suppose you have the expression
   `a = b/c * d`
   Here / and * both have the same priority. b/c first is evaluated because the expression is evaluated from left to right.

2. After evaluating the expression b/c * d, the value is assigned to a because the assignment operator has an order of evaluation from right to left, that is, the right expression is evaluated first.

RELATIONAL OPERATOR

Introduction

Relational operators are used in Boolean conditions or expressions, that is, the expressions that return either true or false. The relational operator returns zero values or nonzero values. The zero value is taken as false while the nonzero value is taken as true.

Program

Th relational operators are as follows:

```
<, <=, >, >=, ==, !=
```

The priority of the first four operators is higher than that of the later two operators. These operators are used in relational expressions such as:

```
7 > 12          // false
20.1 < 20.2       // true
'b' < 'c'    // true
"abb" < "abc"  // true
```

The strings are compared according to dictionary comparison, so if the first characters are equal, the condition is checked for the second characters. If they are also equal then it is checked for the third character, etc. The relational operators return integer values of either zero or non zero.

Note that the equality operator is == and not =. '=' is an assignment operator.

If you want to compare a and b for equality then you should write a == b, not a = b because a = b means you are assigning the value of b to a, as shown in Table 3.1.

TABLE 3.1 Comparing the equality operator (= =) with the '=' assignment operator.

Case	a	b	a = b	a == b
1	5	3	a = 3 (true)	false
2	7	0	a = 0 (false)	false
3	0	0	a = 0 (false)	true

In case 1, the value of a = 5 and b = 3. The assignment expression assigns the value of b to a, so a will be 3. The expression returns a true value because 3 is not zero. For the same case a == b does not make any assignment and returns a false value because in the value of a does not equal that of b.

In case 2, the value of a = 7 and b = 0. The assignment expression assigns the value of b to a, so a will be 0. The expression returns a false value of zero. For the same case, a == b does not make any assignment and returns a false value because the value of a does not equal that of b.

In case 3, the values of a and b are both 0. The assignment expression assigns the value of b to a, so a will be 0. The expression returns a false value of zero. For the same case, a == b does not make any assignment and returns a true value because the value of a equals that of b.

LOGICAL OPERATOR

Introduction

You can combine results of multiple relations or logical operations by using logical operation. The logical operators are negation (!), logical AND (&&), and logical OR (||), in the same order of preference.

Program

```
#include<stdio.h>

main( )
{
  int c1,c2,c3;

printf("ENTER VALUES OF c1, c2 AND c3");
```

```
 scanf("%d%d%d",&c1.&c2,&c3);
  if((c1 < c2)&&(c1<c3))
   printf("\n c1 is less than c2 and c3");
 if (!(c1< c2))
   printf("\n c1 is greater than c2");
 if ((c1 < c2)||(c1 < c3))
   printf("\n c1 is less than c2 or c3 or both");
 }
```

Input

```
c1= 2;
c2= 3;
c3= 4;
```

Output

```
ENTER VALUES OF c1, c2 AND c3
2
3
4
c1 is less than c2 and c3
c1 is less than c2 or c3 or both
```

Explanation

1. Logical AND returns a true value if both relational expressions are true. Logical OR returns true if any of the expressions are true. Negations return complements of values of relational expressions, as shown in Table 3.2.

TABLE 3.2 Results of AND, OR, and Negation.

R1	R2	R1 && R2	R1 || R2	! R1
T	T	T	T	F
T	F	F	T	F
F	T	F	T	T
F	F	F	F	T

2. Logical operators AND, and OR have higher priority than assignment operators, but less than relational operators. Negation operators have the same priority as unary operators, that is, the highest priority.

3. While evaluating logical expressions, C uses the technique of short circuiting. So if the expression is:

   `C1 && C2 && C3 && C4 if C1 is true`

   then only C2 is evaluated. If C1 is false, the expression returns false even if C2, C3, and C4 are true. So if C1 is false C2, C3, and C4 are not evaluated. Remember this when you are doing something such as searching in an array. For example, if you want to search for K in an array, the last value of which is subscript N, you can write the search condition in two ways:

   ```
   I  - (a [i] == K) && (i <= N)
   II -  (i <= N) && (a[i] == K)
   ```

4. In case I you compare the array limit with K and check the bound. This is not correct because if the value of i is more than N you will get the array index out-of-bounds error.
5. In case II, you first check the bound and then compare the array element. This is correct because you will never compare the array element if value of i is more than N.

   The technique of short-circuiting is applicable to the OR operator also. Thus if the expression is:

   `C1 || C2 || C3 || C4 if C1 is True`

   then the expression returns true and C2, C3 and C4 are not evaluated.

TERNARY OPERATOR

Introduction

Ternary operators return values based on the outcomes of relational expressions. For example, if you want to return the value of 1 if the expression is true and 2 if it is false, you can use the ternary operator.

Program/Example

If you want to assign the maximum values of i and j to k then you can write the statement

```
k = ( i>j ) ? i : j;
```

If i > j then k will get the value equal to i, otherwise it will get the value equal to j.

The general form of the ternary operator is:

```
(expr 1) ? expr2 : expr3
```

If expr1 returns true then the value of expr2 is returned as a result; otherwise the value of expr3 is returned.

INCREMENT OPERATOR

Introduction

You can increment or decrement the value of variable using the increment or decrement operator. These operators can be applied only to variables and they can be applied using prefix form or postfix form.

Program

```
#include<stdio.h>

main( )
{
int I,j,k;
i = 3;
j =4;
k = i++ +  -j;
printf("i = %d, j = %d, k = %d",i,j,k);
}
```

Input

```
i =3, j = 4.
```

Output

```
i = 4, j = 3, k = 6.
```

Explanation

When the prefix form is used, the value of the variable is incremented/ decremented first and then applied. In the postfix form, the value is applied and only after the assignment operator is done is the value incremented or decremented.

1. Suppose you write

```
i = 3;
j =4;
k = i++ + --j;
```

   you will get the value of k as 6, i as 4 and j as 3. The order of evaluation is as follows:

   1. i gets the value 3.
   2. j is decremented to 3.
   3. k gets the value 3 + 3.
   4. i is incremented.

2. Suppose you write

```
i = 5;
i = i++ * i++
```

Then you will get the value of i as 27. This is because first the value 5 is used as to make i = 25 and then i is incremented twice. The increment and decrement operators have higher priority than the arithmetic operators.

COMMA OPERATOR

Introduction

You can combine multiple expressions in a single expression using the comma operator.

Program

```
#include<stdio.h>

main()
{
  int i,j,k;
  k = (i = 4, j = 5);
  printf("k = %d",k);
 }
```

Input

i = 4,j = 5.

Output

k = 5.

Explanation

For example, you can write: k = (i = 4, j = 5)

Here the expression is evaluated from left to right, that is, i = 4 is evaluated first then j = 5 is evaluated. The value of the rightmost expression is specified as output, thus k will get the value 5.

BITWISE OPERATOR

Introduction

Bitwise operators interpret operands as strings of bits. Bit operations are performed on this data to get the bit strings. These bit strings are then interpreted according to data type. There are six bit operators: bitwise AND(&), bitwise OR(|), bitwise XOR(^), bitwise complement(~), left shift(<<), and right shift(>>).

Program

```
# include<stdio.h>

main()
{
  char c1,c2,c3;
  printf("ENTER VAULES OF  c1 and c2");
  scanf("%c,%c",&c1,&c2);
  c3 = c1 & c2;
  printf("\n Bitwise AND  i.e. c1 & c2 = %c",c3);
  c3 = c1 | c2;
  printf("\n Bitwise OR  i.e. c1 | c2 = %c",c3);
  c3 = c1 ^ c2;
```

```
    printf("\n Bitwise  XOR i.e. c1 ^ c2 = %c",c3);
    c3 = ~c1;
    printf("\n ones complement  of  c1 = %c",c3);
    c3 =  c1<<2;
    printf("\n left shift by 2 bits c1 << 2 = %c",c3);
    c3 =  c1>>2;
    printf("\n right shift by 2 bits c1 >> 2 = %c",c3);
}
```

Input

```
c1 = 4;
c2 = 6;
```

Output

```
ENTER VALUES OF  c1 and c2
 4
 6

 Bitwise AND  i.e. c1 & c2 = 4
 Bitwise OR  i.e. c1 | c2 =  6
 Bitwise  XOR i.e. c1 ^ c2 = 2
 ones compliment  of  c1 = -4
 left shift by 2 bits c1 << 2 = 16
 right shift by 2 bits c1 >> 2 = 1
```

Explanation

1. Suppose you write
   ```
   char c1, c2, c3;
   c1 = 4;
   c2 = 6;
   ```
   The binary values are
   ```
   c1 = 0000 0100
   c2 = 0000 0110
   ```
2. Suppose you write
   ```
   c3 = c1 & c2;
   ```
   The value of c3 is interpreted as follows:
   ```
       0000 0100
     & 0000 0110
      ----------.
       0000 0100
   ```

Each bit of c1 is compared with the corresponding bit of c2. If both bits are 1 then the corresponding bit is set as 1, otherwise it is set as 0. Thus the value of c3 is 4.

3. `c3 = c1 | c2`

The value of c3 is interpreted as follows:

```
  0000 0100
| 0000 0110
 -----------.
  0000 0110
```

Each bit of c1 is compared with the corresponding bit of c2. If any of the bits are 1 then the corresponding bit is set as 1; otherwise it is set as 0. Thus the value of c3 is 6.

4. `c3 = c1 ^ c2`

The value of c3 is interpreted as follows:

```
  0000 0100
^ 0000 0110
 -----------.
  0000 0010
```

Each bit of c1 is compared with the corresponding bit of c2. If only one bit is 1, the corresponding bit is set to 1; otherwise it is set to 0. Thus you will get the value of c3 as 2 because in the second position for c1, the bit is 0 and for c2, the bit is 1. So only one bit is set.

5. `c3 = ~ c1`

The value of c3 is interpreted as follows:

```
~   0000 0100
    ----------.
    1111 1011
```

Each bit of c1 is complemented; for 1 the complement is 0. Thus you will get the value of c3 as –4, because the leftmost bit is set as 1.

6. `c3 = c1 << 2;`

This is a left-shift operation. The bits are shifted left by two places. c1 indicates the operand that should be an expression returning a whole number. 2 indicates a shift that should not be negative, and its value must be less than the number of bits allocated to that data type.

It is evaluated as follows:

`c1 is 0000 0100`

It is shifted 2 bits to the left to produce

0001 00**

While shifting, the high-order (left) bits are discarded. Since a vacuum is created on the right side, it is filled with 0s to get 0001 0000. Thus the value is 16.

7. `c3 = c1 >> 2;`

   This is a right shift operation. The bits are shifted right by two places. c1 indicates the operand that should be an expression returning a whole number. 2 indicates a shift that should not be negative, and its value must be less than the number of bits allocated to that data type.

   It is evaluated as follows:

   c1 is 0000 0100

   It is shifted 2 bits to the right to produce

   **00 0001

   While shifting, the low-order (right) bits are discarded. The asterisks are replaced using one of the following strategies:

   Logical shift: In this case, the high-order bits are filled with 0s, thus you get 0000 0001.

   Arithmetic shift: In this case the high-order bits are filled with the original sign bits, so if the sign bit is 1, then all bits are filled with 1s; otherwise, they are filled with 0s.

For unsigned data types, logical shift is used, whereas for signed data types arithmetic shift is used.

In these examples, the char data type which is signed is used. In number 4, the sign bit is 0, so you will get the bit pattern 0000 0001 (decimal 1).

OPERATOR PRECEDENCE

Introduction

Since C has various types of operators, it also sets precedence rules so that the value of expressions that involve multiple operators should be deterministic.

Program

The precedence of operators is given in Table 3.3.

TABLE 3.3 Operator precedence rules

Operators	Order of evaluation	Remarks
[] ( ) ->	Left to right	Array subscript, function call
- + sizeof( ) ! ++ --		
& * ~ (cast)	Right to left	Unary
* / %	Left to right	Binary Multiplicative
+ -	Left to right	Binary Additive
>> <<	Left to right	Shift operators
< <= > >=	Left to right	Relational operators
== !=	Left to right	Equality operators
&	Left to right	Bitwise And operator
^	Left to right	Bitwise Xor operator
|	Left to right	Bitwise Or operator
&&	Left to right	Logical And operator
||	Left to right	Logical Or operator
?:	Left to right	Conditional operator
= += -= *= /= %= &= -= |= <<= >>=	Right to left	Assignment
,	Right to left	Comma

Point to Remember

The operators are evaluated according to the precedence as shown in Table 3.3.

4 Control Structures

CONTROL STRUCTURES

Introduction

C provides four general categories of control structures: sequential, selection, iteration and encapsulation.

Program/Example

A *sequential* structure is one in which instructions are executed in sequence. For example,

```
i = i + 1;
j = j + 1;
```

In the selection structure, the sequence of the instruction is determined by using the result of the condition. The statements that can be used in this category are `if` and `switch`. For example:

```
if (a > b)
    i = i + 1;
else
    j = j +1;
```

If the condition is true then the statement `i = i +1` is executed; otherwise `j = j + 1` is executed.7

The iteration structure is one in which statements are repeatedly executed. The iteration structure forms program loops. The number of iterations generally depends on the values of particular variables.

```
for (i=0; i<5; i++)
{
    j = j + 1;
}
```

The statement `j = j + 1` is executed 5 times and the value of `i` changes from 0 to 1, 2, 3, and 4.

Encapsulation structure is the structure in which the other component structures are included. For example, you can include an `if` statement in a `for` loop or a `for` loop in an `if` statement.

Explanation

C provides all the standard control structures that are available in programming languages. These structures are capable of processing any information.

THE `if` STATEMENT

Introduction

The `if` statement is the first selection structure. `if` is used when a question requires a yes or no answer. If you want to choose an answer from several possibilities then use the `switch` statement.

Program/Example

The general format for an `if` statement is:

`if` (condition)

simple or compound statement.

Following are the properties of an `if` statement:

1. If the condition is true then the simple or compound statements are executed.
2. If the condition is false it does not do anything.
3. The condition is given in parentheses and must be evaluated as true (nonzero value) or false (zero value).
4. If a compound statement is provided, it must be enclosed in opening and closing braces.

   Following are the test conditions:

```
    (7)// a non-zero value returns True.
(0)// zero value returns False.
```

```
(i==0)    // True if i=0 otherwise False.
(i = 0)    //  False because value of the expression is
zero.
```

SCOPE OF AN `if` CLAUSE

The scope of an `if` clause determines a range over which the result of the condition affects. The scope of an `if` clause is on the statement which immediately follows the `if` statement. It can be a simple statement or compound statement.

Case 1:

```
if (a>b)
i = i + 1;     // s1
j = j + 1;     // s2
```

Case 2:

```
if (a>b)
{
   i = i + 1; // s1
   j = j + 1; // s2
}
```

If in Case 1 the `if` condition is true, then `s1` is executed because `s1` is a simple statement.

If in Case 2 the `if` condition is true, then both statements `s1` and `s2` are executed because `s1` and `s2` are enclosed in a compound statement.

THE `if-else` STATEMENT

Introduction

When you want to take actions based on the outcome of the conditions, (true or false), then you can use the `if-else` statement.

Program/Example

The general format for an if-else statement is

```
if (condition)
simple or compound statement      // s1
else
simple or compound statement.     // s2
```

If the condition is true then the s1 part is executed and if the condition is false then the s2 part is executed. For example,

```
if (a>b)
printf (" big number is %d", a);    // s1
else
printf (" big number is %d", b);    // s2
```

if a is greater than (b) then s1 is executed. Otherwise s2 is executed.

THE if-else if STATEMENT

Introduction

If you want to make many decisions, then you can use the if-else if statement.

Program

The general format for the if-else if statement is:

```
if (condition 1)
simple or compound statement      // s1
else if (condition 2)
simple or compound statement      // s2
else if ( condition 3)
simple or compound statement      // s3
.....
else if ( conditon n )
simple or compound statement      // sn
```

If condition 1 is true then s1 is executed. If condition 1 is false and condition 2 is true then s2 is executed.

The `else` clause is always associated with the nearest unresolved `if` statement.

```
if (a==5)              // A
if (a==7)              // B
     i = 10;   // C
else           // D
 if (a == 7)   // E
     i = 15;   // F
else           // G
     i = 20;   // H
```

For the `else` statement at position D, the nearest `if` statement is specified at B. So, the `else` statement is associated with `if` at B and not at A.

For the `else` statement at G, the nearest `if` statement is specified at E. So, it is associated with the `if` statement at E and not at A.

```
if (a==5)              // A
if (a==7)              // B
     i = 10;   // C
else           // D
 if (a == 7)   // E
     i = 15;        // F1
     j = 20;        // F2
else           // G
     i = 20;   // H
```

In this case, the `else` statement at G cannot be associated with the `if` statement at E because the `if` statement at E is already resolved. So, it is associated with the `if` statement at A.

Points to Remember

1. You can use `if-else if` when you want to check several conditions but still execute one statement.
2. When writing an `if-else if` statement, be careful to associate your `else` statement to the appropriate `if` statement.
3. You must have parentheses around the condition.
4. You must have a semicolon or right brace before the `else` statement.

THE switch STATEMENT

Introduction

You can use a switch statement when you want to check multiple conditions. It can also be done using an if statement but it will be too lengthy and difficult to debug.

Program/Example

The general format for a switch statement is

```
switch (expressions)
{
case constant  expressions
}
```

Example of a case constant expression and column:

```
switch (i/10)
{
    case 0:   printf ("Number less than 10");      // A
                  break;
    case 1:   printf ("Number less than 20");      // B
                  break;
    case 2:   printf ("Number less than 30");      // C
                  break;
    default:  printf ("Number greater than or equal to 40");    // D
                  break;
}
```

Explanation

1. The switch expression should be an integer expression and, when evaluated, it must have an integer value.
2. The case constant expression must represent a particular integer value and no two case expressions should have the same value.
3. The value of the switch expression is compared with the case constant expression in the order specified, that is, from the top down.
4. The execution begins from the case where the switch expression is matched and it flows downward.

5. In the absence of a break statement, all statements that are followed by matched cases are executed. So, if you don't include a break statement and the number is 5, then all the statements A, B, C, and D are executed.
6. If there is no matched case then the default is executed. You can have either zero or one default statement.
7. In the case of a nested switch statement, the break statements break the inner switch statement.

Point to Remember

The switch statement is preferable to multiple if statements.

THE while LOOP

Introduction

The while loop is used when you want to repeat the execution of a certain statement or a set of statements (compound statement).

Program/Example

The general format for a while loop is

while (condition)

simple or compound statement (body of the loop)

For example,

```
i = 0;
while (i<5)
{
    printf(" the value of i is %d\n", i);
    i = i + 1;
}
```

Explanation

1. Before entering into the loop, the while condition is evaluated. If it is true then only the loop body is executed.

2. Before making an iteration, the `while` condition is checked. If it is true then the loop body is executed.
3. It is the responsibility of the programmer to ensure that the condition is false after certain iterations; otherwise, the loop will make infinite iterations and it will not terminate.
4. The programmer should be aware of the final value of the looping variable. For example, in this case, the final value of the looping variable is 5.
5. While writing the loop body, you have to be careful to decide whether the loop variable is updated at the start of the body or at the end of the body.

THE `do-while` LOOP

Introduction

The `do-while` loop is used when you want to execute the loop body at least once. The `do-while` loop executes the loop body and then traces the condition.

Program/Example

The general format for a `do-while` loop is

do

  simple or compound statement

  `while` (condition)

  For example,

```
i = 0;
do
{
    printf(" the value of i is %d\n", i);
    i = i + 1;
}
while (i<5)
```

Explanation

1. The loop body is executed at least once.
2. The condition is checked after executing the loop body once.

3. If the condition is false then the loop is terminated.
4. In this example, the last value of `i` is printed as 5.

THE `for` LOOP

Introduction

The `for` loop is used only when the number of iterations is predetermined, for example, 10 iterations or 100 iterations.

Program/Example

The general format for the `for` loop is

`for` (initializing; continuation condition; update)

simple or compound statement

For example,

```
for (i = 0; i < 5; i++)
{
    printf("value of i");
}
```

Explanation

1. The `for` loop has four components; three are given in parentheses and one in the loop body.
2. All three components between the parentheses are optional.
3. The initialization part is executed first and only once.
4. The condition is evaluated before the loop body is executed. If the condition is false then the loop body is not executed.
5. The update part is executed only after the loop body is executed and is generally used for updating the loop variables.
6. The absence of a condition is taken as true.
7. It is the responsibility of the programmer to make sure the condition is false after certain iterations.

THE for LOOP WITH A COMMA OPERATOR

Introduction

You may want to control the loop variables in the same for loop. You can use one for loop with a comma operator in such situations.

Program/Example

```
for (i = 0, j = 10; i < 3 && j > 8; i++, j–)
printf (" the value of i and j %d %d\n",i, j);
```

Explanation

1. First i is initialized to 0, and j is initialized to 10.
2. The conditions i<3 and j>8 are evaluated and the result is printed only if both conditions are true.
3. After executing the loop body, i is incremented by 1 and j is decremented by 1.
4. The comma operator also returns a value. It returns the value of the rightmost operand. The value of (i = 0, j = 10) is 10.

THE break STATEMENT

Introduction

Just like the switch statement, break is used to break any type of loop. Breaking a loop means terminating it. A break terminates the loop in which the loop body is written.

Program/Example

For example,

```
i = 0;
while (1)
{
```

```
    i = i + 1;
    printf(" the value of i is %d\n");
    if (i>5) break;
}
```

Explanation

1. The `while (1)` here means the `while` condition is always true.
2. When `i` reaches 6, the `if` condition becomes true and `break` is executed, which terminates the loop.

THE `continue` STATEMENT

Introduction

The `break` statement breaks the entire loop, but a `continue` statement breaks the current iteration. After a `continue` statement, the control returns to top of the loop, that is, to the test conditions. `Switch` doesn't have a `continue` statement.

Program/Example

Suppose you want to print numbers 1 to 10 except 4 and 7. You can write:

```
for(i = 0, i < 11, i++)
{
   if ((i == 4) || (i == 7)) continue;
   printf(" the value of i is %d\n", i);
}
```

Explanation

1. If `i` is 1 then the `if` condition is not satisfied and `continue` is not executed. The value of `i` is printed as 1.
2. When `i` is 4 then the `if` condition is satisfied and `continue` is executed.
3. After executing the `continue` statement, the next statement, (`printf`), is not executed; instead, the updated part of the `for` statement (`i++`) is executed.

5 The printf Function

printf

Introduction

printf is used to display information on screen, that is, standard output. printf is a function that returns the number of characters printed by the printf function.

Program

```
#include <stdio.h>
main()
{
int i = 0;
i=printf("abcde");    // A
printf("total characters printed %d\n",i);  //B
}
```

Explanation

1. Here, five characters are printed by statement A. So, i will get the value 5.
2. Statement B prints the value of i as 5.
3. The general format for the printf statement has a first string argument followed by any additional arguments.
4. In statement B, "total characters printed %d\n" is the first string argument.
5. i is the second argument. You may have multiple arguments, but that depends on what value you have to print. For each additional argument you will have to include a placeholder. Each placeholder begins with %. In statement B, %d is the placeholder.

6. For the second argument i, the placeholder is %d. So when you need an integer value, you have to use %d. The placeholders are given for each data type.
7. For example, if you want to print i and j, you may have to use two placeholders. Any material in the first string argument, other than the placeholder and characters, represents the escape sequence. In this example, the escape sequence character is \n, which is not printed but acts as a directive. For example, the \n directive indicates that the next printing should be done on a new line.

Points to Remember

1. printf is used to direct output to standard output format.
2. printf is a function that returns the number of characters printed.

PLACEHOLDERS

Introduction

Placeholders are used to print values of arguments supplied in print. The directives in the placeholders control printing.

Program/Example

The general form of a placeholder is:

% flags field-width precision prefix type-identifier.

Type-identifiers

The type-identifiers are as follows:

d, i Signed integers

o Unsigned integers displayed in octal form.

u Unsigned integers in decimal form.

x Unsigned integers in hexadecimal form, and the hexadecimal characters a, b, c, d, e, and f printed in lowercase.

X Unsigned integer in hexadecimal form, and the hexadecimal characters A, B, C, D, E, and F printed in uppercase.

c Any value converted to unsigned char and displayed; c is used mainly for printing characters.

s The argument is converted to a character array and is printed; the last null in the string is not printed.

f Floating point.

e, E Floating point displayed in exponential form. It will have one digit to the left of the decimal point; the number of digits on the right side of the decimal point depends on the required precision.

g, G The value can be printed in floating point or exponential form. The exponential form is used if the exponent is less than -1 or if the exponent causes more places than required by the specified precision; the decimal point appears only if it is followed by a digit.

n This indicates to print the number of characters that are printed so far by printf.

p It indicates an additional argument pointer to void; the value of the pointer is converted to a sequence of characters.

Type prefixes

h It can appear before type indicators d, i, o, u, x, and X. It indicates that the value to be displayed should be interpreted as short; for example, short integer (hd) and short unsigned integer (hu).

l It can appear before type-identifiers d, i, o, u, x, and X. It indicates that the value to be displayed should be interpreted as long; for example, long integer (hd) and long unsigned integer (hu).

l, L Available for type-identifiers e, E, f, g, and G. It indicates that a value should be indicated as long double.

Field-width

1. *Field-width* indicates the least number of columns that will be allocated to the output. For example, if you write %4d to i and the value of i is 10, then 4 columns are allocated for i and 2 blank are added on left side of value of i. So the output is bb10. Here, b indicates blank.
2. If the value is more than the specified column, field-width is ignored and the number of columns used is equal to the number of columns required by the arguments. So if i is 12345 then 5 columns are used, even if %4d is specified.

3. In any circumstance, the output width is not shortened, because of field-width.
4. If you specify * instead of field-width then you have to specify additional arguments. For example,

```
printf ("%*d\n", 5, 20);        // A
printf ("%*d\n", 20, 5);        // B
```

In A, 5 is substituted for * and it indicates putting the value 20 in 5 columns.

In B, 20 is substituted for * and it indicates putting the value 5 in 20 columns.

Precision

1. *Precision* indicates the minimum number of digits printed for type integers d, i, o, u, x, and X. For example,

```
i.   printf("%10.4d\n", 35)
```

2. Here 10 is the field-width and 4 is the precision, so 10 columns are used for the 4-digit output. To make 35 into 4 digits, two 0s are added to the left side to make it 0035. To print 0035 in 10 columns, blanks are added to make the output bbbbbb0035.
3. For floating arguments, precision indicates how many digits are printed after decimal points. If precision is more than the number of digits on the right side of the decimal point, 0s are added to the right side.
4. If precision indicates too few digits, then it is ignored and the number of digits are printed as necessary.

Flags

1. *Flag* characters are used to give directives for the output. You can use multiple flag characters in any order.
2. The flag characters are as follows:

   – Indicates that output is left justified.

   ```
   printf("%-10.4d\n", 25)
   ```

   It causes the number to be printed as 0025bbbbbb. Thus, blanks are added to the right side.

   In the absence of a flag, it is printed as bbbbbb0025.

+ Indicates that i number is printed using a sign character (+ or –).

```
printf("%+d\n", -25);
printf("%+d\n", 25);
```

It causes printing as

```
-25
+25
```

<space> Indicates a space for positive values so that positive values and negative values are aligned. For example,

```
printf("% d\n", 25);
printf("% d", 25);
```

It causes printing in the form of

```
b25
25
```

In the first case, blank is displayed.

– # Indicates that the value should be converted to another form before displaying. For example, for hexadecimal values you can indicate 0X; for the floating data type, # indicates that the decimal point should always be included in the output.

0 Used with whole and real numbers, 0 causes 0s to be padded to complete the field width. If the precision is specified as 0, then this flag is ignored; if the 0 and – flags are both specified, the 0 flag is ignored.

Escape Sequence

Escape sequences are the special directives used to format printing. For example, \n indicates that the next printing should start from the first column of the next line. Following are the escape sequences:

\a Alert

Produces a beep or flash; the cursor position is not changed.

\b Backspace

Moves the cursor to the last column of the previous line.

\f Form feed

Moves the cursor to start of next page.

\n New line

Moves the cursor to the first column of the next line.

\r Carriage Return

Moves the cursor to the first column of the current line.

\t Horizontal Tab

Moves the cursor to the next horizontal tab stop on the line.

\v Vertical Tab

Moves the cursor to the next vertical tab stop on the line.

\\

Prints \\.

\”

Prints “

%%

Prints %.

Points to Remember

1. `printf` returns the number of characters printed; if some error occurs then it returns a negative value.
2. Formating of `printf` can be controlled by using flags, field-width, etc.

6 Address and Pointers

ADDRESS

Introduction

For every variable declared in a program there is some memory allocation. Memory is specified in arrays of bytes, the size of which depending on the type of variable. For the integer type, 2 bytes are allocated, for floats, 4 bytes are allocated, etc. For every variable there are two attributes: *address* and *value*, described as follows:

Program

```
#include <stdio.h>
main ()
{
    int i, j, k;        //A
            i = 10;             //B
    j = 20;             //C
    k = i + j; //D

    printf ("Value of k is %d\n", k);
}
```

Explanation

1. Memory allocations to the variables can be explained using the following variables:

   100,i   10

   200, j   20

   300,k   30

When you declare variables i, j, k, memory is allocated for storing the values of the variables. For example, 2 bytes are allocated for i, at location 100, 2 bytes are allocated for j at location 200, and 2 bytes allocated for k at location 300. Here 100 is called the address of i, 200 is called address of j, and 300 is called the address of k.

2. When you execute the statement i = 10, the value 10 is written at location 100, which is specified in the figure. Now, the address of i is 100 and the value is 10. During the lifetime of variables, the address will remain fixed and the value may be changed. Similarly, value 20 is written at address 200 for j.
3. During execution, addresses of the variables are taken according to the type of variable, that is, local or global. Local variables usually have allocation in stack while global variables are stored in runtime storage.

Points to Remember

1. Each variable has two attributes: address and value.
2. The address is the location in memory where the value of the variable is stored.
3. During the lifetime of the variable, the address is not changed but the value may change.

POINTERS

Introduction

A *pointer* is a variable whose value is also an address. As described earlier, each variable has two attributes: address and value. A variable can take any value specified by its data type. For example, if the variable i is of the integer type, it can take any value permitted in the range specified by the integer data type. A pointer to an integer is a variable that can store the address of that integer.

Program

```
#include <stdio.h>
main ()
{
```

```
    int i;              //A
    int * ia;           //B
    i = 10;             //C
    ia = &i;            //D

      printf (" The address of i is %8u \n", ia);         //E
      printf (" The value at that location is %d\n", i);  //F
     printf (" The value at that location is %d\n", *ia); //G
   *ia = 50;                                //H
      printf ("The value of i is %d\n", i);               //I
}
```

Explanation

1. The program declares two variables, so memory is allocated for two variables. `i` is of the type of `int`, and `ia` can store the address of an integer, so it is a pointer to an integer.
2. The memory allocation is as follows:

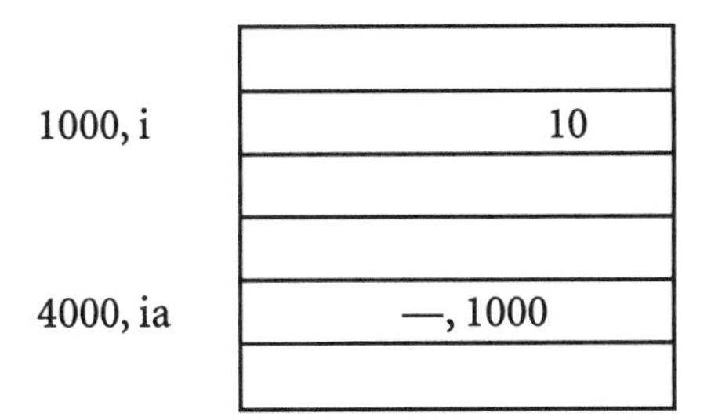

3. `i` gets the address 1000, and `ia` gets address 4000.
4. When you execute `i = 10`, 10 is written at location 1000.
5. When you execute `ia = &i` then the address and value are assigned to `i`, thus `i` has the address of 4000 and value is 1000.
6. You can print the value of `i` by using the format `%au` because addresses are usually in the format unsigned long, as given in statement E.
7. Statement F prints the value of `i`, (at the location 1000).
8. Alternatively, you can print the value at location 1000 using statement G. `*ia` means you are printing the value at the location specified by `ia`. Since `i` has the value for 1000, it will print the value at location 1000.

9. When you execute `*ia = 50`, which is specified by statement H, the value 50 is written at the location by `ia`. Since `ia` specifies the location 1000, the value at the location 1000 is written as 50.
10. Since `i` also has the location 1000, the value of `i` gets changed automatically from 10 to 50, which is confirmed from the `printf` statement written at position `i`.

Points to Remember

1. Pointers give a facility to access the value of a variable indirectly.
2. You can define a pointer by including a* before the name of the variable.
3. You can get the address where a variable is stored by using &.

7 The scanf Function

scanf

Introduction

The scanf function is used to read information from a standard input device (keyboard). scanf starts with a string argument and may contain additional arguments. Any additional arguments must be pointers (to implement calls by reference).

Program

```
#include <stdio.h>
main()
{
    int i = 0;
    int k,j=10;
    i=scanf("%d%d%d",&j,&k,&i);
    printf("total values inputted %d\n",i);
    printf("The input values %d %d\n",j,k);

}
```

Explanation

1. Statement A indicates scanf; it is used for inputting values for i, j, k.
2. You have to use the address of the variable as an additional variable, for example, &i.
3. The first argument is always a string argument with placeholders.
4. During execution of scanf, the input is processed and it is matched against the string argument. The process is continued until the matching is complete.

5. When the first placeholder is encountered in the string argument, a value of the specific type of the first element constitutes a match. It is repeated for each placeholder.
6. If there are one or more whitespace characters in the first string argument between the placeholders, any sequence of one or more whitespace characters in the input completes a match.
7. If there are other characters in a string argument, the input should have the same character in the same sequence in order to constitute a match.
8. Once a match is not found, the function is terminated. Matching fails if the expected input is missing.
9. `scanf` returns the number of values that have been succesfully input. In this example, if you type A instead of an integer when you are giving the value for `k`, then `scanf` returns only 1, although `k` gets the value 65 (the ASCII value for A).

Points to Remember

1. Input can be done using `scanf`.
2. For `scanf`, the address of the variable should be passed.
3. `scanf` returns the number of successful inputs.

THE `scanf` PLACEHOLDERS

Introduction

The `scanf` placeholder consists of % at the beginning and a type indicator at the end. Apart from that it can have *, a maximum field-width indicator, and a type indicator modifier, for example,

`%10.2f,%10d`

Program/Example

Type indicators

`d, i` Used for signed integers; the expected argument should be a pointer to `int`.

o Used for unsigned int expected's value. It should be an integer in octal form.

U Unsigned integer in decimal form.

X, X Unsigned integer in hexadecimal form.

E, E, f, g, G Floating-point values.

S Character string. It matches a sequence of non-whitespace characters terminated by an end-of-line or end-of-file character. The additional argument should be a pointer to char and should point to an area that is large enough to hold the input string as well as the NULL terminator.

C Matches the number of characters according to a specified field-width. If no width is specified then a single character is assumed. The additional argument must be a pointer to char; the area pointed to should be large enough to hold the specified number of characters.

N Does not read any input but writes the number of characters so far in the target variable.

Use of *

The * is used to suppress input. For example, with %*d, if your input consists of 5 values and you want to ignore the middle 3 values, you can write:

```
scanf(" %d %*d %*d%*d %d ", &i, &j)
```

So, if your input is

10 20 30 40 50

it will get the value 10 and j will get the value 50. This is useful when you are getting the input from a file.

Field-width

It indicates the maximum number of characters that are read into the variables.

Explanation

1. scanf requires two inputs: the first is a string argument and the second is a set of additional arguments.
2. You can define how the input is to be taken by using placeholders.

8 Preprocessing

PREPROCESSOR

Introduction

C's preprocessor provides a facility for defining constant and substitution, which are commonly called macros. It is used when you either want the make to program more readable or when you don't have enough information about certain values. For example, if your input is in U.S. dollars, and your processing is done in terms of rupees, then your program may have the expression

```
Rs = usd * 46;
```

where 46 is the currency rate. You can write the expression as:

```
# define currency_rate 46
rs = usd * currency_rate;
```

So, if the currency rate is changed, you can make the necessary change only in one place. The preprocessor directive is defined here.

Program

```
# include <stdio.h>

#define VAL 35          // A
#define HELLO "HELLO";       // B

main ()
{
    int res;

    res = VAL-5;              // C
    printf ("res = VAL-5: res == %d\n", res);
```

```
    printf ( HELLO);          //D
}
```

Statements A and B indicate preprocessor directives. VAL is defined as integer 35 and HELLO is a string as "HELLO". Whenever VAL and HELLO appear, they are replaced by the specified values.

In statement C, VAL 35 is replaced by VAL -5. So the statement becomes

```
res = 35-5;
```

Statement D, after replacement, becomes

```
printf ("HELLO")
```

The preprocessor directives are not C statements, so they do not end with semicolons.

The include directive tells the compiler to include all the contents of a specified file in the source file before giving the source file for compiling.

Explanation

1. The preprocessor substitutes strings that are specified by using define directive

   ```
   #define constant identifer "value"
   ```

2. Following are valid define expressions:

   ```
   #define TRUE              1
   #define FALSE             0
   #define BS         '\b'
   #define TAB               '\011'
   ```

undef

Introduction

If you want to nullify the effect of the define directive and specify a new value, then you can use the undef directive.

Program

```
#include <stdio.h>
#define VAL 40;         //A
#undef  VAL             //B
#define  VAL 40         //C
main()
{
   printf ("%d\n", VAL);    //D
}
```

Explanation

1. Statement A defines VAL as 40, that is, an erroneous definition.
2. Statement B indicates that the afore mentioned definition no longer exists.
3. Statement C allows a new definition.
4. Statement D uses new definition of 40.

Point to Remember

The undef directive nullifies the effect of an earlier definition.

ifdef

Introduction

The ifdef directive makes substitutions based on whether a particular identifier is defined.

Program

Suppose we have three files:

```
file1.h
#define USD 1
file2.h
#define UKP 1
file3
#include <stdio.h>
#include <file1.h>              //A
```

```
#ifdef USD                      // B
      #define currency_rate 46       //C
#endif                  //D

#ifdef UKP                      // E
   #define currency_rate 100 //F
#endif                  //G
main()
{
   int rs;
   rs = 10 * currency_rate;  //H
   printf ("%d\n", rs);
}
```

Explanation

1. Statement A includes file1.h, so the content of the file is substituted in that position. If the file name is given in angle brackets, it means the file is searched in the default search path. If the file name is specified within " ", like include "file1.h", then file1.h is searched only in the current directory.
2. Statement B is an ifdef directive and it checks whether the identifier USD is defined. Since it is defined in file1.h, the condition is true and the currency rate is defined as 46. You can include multiple directives in if, def and endif.
3. Statement E checks whether the identifier UKP is defined. Since it is not defined, because file2.h, is not included in the file, the condition is false and its defined directive is not processed.
4. The currency rate in file3 is taken as 46.
5. In the expression in statement H, the currency rate is substituted as 46.
6. If, instead of file1.h, you include file2.h, then the currency rate will be 100.

Points to Remember

1. ifdef is used to make a substitution depending on whether a certain identifier is defined.
2. If the identifier is defined, it returns true; otherwise, it is false.

ifndef

Introduction

#ifndef is used to check whether a particular symbol is defined.

Program

Suppose wse have three files:

```
file1.h
#define USD 1

file2.h
#define UKP 1

file3
#include <stdio.h>
#include <file1.h>              //A

#ifndef USD                     // B
       #define currency_rate 100      //C
#endif                 //D

#ifndef UKP                    // E
   #define currency_rate 46 //F
#endif                 //G
main()
{
    int rs;
    rs = 10 * currency_rate;  //H
    printf ("%d\n", rs);
}
```

Explanation

1. ifndef is a complement of ifdef. That is, if the symbol is defined, ifndef returns false.
2. Statement B is an ifndef directive and it checks whether the identifier USD is defined. Since it is defined in file1.h, the condition is false and further processing is not done.

3. Statement E checks whether the identifier UKP is defined. Since it is not defined, because file2.h is not included in the file, the condition is true and the currency rate is defined as 46.
4. The currency rate in file3 is taken as 46.
5. In the expression in statement H, the currency rate is substituted as 46.
6. If, instead of file1, you include file2.h, then the currency rate will be 100.
7. ifndef is a complement of ifdef. So, when ifdef returns true, ifndef returns false.

#if

Introduction

#if allows you to define more generalized conditions. Multiple conditions, which are connected by relational operators such as AND(&&), OR(||), are allowed.

Program

Suppose we have three files:

```
file1.h
#define USD 1

file2.h
#define UKP 1

file3
#include <stdio.h>
#include <file1.h>                    //A

#if ((1>0) &&(defined (USD))          // B
       #define currency_rate 46               //C
#endif                          //D

#if (defined (UKP))                   // E
   #define currency_rate 100 //F
#endif                          //G
```

```
main()
{
    int rs;
    rs = 10 * currency_rate; //H
    printf ("%d\n", rs);
}
```

Explanation

1. Statement B indicates the `if` directive.
2. The generalized form of the `if` directive is
   ```
   #if <condition>
   #endif
   ```
3. The condition (1>0) is absolutely not necessary here. It is given just to indicate how you can concatanate multiple conditions.
4. The condition defined (USD) is true only if the identifier USD is defined.
5. Since `file1.h` is included and USD gets defined, the condition is evaluated as true and the currency rate is defined as 46.
6. In statement E, the condition is false because UKP is not defined.
7. In the statement H, the currency rate is 46.

Points to Remember

1. The `if` directive allows us to use a condition more generalized than `ifdef`.
2. The `defined()` predicate returns true if the symbol is defined; otherwise, it is false.

ifelse

Introduction

The `ifelse` directive lets us specify the action if the condition is not true.

Program

Suppose we have three files:

```
file1.h
#define USD 1
file2.h
#define UKP 1

file3
#include <stdio.h>
#include <file1.h>                        //A

#if (defined (USD))             // B
        #define currency_rate 46
#else
       #define currency_rate 100               //C
#endif                          //D

 main()
{
   int rs;
   rs = 10 * currency_rate;     //H
   printf ("%d\n", rs);
}
```

Explanation

1. Statement B indicates the `ifelse` directive.
2. If the identifier `USD` is defined, the currency rate is taken as 46; otherwise, the currency rate is taken as 100.
3. Since `USD` is defined in `file1.h`, the currency rate is taken as 46.

Point to Remember

The `ifelse` directive allows us to take action if the condition is not satisfied.

`ifelif`

Introduction

`ifelif` allows us to take one action if there are multiple decision points. For example, if you want to take the currency rate of 1 if USD and UKP are not defined, you can write the following program.

Program

Suppose we have three files:

file1.h

```
#define USD 1
```

file2.h

```
#define UKP 1
```

file3

```
#include <stdio.h>
#include <file1.h>                    //A

#if (defined (USD))                   // B
     #define currency_rate 46
#elif (defined (UKP))
     #define currency_rate 100        //C
#else
     # define currency_rate 1         //D
#endif

 main()
{
   int rs;
   rs = 10 * currency_rate;           //H
   printf ("%d\n", rs);
}
```

Explanation

1. Statement B includes the `ifelif` directive. It is similar to the `else` directive.
2. `#elif` appears only after `#if`, `#ifdef`, `#ifndef`, and `#elif`.
3. `#elif` is similar to `#else` but it is followed by a condition.
4. You can have as many `#elif` directives as you want.
5. If USD is defined, then the currency rate is 46; otherwise, if UKP is defined, then the currency rate is 100; otherwise, the currency rate is 1.
6. In this case, if you remove the statement `include file1.h` at position A, then USD and UKP are not defined and currency rate is taken as 1.

Points to Remember

1. `#elif` is similar to `#else` but it is followed by a condition.
2. `#elif` allows taking action in the case of multiple decision points.

ERROR DIRECTIVE

Introduction

The *error directive* is used to specify an error message for a specific situation. In the following program, the error message is displayed if USD and UKP are not defined.

Program

Suppose we have three files:

file1.h

```
#define USD 1
```

file2.h

```
#define UKP 1
```

file3

```
#include <stdio.h>
#include <file1.h>                          //A

#if !defined (USD) || !defined (UKP)      // B
#error "ERROR: NO_CURRENCY rate is specified."  //C
#endif

main()
{
    int rs;
    rs = 10 * currency_rate;               //D
    printf ("%d\n", rs);
}
```

Explanation

1. Statement B checks whether UKP or USD is defined.
2. If both are not defined then the preprocessor displays an error.

Points to Remember

1. The #error directive allows us to specify an error message.
2. The error message is generated by the preprocessor.

#line

Introduction

The #line directive allows you to define arbitrary line numbers for the source lines. Normally, the compiler counts lines starting at line number 1; using the #line directive, you can specify an arbitrary line number at any point. The compiler then uses that line number for subsequent counts.

Program

```
#include <stdio.h>
main()
{
    printf("A\n");              //A

     #line100                   //H
    printf("B\n");              //B
    printf("C FILE %s  LINE %d\n", __FILE__, __LINE__ );//C
     #line200                   //K

    printf("D\n");              //D
    printf("E\n");              //E
}
```

Explanation

1. The statement H indicates the #line directive.

2. The #line number in statement B is taken as 100 and for statement C, it is taken as 101.
3. The #line number in statement D is taken as 200 and for statement E, it is taken as 201.
4. If you introduce any error in statement B then the compiler will display the error at #line number 100.
5. C has provided two special identifiers: __FILE__ and __LINE__, which indicate the file name of the source file and the current line number, respectively.

Point to Remember

#line is used to indicate line numbers which can be used for debugging.

MACRO

Introduction

Macros allow replacement of the identifier by using a statement or expression. The replacement text is called the macro body. C uses macros extensively in its standard header files, such as in getchar(), getc().

Program

```
#define CUBE(x)  x*x*x        //A
#include <stdio.h>
main ()
{
    int k = 5;
    int j = 0;
    j = CUBE(k);              //B  j = k*k*k

    printf ("value of j is %d\n", j);     //C
}
```

Explanation

1. You can define the macro CUBE as in statement A.

2. The macro can be defined by using parameters, but that is not mandatory.
3. The parameter name that is used in a macro definition is called the *formal* parameter. In this example, it is x.
4. x*x*x is called the macro body.
5. There should not be any spaces between the macro name and the left parenthesis.
6. CUBE(k) in statement B indicates a macro call.
7. An argument such as k, which is used for calling a macro, is called an *actual* parameter.
8. While expanding the macro, the actual parameter is substituted in the formal parameter and the macro is expanded. So you will get the expansion as j = k*k*k.
9. The value of j is calculated as 125.
10. Since macro expansion is mainly a replacement, you can use any data type for the actual parameter. So, the above macro works well for the float data type.

Points to Remember

1. A macro is used when you want to replace a symbol with an expression or a statement.
2. You can define macros by using parameters.

MACRO AND FUNCTION

Introduction

While writing the macro, you have to write the macro body carefully because the macro just indicates replacement, not the function call.

Program

```
#include <stdio.h>
#define add(x1, y1)  x1+y1   //E
#define mult(x1,y2) x2*y2    //F
```

```
main ()
{
    int a,b,c,d,e;
    a = 2;
    b = 3;
    c = 4;
    d = 5;
    e = mult(add(a, b), add(c, d)); //A

    // mult(a+b, c+d)                //B
    // a+b * c+d                     //C

    printf ("The value of e is %d\n", e);
}
```

Explanation

1. Statement E indicates a macro for adding two numbers.
2. Statement F indicates a macro for multiplying two numbers.
3. Statement A indicates a macro that is supposed to add two numbers and then multiply two numbers. In this case, it is supposed to perform the calculation (2+3) * (4+5).
4. The actual expansion of macro adds is given in statement B.
5. The final expansion of `mult` gives the expansion `a+b * c+d`, which is erroneous.
6. The final value of `e` is 17, which is not correct.
7. To get the correct value, use the following definition:
   ```
   #define add(x1, y1) (x1+y1)
   #define mult(x2, y2) (x2*y2)
   ```

Point to Remember

While using the macro, you have to write the expression correctly. You can use parentheses to give the correct meaning to the expression.

9 Arrays

ARRAYS

Introduction

An *array* is a data structure used to process multiple elements with the same data type when a number of such elements are known. You would use an array when, for example, you want to find out the average grades of a class based on the grades of 50 students in the class. Here you cannot define 50 variables and add their grades. This is not practical. Using an array, you can store grades of 50 students in one entity, say grades, and you can access each entity by using subscript as grades[1], grades[2]. Thus you have to define the array of grades of the float data type and a size of 50. An array is a composite data structure; that means it had to be constructed from basic data types such as array integers.

Program

```
#include <stdio.h>
main()
{
    int a[5];  \\A
    for(int i = 0;i<5;i++)
    {
        a[i]=i;\\B
    }
    printarr(a);
}
void printarr(int a[])
{
    for(int i = 0;i<5;i++)
    {
        printf("value in array %d\n",a[i]);
    }
}
```

Explanation

1. Statement A defines an array of integers. The array is of the size 5—that means you can store 5 integers.
2. Array elements are referred to using subscript; the lowest subscript is always 0 and the highest subscript is (size –1). If you refer to an array element by using an out-of-range subscript, you will get an error. You can refer to any element as `a[0]`, `a[1]`, `a[2]`, etc.
3. Generally, you can use a `for` loop for processing an array. For the array, consecutive memory locations are allocated and the size of each element is same.
4. The array name, for example, `a`, is a pointer constant, and you can pass the array name to the function and manipulate array elements in the function. An array is always processed element by element.
5. When defining the array, the size should be known.

*The array subscript has the highest precedence among all operators thus a[1] * a[2] gives the multiplication of array elements at position 1 and position 2.*

Points to Remember

1. An array is a composite data structure in which you can store multiple values. Array elements are accessed using subscript.
2. The subscript operator has the highest precedence. Thus if you write `a[2]++`, it increments the value at location 2 in the array.
3. The valid range of subscript is 0 to size –1.

ADDRESS OF EACH ELEMENT IN AN ARRAY

Introduction

Each element of the array has a memory address. The following program prints an array limit value and an array element address.

Program

```
#include <stdio.h>
void printarr(int a[]);
main()
{
    int a[5];
    for(int i = 0;i<5;i++)
    {
        a[i]=i;
    }
    printarr(a);
}
void printarr(int a[])
{
    for(int i = 0;i<5;i++)
    {
        printf("value in array %d\n",a[i]);
    }
}
void printdetail(int a[])
{
    for(int i = 0;i<5;i++)
    {
        printf("value in array %d and address is %16lu\n",a[i],&a[i]);
\\ A
    }
}
```

Explanation

1. The function printarr prints the value of each element in arr.
2. The function printdetail prints the value and address of each element as given in statement A. Since each element is of the integer type, the difference between addresses is 2.
3. Each array element occupies consecutive memory locations.
4. You can print addresses using place holders %16lu or %p.

Point to Remember

For array elements, consecutive memory locations are allocated.

ACCESSING AN ARRAY USING POINTERS

Introduction

You can access an array element by using a pointer. For example, if an array stores integers, then you can use a pointer to integer to access array elements.

Program

```
#include <stdio.h>
void printarr(int a[]);
void printdetail(int a[]);
main()
{
    int a[5];
    for(int i = 0;i<5;i++)
    {
        a[i]=i;
    }
    printdetail(a);
}
void printarr(int a[])
{
    for(int i = 0;i<5;i++)
    {
        printf("value in array %d\n",a[i]);
    }
}
void printdetail(int a[])
{
    for(int i = 0;i<5;i++)
    {
        printf("value in array %d and address is %8u\n",a[i],&a[i]);
    }
}
void print_usingptr(int a[]) \\ A
{
    int *b;     \\ B
    b=a;              \\ C
    for(int i = 0;i<5;i++)
    {
```

```
        printf("value in array %d and address is %16lu\n",*b,b); \\ D
        b=b+2; \\E
    }
}
```

Explanation

1. The function print_using pointer given at statement A accesses elements of the array using pointers.
2. Statement B defines variable b as a pointer to an integer.
3. Statement C assigns the base address of the array to b, thus the array's first location (a[0]) is at 100; then b will get the value 100. Other elements of the array will add 102,104, etc.
4. Statement D prints two values: *b means the value at the location specified by b, that is, the value at the location 100. The second value is the address itself, that is, the value of b or the address of the first location.
5. For each iteration, b is incremented by 2 so it will point to the next array location. It is incremented by 2 because each integer occupies 2 bytes. If the array is long then you may increment it by 4.

Points to Remember

1. Array elements can be accessed using pointers.
2. The array name is the pointer constant which can be assigned to any pointer variable.

MANIPULATING ARRAYS USING POINTERS

Introduction

When the pointer is incremented by an increment operator, it is always right incremented. That is, if the pointer points to an integer, the pointer is incremented by 2, and, if it is long, it is incremented by 4.

Program

```
#include <stdio.h>
void printarr(int a[]);
void printdetail(int a[]);
void print_usingptr(int a[]);
main()
{
    int a[5];
    for(int i = 0;i<5;i++)
    {
        a[i]=i;
    }
    print_usingptr(a);
}
void printarr(int a[])
{
    for(int i = 0;i<5;i++)
    {
        printf("value in array %d\n",a[i]);
    }
}
void printdetail(int a[])
{
    for(int i = 0;i<5;i++)
    {
        printf("value in array %d and address is %8u\n",a[i],&a[i]);
    }
}
void print_usingptr(int a[])
{
    int *b;
    b=a;
    for(int i = 0;i<5;i++)
    {
        printf("value in array %d and address is %16lu\n",*b,b);
        b++;                    // A
    }
}
```

Explanation

1. This function is similar to the preceding function except for the difference at statement A. In the previous version, b = b+2 is used. Here b++ is used to increment the pointer.
2. Since the pointer is a pointer to an integer, it is always incremented by 2.

Point to Remember

The increment operator increments the pointer according to the size of the data type.

ANOTHER CASE OF MANIPULATING AN ARRAY USING POINTERS

Introduction

You can put values in the memory locations by using pointers, but you cannot assign the memory location to an array to access those values because an array is a pointer constant.

Program

```
#include <stdio.h>
void printarr(int a[]);
void printdetail(int a[]);
void print_usingptr_a(int a[]);
main()
{
    int a[5];
    int *b;
    int *c;
    for(int i = 0;i<5;i++)
    {
       a[i]=i;
    }
    printarr(a);
    *b=2;            \\ A
```

```
    b++;                \\ B
    *b=4;               \\ C
    b++;
    *b=6;               \\ D
    b++;
    *b=8;               \\ E
    b++;
    *b=10;
    b++;
    *b=12;
    b++;
    a=c; //error        \\F
    printarr(a);


}
void printarr(int a[])
{
    for(int i = 0;i<5;i++)
    {
        printf("value in array %d\n",a[i]);
    }
}
void printdetail(int a[])
{
    for(int i = 0;i<5;i++)
    {
        printf("value in array %d and address is %16lu\n",a[i],&a[i]);
    }
}

void print_usingptr_a(int a[])
{

    for(int i = 0;i<5;i++)
    {
        printf("value in array %d and address is %16lu\n",*a,a); \\ F
        a++; // increase by 2 bytes          \\ G
    }
}
```

Explanation

1. You can assign a value at the location specified by `b` using statement A.
2. Using statement B, you can point to the next location so that you can specify a value at that location using statement C. Using this procedure, you can initialize 5 locations.
3. You cannot assign the starting memory location as given by statement F to access those elements because `a` is a pointer constant and you cannot change its value.
4. The function `print_usingptr_a` works correctly even though you are writing `a++`. This is because when you pass `a` as a pointer in an actual parameter, only the value of `a` is passed and this value is copied to the local variable. So changing the value in the local variable will not have any effect on the outside function.

Point to Remember

The array limit is a pointer constant and you cannot change its value in the program.

TWO-DIMENSIONAL ARRAY

Introduction

You can define two- or multi-dimensional arrays. It is taken as an array of an array. Logically, the two-dimensional array 3 X 2 is taken as

```
3    1
5    2
8    7
```

Here there are three arrays, i.e. one array in each row. The values are stored as

```
3 1 5 2   8 7
```

This style is called *row measure form*. Each row array is represented as `a[0]`, which consists of elements 3 and 1. `a[1]` consists of 5 2 and `a[2]` consists of 8 7. Each element of `a[0]` is accessed as `a[0] [0]` and `a[0] [1]`, thus the value of `a[0][0]` and `a[0][1]` is 1.

Program

```
#include <stdio.h>
void printarr(int a[][]);
void printdetail(int a[][]);
void print_usingptr(int a[][]);
main()
{
    int a[3][2];      \\ A
    for(int i = 0;i<3;i++)
        for(int j=0;j<2 ;j++)
        {
            {
                a[i]=i;
            }
        }
    printdetail(a);
}
void printarr(int a[][])
{
    for(int i = 0;i<3;i++)
        for(int j=0;j<2;j++)
        {
            {
                printf("value in array %d\n",a[i][j]);
            }
        }
}
void printdetail(int a[][])
{
    for(int i = 0;i<3;i++)
        for(int j=0;j<2;j++)
        {
            {
                printf(
                "value in array %d and address is %8u\n",
                a[i][j],&a[i][j]);
            }
        }
}
void print_usingptr(int a[][])
{
    int *b;    \\ B
```

```
    b=a;                \\ C
    for(int i = 0;i<6;i++)   \\ D
    {
        printf("value in array %d and address is %16lu\n",*b,b);
        b++; // increase by 2 bytes \\ E
    }
}
```

Explanation

1. Statement A declares a two-dimensional array of the size 3 × 2.
2. The size of the array is 3 × 2, or 6.
3. Each array element is accessed using two subscripts.
4. You can use two `for` loops to access the array. Since `i` is used for accessing a row, the outer loop prints elements row-wise, that is, for each value of `i`, all the column values are printed.
5. You can access the element of the array by using a pointer.
6. Statement B assigns the base address of the array to the pointer.
7. The `for` loop at statement C increments the pointer and prints the value that is pointed to by the pointer. The number of iterations done by the `for` loop, 6, is equal to the array.
8. Using the output, you can verify that C is using row measure form for storing a two-dimensional array.

Points to Remember

1. You can define a multi-dimensional array in C.
2. You have to provide multiple subscripts for accessing array elements.
3. You can access array elements by using a pointer.

THREE-DIMENSIONAL ARRAY

Introduction

Just as with a two-dimensional array, you can define three-dimensional arrays and processing in a similar way. Each three-dimensional array is taken as an array of two-dimensional arrays.

Program

```
#include <stdio.h>
main()
{
    int a[2][3][4];   \\ A
    int b[3][4];      \\ B
    int c[4];         \\ C
    int cnt=0;
    for(int i=0;i<2;i++)
       for(int j=0;j<3;j++)
             for(int k=0;k<4;k++)
             {
                    a[i][j][k] = cnt;
                    cnt;
             }
}
void print_onedim(int a[])   \\ D
{
    for(int i=0;i<4;i++)
       printf("%d ",a[i]);
}
void print_twodim(int a[][4])                \\ E
{
    for(int j=0;j<3;j++)
       print_onedim(a[j]);
    printf("\n");
}
void print_threedim(int a[][3][4])           \\ F
{
    printf("Each two dimension matrix\n");
    for(int j=0;j<2;j++)
       print_twodim(a[j]);;

}
```

Explanation

1. The three-dimensional array consists of two arrays of the size 3 × 4. Each is referred to as a[0] a[1]. Thus a[0] consists of 12 elements and a[1] also consists of 12 elements.
2. Each two-dimensional array is taken as three arrays of the size 4.

3. The function `print_onedim` prints a single-dimensional array.
4. The function `print_twodim` prints a two-dimensional array and it calls the function for printing a single-dimensional value.
5. The function `print_threedim` prints a three-dimensional array and calls the function for printing a three-dimensional value.
6. Dimension two, which is closer to the array name, is called the outermost dimension, and dimension four, which is far from the array name declaration, is called the innermost dimension.
7. When you pass an array to the function, you have to specify the inner dimension. For example, to print `two_dim` you have to specify the inner dimension, i.e. 4, and for `printthree_dim` you have to pass 3 and 4 as inner dimensions.
8. When you pass a single-dimension array, you need not pass a dimension because the function knows what the best address of the array is.
9. In a case of a two-dimensional array, we have to pass the inner dimension because only then does the function know the base address of each array. For example, if the declaration is

   ```
   int a[3][4]
   ```

   it is considered as three arrays of the size 4. So the base address of `a[0]` is `a` itself. The base address of `a[1]` is `a+8` because the first row has 4 elements of size 2 bytes each; thus we can get the base address of `a[1]`. Similarly, the base address of `a[2]` is `a+16`. Thus, to calculate the base address, you should know the inner dimension, 4.

Points to Remember

1. You can declare a multi-dimensional array and access it in a similar way to accessing a two-dimensional array.
2. While passing an array to the function, you have to specify the inner dimension to facilitate calculations of the base addresses.

POINTER ARRAYS

Introduction

You can define a *pointer array* similarly to an array of integers. In the pointer array, the array elements store the pointer that points to integer values.

Program

```
#include <stdio.h>
void printarr(int *a[]);
void printarr_usingptr(int *a[]);
int *a[5];      \\ A
main()
{

    int i1=4,i2=3,i3=2,i4=1,i5=0;           \\ B
    a[0]=&i1;                               \\ C
    a[1]=&i2;
    a[2]=&i3;
    a[3]=&i4;
    a[4]=&i5;

    printarr(a);
    printarr_usingptr(a);
}
void printarr(int *a[])                     \\ D
{
    printf("Address          Address in array           Value\n");
    for(int  j=0;j<5;j++)
    {
       printf("%16u             %16u                         %d\n",
       a[j],a[j],a[j]);             \\E
    }

}
void printarr_usingptr(int *a[])
{
    int j=0;
    printf("using pointer\n");
    for( j=0;j<5;j++)
    {
       printf("value of elements   %d %16lu %16lu\n",**a,*a,a); \\ F
       a++;
    }

}
```

Explanation

1. Statement A declares an array of pointers so each element stores the address.
2. Statement B declares integer variables and assigns values to these variables.
3. Statement C assigns the address of i1 to element a[0] of the array. All the array elements are given values in a similar way.
4. The function print_arr prints the address of each array element and the value of each array element (the pointers and values that are pointed to by these pointers by using the notations &a[i], a[i] and *a[i]).
5. You can use the function printarr_usingptr to access array elements by using an integer pointer, thus a is the address of the array element, *a is the value of the array element, and **a is the value pointed to by this array element.

Point to Remember

You can store pointers in arrays. You can access values specified by these values by using the * notations.

10 Function

FUNCTION

Introduction

Functions are used to provide modularity to the software. By using functions, you can divide complex tasks into small manageable tasks. The use of functions can also help avoid duplication of work. For example, if you have written the function for calculating the square root, you can use that function in multiple programs.

Program

```
#include <stdio.h>
int add (int x, int y)        //A
{
    int z;              //B
    z = x + y;
    return (z);         //C
}
main ()
{
    int i, j, k;
    i = 10;
    j = 20;
    k = add(i, j);              //D
    printf ("The value of k is%d\n", k);   //E
}
```

Explanation

1. This function adds two integers and returns their sum.

2. When defining the name of the function, its return data type and parameters must also be defined. For example, when you write

   ```
   int add (int x, int y)
   ```

   `int` is the type of data to be returned, `add` is the name of the function, and `x` and `y` are the parameters of the type `int`. These are called *formal parameters.*
3. The body of a function is just like the body of main. That means you can have variable declarations and executable statements.
4. A function should contain statements that return values compatible with the function's return type.
5. The variables within the function are called *local variables.*
6. After executing the return statement, no further statements in the function body are executed.
7. The name of the function can come from the arguments that are compatible with the formal parameters, as indicated in statement D.
8. The arguments that are used in the call of the function are called *actual parameters.*
9. During the call, the value of the actual parameter is copied into the formal parameter and the function body is executed.
10. After the return statement, control returns to the next statement which is after the call of the function.

Points to Remember

1. A function provides modularity and readability to the software.
2. To define the function, you have to define the function name, the return data type and the formal parameters.
3. Functions do not require formal parameters.
4. If the function does not return any value, then you have to set the return data type as void.
5. A call to a function should be compatible with the function definition.

THE CONCEPT OF STACK

Introduction

A *stack* is memory in which values are stored and retrieved in "last in first out" manner by using operations called *push* and *pop*.

Program

Suppose you want to insert values in a stack and retrieve values from the stack. The operations would proceed in the following manner:

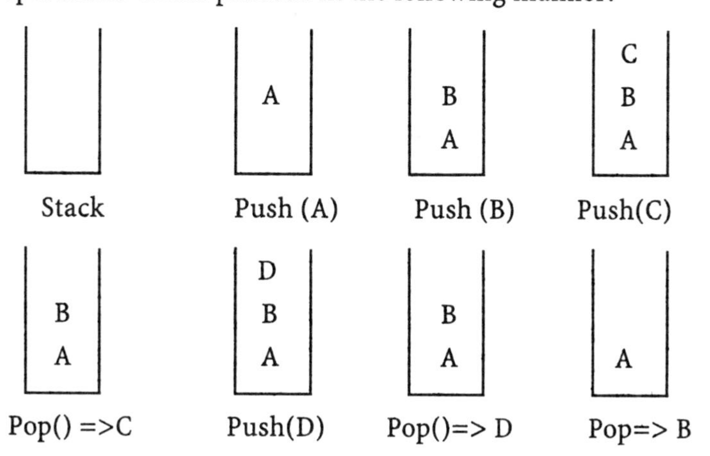

Explanation

1. Initially, the stack is empty. When you start push A, A is placed in the stack.
2. Similarly, push B and push C put these elements in the stack; the last element pushed is C.
3. The pop operation takes the topmost element from the stack. Thus the element C, which was put in last, is retrieved first. This method is called *last-in first-out (LIFO)*.
4. The push D operation puts element D in the stack above B.
5. Thus push puts the element on the top of the stack and pop takes the element from the top of the stack. The element A which is pushed is the last element taken from the stack.

Point to Remember

The last-in first-out retrieval from the stack is useful for controlling the flow of execution during the function call.

THE SEQUENCE OF EXECUTION DURING A FUNCTION CALL

Introduction

When the function is called, the current execution is temporarily stopped and the control goes to the called function. After the call, the execution resumes from the point at which the execution is stopped.

To get the exact point at which execution is resumed, the address of the next instruction is stored in the stack. When the function call completes, the address at the top of the stack is taken.

Program

```
main ( )
{
    printf ("1 \n");  // 1
    printf ("2 \n");  // 2
    printf ("3 \n");  // 3
    printf ("4 \n");  // 4
    printf ("5 \n");  // 5
   f1 ( );
    printf ("6 \n");  // 6
    printf ("7 \n");  // 7
    printf ("8 \n");  // 8
}
void f1 (void)
{
    printf ("f1-9 \n");      // 9
    printf ("f1-10 \n");     // 10
    f2 ( );
    printf ("f1-11 \n");     // 11
    printf ("f1-12 \n");     // 12
}
void f2 (void)
```

```
{
    printf ("f2-13 \n");      // 13
    printf ("f2-14 \n");      // 14
    printf ("f3-15 \n");      // 15
}
```

Explanation

1. Statements 1 to 5 are executed and function `f1( )` is called.
2. The address of the next instruction is pushed into the stack.
3. Control goes to function `f1( )`, which starts executing.
4. After the 10th statement, fuction `f2` is called and address of the next instruction, 11, is pushed into the stack.
5. Execution begins for function `f2` and statements 13, 14, and 15 are executed.
6. When `f2` is finished, the address is popped from the stack. So address 11 is popped.
7. Control resumes from statement 11.
8. Statements 11 and 12 are executed.
9. After finishing the `f1` address is popped from the stack, i.e. 6.
10. Statements 6, 7, and 8 are executed.
11. The execution sequence is

```
1 2 3 4 5   f1_9  f1_10   f2_13  f2_14  f2_15    f1_11  f1_12  6 7  8.
```

Points to Remember

1. Functions or sub-programs are implemented using a stack.
2. When a function is called, the address of the next instruction is pushed into the stack.
3. When the function is finished, the address for execution is taken by using the pop operation.

PARAMETER PASSING

Introduction

Information can be passed from one function to another using parameters.

Program

```
main ( )
{
    int i;
    i = 0;
    printf (" The value of i before call %d \n", i);
    f1 (i);
    printf (" The value of i after call %d \n", i);
}
void f1 (int k)
{
    k = k + 10;
}
```

Explanation

1. The parameter used for writing the function is called the formal parameter, k in this case.
2. The argument used for calling the function is called the actual parameter.
3. The actual and formal parameters may have the same name.
4. When the function is called, the value of the actual parameter is copied into the formal parameter. Thus k gets the value 0. This method is called *parameter passing by value*.
5. Since only the value of i is passed to the formal parameter k, and k is changed within the function, the changes are done in k and the value of i remains unaffected.
6. Thus i will equal 0 after the call; the value of i before and after the function call remains the same.

Points to Remember

1. C uses the method of parameter passing by value.
2. In parameter passing by value, the value before and after the call remains the same.

CALL BY REFERENCE

Introduction

Suppose you want to pass a parameter under the following conditions:

1. You need to change the value of the parameter inside the function.
2. You are interested in the changed value after the function completes.

In languages such as Pascal, you have the option of passing the parameter by reference. C, however, does not support this. As explained in the previous example, you cannot have a changed value after the function call because C uses the method of parameter passing by value. Instead, you'll have to implement the function indirectly. This is done by passing the address of the variable and changing the value of the variable through its address.

Program

```
main ( )
{
    int i;
    i = 0;
   printf (" The value of i before call %d \n", i);
    f1 (&i);          // A
   printf (" The value of i after call %d \n", i);
}
void  (int *k)        // B
{
    *k = *k + 10;     // C
}
```

Explanation

1. This example is similar to the previous example, except that the function is written using a pointer to an integer as a parameter.
2. Statement C changes the value at the location specified by *k.
3. The function is called by passing the address of `i` using notation `&i`.
4. When the function is called, the address of `i` is copied to k, which holds the address of the integer.
5. Statement C increments the value at the address specified by k.
6. The value at the address of `i` is changed to 10. It means the value of `i` is changed.
7. The `printf` statements after the function call prints the value 10, that is, the changed value of `i`.

Points to Remember

1. Call by reference is implemented indirectly by passing the address of the variable.
2. In this example, the address of `i` is passed during the function call. It does not change; only the value of the address is changed by the function.

THE CONCEPT OF GLOBAL VARIABLES

Introduction

The various modules can share information by using *global variables*.

Program

```
#include <stdio.h>
int i =0;      //Global variable
main()
{
    int j;            // local variable in main
    void f1(void)  ;
    i =0;
    printf("value of i in main %d\n",i);
```

```
    f1();
    printf("value of i after call%d\n",i);
}
void f1(void)
{
    int k;              // local variable for f1.
    i = 50;
}
```

Explanation

1. When you define a variable inside the function block it is called a local variable.
2. The local variable can be accessed only in the block in which it is declared.
3. `j` is the local variable for main and it can be accessed only in the block main. That means you cannot access it in function `f1`.
4. `k` is the local variable for function `f1` and it cannot be accessed in main.
5. The variable `i`, which is outside main, is called a global variable. It can be accessed from function main as well as function `f1`.
6. Any expression in this function is going to operate on the same `i`.
7. When you call function `f1`, which sets the value of `i` to 50, it is also reflected in main because main and `f1` are referring to the same variable, `i`.

Points to Remember

1. Global variables can be accessed in all the functions in that file.
2. Any update to the global variable also affects the other functions, because all functions refer to the same value of i.
3. When you want to share information between multiple functions, you can use the concept of global variables.

RESOLVING VARIABLE REFERENCES

Introduction

When the same variable is resolved using both local definition and global definition, the local definition is given preference. This is called the rule of inheritance. It says that when you can resolve a reference to the variable by using multiple definitions, the nearest definition is given preference. Since local definition is the nearest, it gets preference.

Program

```
int i =0;              //Global variable  /A
main()
{
    int i ;                    // local variable for main   / B
    void f1(void)  ;           //C
    i =0;                      // D
    printf("value of i in main %d\n",i);   // E
    f1();                                  // F
    printf("value of i after call%d\n",i); // G
}
void f1(void)                       // H
{
    int i=0;           //local variable for f1      // I
    i = 50;                                // J
}
```

Explanation

1. Here i is declared globally and locally in function main and in function f1, respectively, as given in statements A, B and I.
2. Statement D refers to i, which can be resolved by using both local definition and global definition. Local definition is given more preference. So statement D refers to the definition at statement B and all the statements in main refer to the definition at statement B, that is, the local definition.
3. When a function is called, statement i = 50 refers to the local definition in that function (definition at statement I).
4. Using statement G, the value of i is 0 because both main and function f1 refer to their local copies of i. So the changed value of f1 is not reflected in main.
5. Even if you comment local definition of function f1 at statement I the value printed remains the same. This is because main refers to its local copy while f1 refers to the global variable i — the two are different.

Point to Remember

When a variable can be resolved by using multiple references, the local definition is given more preference.

SYNTAX OF FUNCTION DEFINITION

Introduction

When specifying the function, you have to specify the return type, function name, parameter, list, and function body. Within the function body, you can have local definition and return statements.

Program/Example

The general format of a function is

```
<Return type>    <Function name>    <Parameter list>
{
    <local definitions>
     executable statements;
     Return (expression);
}
```

For example,

```
int f1 (int j, float f)
{
   int k;
   k = 1;
   return (k);
}
```

Explanation

1. A function returns a value of the type that is specified by the return type. If you don't specify a written type, it is assumed that it returns an `int` value.
2. When the function does not return a value, you have to specify the return data type as void. When the function returns void, you may not write return in the body or you can write the return statement as `return;` .

3. All functions must be named.
4. You can specify parameters in the parameter list, separated by commas. While specifying the parameters, you have to specify the parameter data type and parameter name.
5. If you don't specify parameters, then you can specify only parentheses as shown here:

   ```
   int f1( )
   ```
6. When you want to use variables only for the function then you can declare them just as in main.
7. A function returns a value to the caller using the return statement. You may have multiple return statements and the return expression should evaluate to a value that is compatible with the return data type.
8. A function returns to the caller after executing the first return statement it encounters during execution.
9. A call to a function should match the definition of the function.
10. The order of parameters in the call is important because the actual parameter value is copied to the formal parameter value according to the order. It means that the first argument in the call is copied to the first parameter, the second argument is copied to the second parameter, etc.
11. When you are using whole numbers as parameters, it is better to declare them by using the data type `int`. Because all your lower data types' actual parameters can be used for passing the value, your function can be useful for multiple data types.
12. When you are using real numbers as parameters, it is better to declare them as double so that the function can be used for both the float and double data types.

Points to Remember

1. While specifying the function you have to specify five main functions: written type, function name, parameter, list, function body and return statement.
2. Function name and function body are necessary, while the others are optional.

CALLING FUNCTION

Introduction

When a function is written before main it can be called in the body of main. If it is written after main then in the declaration of main you have to write the prototype of the function. The prototype can also be written as a global declaration.

Program

```
Case 1:
#include <stdio.h>
main ( )
{
    int i;
    void  (int *k)              // D

    i = 0;
    printf (" The value of i before call %d \n", i);
    f1 (&i);            // A
    printf (" The value of i after call %d \n", i);
}
void  (int *k)          // B
{
    *k = *k + 10;       // C
}

Case 2:

#include <stdio.h>
void  (int *k)          // B
{
    *k = *k + 10;       // C
}

main ( )
{
    int i;
    i = 0;
    printf (" The value of i before call %d \n", i);
```

```
    f1 (&i);          // A
    printf (" The value of i after call %d \n", i);
}

Case 3:
#include <stdio.h>
void  f1(int *k)              // B
{
    *k = *k + 10;     // C
}

main ( )
{
    int i;
    i = 0;
printf ("The value of i before call %d \n", i);
    f1 (&i);          // A
printf ("The value of i after call %d \n", i);
}
```

Explanation

1. In Case 1, the function is written after main, so you have to write the prototype definition in main as given in statement D.
2. In Case 2, the function is written above the function main, so during the compilation of main the reference of function f1 is resolved. So it is not necessary to write the prototype definition in main.
3. In Case 3, the prototype is written as a global declaration. So, during the compilation of main, all the function information is known.

11 Storage of Variables

STORAGE

Introduction

In the variable declaration you can also define lifetime or storage duration of the variable. *Lifetime* indicates the length of time the variable value is guaranteed during execution. For example, if the variable is defined inside the function, its value is kept until the function executes. After completion of the function, the storage allocated for the variable is freed.

Program

```
#include <stdio.h>
int g = 10;    \\ A
main()
{
    int i =0;  \\ B
    void f1(); \\ C
    f1();              \\ D
    printf(" after first call \n");
    f1();              \\ E
    printf("after second  call \n");
    f1();              \\ F
    printf("after third  call \n");

}
void f1()
{
    static int k=0;   \\ G
    int j = 10;                \\ H
```

```
    printf("value of k %d j %d",k,j);
    k=k+10;
}
```

Explanation

1. Variables in C language can have automatic or static lifetimes. Automatic means the variable is in existence until the function in which it is defined executes; static means the variable is retained until the program executes.
2. The variable that is defined outside the function, such as g in statement A, is called a global variable because it is accessible from all the functions. These global variables have static lifetimes, that is, variable return throughout the program execution. The value of the variable, as updated from one function, affects another function that refers to that variable. It means that the updating in this variable is visible to all functions.
3. Variables such as i, defined in main, or j, defined in f1, are of the automatic type; i exists until main is completed and j exists until f1 is completed.
4. You can define the lifetime of a local variable in a function as given in statement G. The variable k has a static lifetime; its value is returned throughout the execution of the program.
5. The function f1 increments the value of k by 10 and prints the values of j and k.
6. When you call the function for the first time using statement D, k is printed as 0, j is printed as 10, k is incremented to 10, the space of j is reallocated, and j ceases to exist.
7. When you call the function the second time, it will give 10 (the previous value of k) because k is a static variable. There are reallocations for j so j is printed as 10.
8. When you call the function the third time, j is still printed as 10.

Points to Remember

1. The variables in C can have static or automatic lifetimes.
2. When a variable has a static lifetime, memory is allocated at the beginning of the program execution and it is reallocated only after the program terminates.

3. When a variable has an automatic lifetime, the memory is allocated to the variable when the function is called and it is deallocated once the function completes its execution.
4. Global variables have static lifetimes.
5. By default, local variables have automatic lifetimes.
6. To make a local variable static, use the storage-class specifier.

EXTERNAL REFERENCES

Introduction

In collaborative software development it is common for multiple users to write programs in different files. For example, one user implements function f1, a second user implements function f2, while a third user implements the main function. C has a provision to compile programs even if function or variable implementation is not available. In such cases, the program is compiled but it is not yet fit for execution. The program is not executable until all the references in the file are available.

Program

```
\\ Program in file external.c

#include <stdio.h>                    \\ A
#include <d:\cbook\storage\f1.cpp>  \\ B
extern int i;                         \\ C
main()
{
    i =0;                                         \\ D
    printf("value of i %d\n",i);
}

\\ Program in file f1.cpp

int i =7;                        \\ E
```

Explanation

1. Here the program is written in two files: `extern1.c` and `f1.cpp`. The file `extern1.c` has the main and reference of variable `i`.
2. The file `f1.cpp` has the declaration of `i`.
3. In the file `extern.c` there is a reference of `i` so the compiler should know the data type of `i`. This is done using the extern definition by statement C. *Extern* means that the variable or function is implemented elsewhere but is referred to in the current file.
4. Statement D refers to `i`.
5. The definition of `i` is given in the file `f1.cpp`, as given by statement E.
6. In the absence of an `include` directive in statement B, you can still compile the file; it will give no errors. Such a file is called an *object file. It is not fit for execution* because the reference of `i` is not resolved.
7. When you write statement B the reference of `i` is re-sorted and the executable file can be made.

Points to Remember

1. Extern definition is used when you have to refer a function or variable that is implemented elsewhere or it will be implemented later on.
2. When all the references are resolved then only the executable file is made.

REGISTER VARIABLES

Introduction

When you want to refer a variable, many times you can allocate fast memory in the form of a register to that variable. For variables such as loop counters, register allocation is done. The processor has memory in the form of register for its temporary storage. The access time of the register is much less than main memory. That is the reason that register allocation provides more speed. But the processor has a limited number of registers. So the register declaration acts as a directive; it does not guarantee the allocation of a register for storing value of that variable.

Program

```
#include <stdio.h>
main()
{
   register int  i = 0;     \\ A

   for( i=0;i<2;i++)
   {
      printf("value of i is %d\n",i);
   }

}
```

Explanation

1. Here the register allocation directive is given for variable `i`. During execution, `i` will be allocated a register if it is available; otherwise, `i` will receive normal memory allocations.
2. You can use a register directive only for variables of the automatic storage class, not for global variables.
3. Generally, you can use register storage for `int` or `char` data types.

You cannot use register allocation for global variables because memory is allocated to the global variable at the beginning of the program execution. At that time, it is not certain which function is invoked and which register is used. Function code may use the register internally, but it also has access to a global variable, which might also use the same register. This leads to contradiction, so global register variables are not allowed.

Points to Remember

1. Register allocation is done for faster access, generally for loop counters.
2. You cannot declare global register variables.

SCOPE OF VARIABLES

Introduction

In C you can define a variable in the block. The blocks are marked using { and } braces. The blocks can also be defined using the for statement. The scope of the variable is in the block in which it is declared, meaning that you can use that variable anywhere in the block. Even if some block is declared in that block, you can use that variable. When the variable is referred in the block and if it can be resolved using two definitions, then the nearest definition has more precedence. So the variable is interpreted according to the nearest definition. Even if the two definitions define two different data types for variables, they are accepted.

Program

```
#include <stdio.h>
main()
{       \\ Block 1
    int  i = 10;       \\ A

    {           \\ Block 2
        int i = 0;           \\ B
        for( i=0;i<2;i++)    \\ C
        {
                printf("value of i is %d\n",i);
        }       \\ End of block 2
    }
    printf("the value of i is %d\n",i);     \\ D
}   \\ End of block 1
```

Explanation

1. The statement `block 1` defines the start of block 1.
2. The statement 'end of block 1' defines the end of block 1.
3. Statement A defines variable `i` which has the scope in the entire block 1.
4. The statement `block 2` defines the start of block 2.
5. The statement 'end of block 2' defines the end of block 2.
6. Statement B defines variable `i` which is entirely in block 2.

7. The for loop refers i, which can be resolved using two definitions: statement A and statement B.
8. Since the definition of statement B is nearest, the variable is referred using that definition, so the for loop modifies the value of i at statement B.
9. Variable i at statement A and variable i at statement B are two independent variables even though they have the same name. Statement D is outside block 2, so it prints the value of variable i in block 1.

Points to Remember

1. In C, you can define variables in the block, which is demarcated by using { and } braces. The variable has the scope inside the block in which it is declared.
2. When the variable is resolved using two definitions, the nearest definition has more precedence.

FURTHER SCOPE OF VARIABLES

Introduction

In C, you can define the counter in the for loop itself; the counter has scope up to the end of the for loop.

Program

```
#include <stdio.h>
main()
{
    int  k = 10;

    {
        for(int  i=0;i<2;i++) \\ A
        {       \\ B
                printf("value of i is %d\n",i);     \\ C
        }       \\ D
    }
    printf("the value of i is %d\n",i);     \\ E
}
```

Explanation

1. The counter variable `i` is defined at statement A.
2. The scope of the `for` loop is up to statement D, which is the end of the `for` loop for statement B.
3. If you do not comment out statement E, you will get an error, because you cannot refer `i` in the outside block.

Point to Remember

You can define counter variable inside the `for` loop.

12 Memory Allocation

DYNAMIC MEMORY ALLOCATIONS

Introduction

You can use an array when you want to process data of the same data type and you know the size of the data. Sometimes you may want to process the data but you don't know what the size of the data is. An example of this is when you are reading from a file or keyboard and you want to process the values. In such a case, an array is not useful because you don't know what the dimension of the array should be. C has the facility of dynamic memory allocations. Using this, you can allocate the memory for your storage. The allocation is done at runtime. When your work is over, you can deallocate the memory. The allocation of memory is done using three functions: `malloc`, `relloc`, and `calloc`. The functions return the pointers to void, so it can be typecast to any data type, thus making the functions generic. These functions take the input as the size of memory requirement.

Program

```
#include <stdio.h>
#include <malloc.h>
main()
{
    int  *base;         \\ A
    int i;
    int cnt=0;
    int sum=0;
    printf("how many integers you have to store \n");
    scanf("%d",&cnt);        \\ B
```

```
    base = (int *)malloc(cnt * sizeof(int));       \\ C
    printf("the base of allocation is %16lu \n",base);    \\ D
    if(!base) \\ E
       printf("unable to allocate size \n");
    else
    {
       for(int j=0;j<cnt;j++)                   \\ F
             *(base+j)=5;
    }
    sum = 0;
    for(int j=0;j<cnt;j++)                      \\ G
       sum = sum + *(base+j);
    printf("total sum is %d\n",sum);
    free(base);                \\ H
    printf("the base of allocation is %16lu \n",base);
    base = (int *)malloc(cnt * sizeof(int));
    printf("the base of allocation is %16lu \n",base);
    base = (int *)malloc(cnt * sizeof(int));            \\ I
    printf("the base of allocation is %16lu \n",base);
    base = (int *)calloc(10,2);                         \\ J
    printf("the base of allocation is %16lu \n",base);
}
```

Explanation

1. This program demonstrates the use of dynamic memory allocation for processing `n` integers where `n` is not defined at compilation time, but the user instead specifies the number of integers to be processed.
2. The processing adds 5 to the value of each integer.
3. Statement B reads how many integers you have to process.
4. Statement C allocates memory for the required integers by using the function `malloc`.
5. `malloc` takes the size in bytes as input.
6. The size of the operator returns how many bytes can be occupied by one unit of the specified data type. The size of `int` returns two bytes. If you give the value `cnt` as 10 then it will allocate 20 bytes.
7. `malloc` returns the pointer to void, which is typecast as a pointer to an integer. The value starts at the address of the memory from where allocations are done. The value is stored in the variable base, which is declared at statement

A. If memory allocations cannot be done, the base will get the value 0, which can be tested using an `if` statement. The `for` loop F puts a value of 5 in the allocated memory. Note that the first value is stored in the location specified by the base and the next value is stored according to base +`j`. If the base is 100 and `j` is 1 then the value of base + 1 is 102, according to pointer arithmetic, and not 101, because this is a pointer to an integer and an integer occupies two bytes. You can retreive the value by using a pointer to an integer as specified by the `for` loop in statement G. After your work is over, you can return the memory using the function `free`. `free` takes a pointer to storage as input.

8. You can again allocate more or less memory by using the function `malloc`. You can again allocate memory without deallocating previous memory as given by statement I. You can allocate the memory similarly to `malloc` by using the function `calloc`. `calloc` takes two arguments: total number of data and the size of each data.

Points to Remember

You can allocate memory at runtime by using the function `malloc`. `malloc` allocates memory specified using an argument in terms of bytes, and returns the pointer to storage from where the memory is allocated. You can deallocate the memory by using the function `free`.

The prototypes of the function are available in the hidden files `malloc.h`.

13 Recursion

RECURSION

Introduction

You can express most of the problems in the following program by using *recursion*. We represent the function add by using recursion.

Program

```
#include <stdio.h>
int add(int pk,int pm);
main()
{
    int k ,i,m;
    m=2;
    k=3;
    i=add(k,m);
    printf("The value of addition is %d\n",i);
}
int add(int pk,int pm)
{
    if(pm==0) return(pk);            \\ A
    else return(1+add(pk,pm-1));     \\ B
}
```

Explanation

1. The add function is recursive as follows:

```
        add (x, y) = 1 + add(x, y-1)          y > 0
                   =  x                    y = 0
```

for example,

```
add(3, 2) = 1 + add(3, 4)
add(3, 1) = 1 + add(3, 0)
add(3, 0) = 3
add(3, 1) = 1+3 = 4
add(3, 2) = 1+4 = 5
```

2. The recursive expression is `1+add(pk, pm-1)`. The terminating condition is `pm = 0` and the recursive condition is `pm > 0`.

STACK OVERHEADS IN RECURSION

Introduction

If you analyze the address of local variables of the recursive function, you will get two important results: the depth of recursion and the stack overheads in recursion. Since local variables of the function are pushed into the stack when the function calls another function, by knowing the address of the variable in repetitive recursive call, you will determine how much information is pushed into the stack. For example, the stack could grow from top to bottom, and the local variable `j` gets the address 100 in the stack in the first column. Suppose stack overheads are 16 bytes; in the next call `j` will have the address 84, in the call after that it will get the address 16. That is a difference of 16 bytes. The following program uses the same principle: the difference of the address in consecutive calls is the stack overhead.

Program

```
#include <stdio.h>
int fact(int n);
long  old=0;  \\E
long current=0;       \\F
main()
{
    int k = 4,i;
    long diff;
    i =fact(k);
    printf("The value of i is %d\n",i);
    diff = old-current;
```

```
    printf("stack overheads are %16lu\n",diff);
}
int fact(int n)
{
    int j;
    static int m=0;
    if(m==0) old =(long) &j; \\A
    if(m==1) current =(long) &j;     \\B
    m++;                 \\C
    printf("the address of j and m is  %16lu %16lu\n",&j,&m);    \\D
    if(n<=0)
        return(1);
    else
        return(n*fact(n-1));
}
```

Explanation

1. The program calculates factorials just as the previous program.
2. The variable to be analyzed is the local variable `j`, which is the automatic variable. It gets its location in the stack.
3. The static variable `m` is used to track the number of recursive calls. Note that the static variables are stored in memory locations known as data segments, and are not stored in stack. Global variables such as `old` and `current` are also stored in data segments.
4. The program usually has a three-segment text: first, storing program instructions or program code, then the data segment for storing global and static variables, and then the stack segment for storing automatic variables.
5. During the first call, `m` is 0 and the value of `j` is assigned to the global varable `old`. The value of `m` is incremented.
6. In the next call, `m` is 1 and the value of `j` is stored in current.
7. Note that the addresses of `j` are stored in long variables of type castings.
8. `old` and `current` store the address of `j` in consecutive calls, and the difference between them gives the stack overheads.
9. You can also check the address of `j` and check how the allocation is done in the stack and how the stack grows.

You can also check whether the address of m is constant.

Points to Remember

1. The recursive program has a stack overhead.
2. You can calculate stack overheads by analyzing the addresses of local variables.

WRITING A RECURSIVE FUNCTION

Introduction

A recursive function is a function that calls itself. Some problems can be easily solved by using recursion, such as when you are dividing a problem into sub-problems with similar natures. Note that recursion is a time-consuming solution that decreases the speed of execution because of stack overheads. In recursion, there is a function call and the number of such calls is large. In each call, data is pushed into the stack and when the call is over, the data is popped from the stack. These push and pop operations are time-consuming operations. If you have the choice of iteration or recursion, it is better to choose iteration because it does not involve stack overheads. You can use recursion only for programming convenience. A sample recursive program for calculating factorials follows.

Program

```
#include <stdio.h>
int fact(int n);
main()
{
    int k = 4,i;
    i =fact(k);         \\ A
    printf("The value of i is %d\n",i);
}
int fact(int n)
{
    if(n<=0)            \\ B
```

```
        return(1);             \\ C
    else
        return(n*fact(n-1)); \\ D
}
```

Explanation

1. You can express factorials by using recursion as shown:

   `fact (5) = 5 * fact (4)`

   In general,

   `fact (N) = N * fact (N-1)`

   `fact` 5 is calculated as follows:

   ```
   fact (5) = 5 * fact (4) i.e. there is call to fact (4) \\ A
       fact (4) = 4 * fact (3)
       fact (3) = 3 * fact (2)
       fact (2) = 2 * fact (1)
       fact (1) = 1 * fact (0)
       fact (0) = 1                 \\ B
   ```

   `fact (1) = 1 * 1`, that is the value of the `fact(0)` is substituted in 1.

   ```
       fact (2) = 2 * 1 = 2
       fact (3) = 3 * 2 = 6
       fact (4) = 4 * 6 = 24
       fact (5) = 5 * 24 = 120             \\ C
   ```

2. The operations from statements B to A are collectivelly called the *winding phase*, while the operations from B to C are called the unwinding phase. The winding phase should be the terminating point at some time because there is no call to function that is given by statement B; the value of the argument that equals 0 is the terminating condition. After the winding phase is over, the unwinding phase starts and finally the unwinding phase ends at statement C. In recursion, three entities are important: recursive expressions, recursive condition, and terminating condition. For example,

   ```
       fact ( N) = N * fact (N-1) N > 0
                 = 1              N = 0
   ```

   `N * fact (N-1)` indicates a recursive expression.

   `N > 0` indicates a recursive condition.

   `N = 0` indicates a terminating condition.

3. You should note that the recursive expression is such that you will get a terminating condition after some time. Otherwise, the program enters into

an infinite recursion and you will get a stack overflow error. Statement B indicates the terminating condition, that is, `N = 0`.

4. The condition `N > 0` indicates a recursive condition that is specified by the `else` statement. The recursive expression is `n * fact(n-1)`, as given by statement D.

Points to Remember

1. Recursion enables us to write a program in a natural way. The speed of a recursive program is slower because of stack overheads.
2. In a recursive program you have to specify recursive conditions, terminating conditions, and recursive expressions.

14 Strings

STRINGS AS AN ARRAY OF CHARACTERS

Introduction

A *string* is defined as an array of characters. Strings are terminated by the special character '\o'; this is called a *null parameter* or *null terminator*. When you declare the string, you should ensure that you should have sufficient room for the null terminator. The null terminator has ASCII value 0.

Program

```
main ( )
{
    char s1[6];          \\ A
    char s2[6];
    char ch;
    int cnt = 0;
    s1 = "Hello";          \\ B
    printf ("%s \n", s1);       \\ C
    s2 = {'H', 'e', 'l', 'l', 'o'}  \\ D
    printf("%s \n", s2);              \\ E
    while (  (ch = getchar() )! = '#' && (cnt < 6-1) )           \\ F
        s1[cnt++] = ch;              \\ G
    s1[cnt] = '\0';          \\ H
}
```

Explanation

1. The size of the string is 6, which is the last element terminator, so you can use only 5 positions.

2. In statement B, the string "Hello" is assigned so that the array elements are
   `H e l l o \0`
3. The null terminator is appended automatically.
4. Statement B puts the data in a string using standard array notation.
5. You can print a string using the placeholder %s; the string is printed until it encounters a null character.
6. The while loop in statement H inputs the string by reading character by character.
7. The function getchar returns the character.
8. Note that the counter is incremented up to 5 so as to accommodate the last null terminator.
9. The null terminator is put in place by statement H.
10. The while loop can be terminated before counter 5 by putting in the # character.

Points to Remember

1. A string is a character array with a null terminator at the end.
2. You can initialize the array using different methods.

STRING DEFINITION

Introduction

A string can be defined using a character array or a pointer to characters. Although the two definitions look similar, they are actually different.

Program

```
main ( )
{
    char * s1 = "abcd";        \\ A
    char  s2[] = "efgh";        \\ B
    printf( "%s %16lu \n, s1, s1);   \\ C
    printf( "%s %16lu \n, s2, s2);    \\ D
```

```
    s1 = s2;                    \\ E
    printf( "%s %16lu \n, s1, s1);     \\ F
    printf( "%s %16lu \n, s2, s2);     \\ G
}
```

Explanation

1. Statement A declares s1 as a pointer to a character. When this definition is encountered, the compiler allocates space for the string abcd; the base address of the string is assigned to s1, which is the pointer variable.
2. Statement B declares s2 as a character array. The size of the array is 5 because of an additional null terminator in this case. Also, a space of 5 characters is allocated and the base address is given to s2, which is the pointer constant. During the lifetime of the program, we cannot change the value of s2.
3. The allocation for s1 is the allocation required by the pointer variable.
4. Statement C prints s1, using two place holders: %s and %16lu. Using %s, you will print the string as "abcd". Using %16lu you will print the base address of the string.
5. Statement E assigns a base address of s2 to s1; that is possible because s1 is a variable.

Point to Remember

When the string is declared as a character pointer, a space is allocated for the pointer variable, which holds the base address of the string.

STRINGS AS PARAMETERS

Introduction

The string can be passed to a function just as in a normal array. The following examples are used for printing the number of characters in the string:

Program

```
main ( )
{
```

```
    char s1[6] = "abcde ";
    int cnt = 0;
    cnt = cnt_str(s1);        \\ A
    printf( " total characters are %d \n", cnt);
}
int cnt_str(char s1[]);       \\ B
{
    int cn = 0;
   while ( (cn < 6) && s1[cn]! = '\0')
    cn++;
    return(cn);
}
```

Explanation

1. A function, cnt_str, calculates the number of characters in a string. The string is passed just as a character array. When the array is passed, the base address of the array is actually what gets passed.
2. Statement B is called to a function in which s1 is passed just as a normal array.

15 Structures

STRUCTURES

Introduction

Structures are used when you want to process data of multiple data types but you still want to refer to the data as a single entity. Structures are similar to records in Cobal or Pascal. For example, you might want to process information on students in the categories of name and marks (grade percentages). Here you can declare the structure 'student' with the fields 'name' and 'marks', and you can assign them appropriate data types. These fields are called members of the structure. A member of the structure is referred to in the form of structurename.membername.

Program

```
struct student       \\ A
{
   char name[30];    \\ B
   float marks;      \\ C
} student1, student2;       \\ D

main ( )
{
   struct student student3; \\ E
   char s1[30];             \\ F
   float  f;                \\ G
   scanf ("%s", name);      \\ H
   scanf (" %f", & f);      \\ I
   student1.name = s1;      \\ J
   student2.marks = f;      \\ K
   printf (" Name is %s \n", student1.name);     \\ L
```

```
    printf (" Marks are %f \n", student2.marks);   \\ M
}
```

Explanation

1. Statement A defines the structure type student. It has two members: name and marks.
2. Statement B defines the structure member name of the type character 30.
3. Statement C defines the structure member marks of the type float.
4. Statement D defines two structure variables: structure1 and structure2. In the program you have to use variables only. Thus struct student is the data type, just as int and student1 is the variable.
5. You can define another variable, student3, by using the notations as specified in statement E.
6. You can define two local variables by using statements F and G.
7. Statement J assigns s1 to a member of the structure. The structure member is referred to as structure variablename.membername. The member student1.name is just like an ordinary string, so all the operations on the string are allowed. Similarly, statement J assigns a value to student1.marks
8. Statement L prints the marks of student1 just as an ordinary string.

Points to Remember

1. Structures are used when you want to process data that may be of multiple data types.
2. Structures have members in the form: structurename.membername.

COMPLEX STRUCTURE DEFINITIONS

Introduction

You can define structures of arrays or arrays of structures, etc. The following section gives definitions of complex structures.

Program

```
Struct address        \\ A
{
   plot char [30], struc char[30];
   city char[30]
}
struct student        \\ B
{
   name char[30];
   marks float;
   struct address adr;       \\ C
}
main ( )
{
   struct student student1; \\ D
   struct student class[20];       \\ E
   class[1].marks = 70;      \\ F
   class[1].name = " Anil ";
   class[1].adr.plot = "7 ";          \\ G
   class[1].adr.street = " Mg Road";
   class[1].adr.city = "mumbai";

   printf( " Marks are %d\n", class[1].marks);
   printf( " name are %s\n", class[1].name);
   printf( " adr.plot is %s\n", class[1].adr.plot);
   printf( " adr.street is %s\n", class[1].adr.stret);
   printf( " adr.city is %s\n", class[1].adr.city);
}
```

Explanation

1. Statement A declares the address of a structure containing the members `plot`, `street` and `city`.
2. Statement B declares a structure having 3 members: `name`, `marks`, and `adr`. The data type of `adr` is structure address, which is given by statement C.
3. Statement D defines the variable `student1` of the data type `struct student`.
4. Statement E defines an array class with 20 elements. Each element is a structure.
5. You can refer to marks of the students of `class[1]` using the notation `class[1].marks`. `class[1]` indicates the first element of the array, and

since each element is a structure, a member can be accessed using dot notation.

6. You can refer to the plot of a student of `class[1]` using the notation `class[1].adr.plot`. Since the third element of the structure is adr, and `plot` is a member of adr, you can refer to members of the nested structures.
7. If you want to refer to the first character of the character array `plot`, then you can refer it as

   `Class[1].adr.plot[0]`

   because `plot` is a character array.

Points to Remember

1. When a structure is a member of another structure it is called a *nested structure*.
2. You can define structures of arrays or arrays of structures, and the members are referred to using dot notations.

MEMORY ALLOCATION TO STRUCTURE

Introduction

For each structure, variable memory is allocated. The following sections give the memory layout of the structure `student1`.

Program/Example

student1

	student1	0	name
30	marks		
34	adr		plot
64	street		
94			city

Explanation

1. Suppose the base address of the allocations is 0; then the first member name starts from 0.
2. Since name has 30 characters, the second member, marks, starts from location 30; marks occupies 4 bytes.
3. The third member, adr, starts from location 34, so the first member of adr starts from location 34. Period plot occupies 30 bytes, so street starts at 64.
4. city starts at 94.
5. You can print the addresses of the members using the following printf statements:
   ```
   printf( "16lu\n", &student1.marks);
   printf( "16lu\n", &student1.adr.plot);
   ```

Point to Remember

The structure members are allocated consecutive memory locations.

PROGRAMMING WITH STRUCTURES

Introduction

You can write programs with structures by using modular programming.

Program

```
struct student
{
    name char[30];
    marks float;
}
main ( )
{
    struct student student1;
    student1 = read_student ( )
    print_student( student1);
    read_student_p(student1);
```

```
    print_student (student1);
}
struct student read_student( )      \\ A
{
    struct student student2;
    gets(student2.name);
    scanf("%d",&student2.marks);
    return (student2);
}
void print_student (struct student student2)      \\ B
{
    printf( "name is %s\n", student2.name);
    printf( "marks are%d\n", student2.marks);
}
void read_student_p(struct student student2)      \\ C
{
    gets(student2.name);
    scanf("%d",&student2.marks);

}
```

Explanation

1. The function read_student reads values in structures and returns the structure.
2. The function print_student takes the structure variable as input and prints the content in the structure.
3. The function read_student_p reads the data in the structure similarly to read_student. It takes the structure student as an argument and puts the data in the structure. Since the data of a member of the structure is modified, you need not pass the structure as a pointer even though structure members are modified. Here you are not modifying the structure, but you are modifying the structure members through the structure.

Points to Remember

1. You can write a function that returns the structure. While writing the function, you should indicate the type of structure that is returned by the function. The return statement should return the structure using a variable.

2. You can pass a structure as an argument. You can modify a member of the structure by passing the structure of an argument. The changes in the member made by the function are retained in the called module. This is not against the principle of call by value because you are not modifying the structure variable, but are instead modifying the members of the structure.

STRUCTURE POINTERS

Introduction

You can process the structure using a *structure pointer*.

Program

```
struct student         \\ A
{
    char name[30];     \\ B
    float marks;       \\ C
};                \\ D

main ( )
{
    struct student *student1;         \\ E
    struct student student2; \\ F
    char s1[30];
    float  f;
    student1 = &student2;      \\ G
    scanf (“%s”, name);        \\ H
    scanf (“ %f”, & f);        \\ I
    *student1.name = s1;       \\ J  student1-> name = f;
    *student2.marks = f;       \\ K  student1-> marks = s1;

    printf (“ Name is %s \n”, *student1.name);     \\ L
    printf (“ Marks are %f \n”, *student2.marks); \\ M
}
```

Explanation

1. Statement E indicates that student1 is the pointer to the structure.

2. Statement F defines the structure variable student2 so that memory is allocated to the structure.
3. Statement G assigns the address of the structure student2 to the pointer variable structure student1.
4. In the absence of statement G, you cannot refer to the structure using a pointer. This is because when you define the pointer to the structure, the memory allocation is done only for pointers; the memory is not allocated for structure. That is the reason you have to declare a variable of the structure type so that memory is allocated to the structure and the address of the variable is given to the point.
5. Statement J modifies a member of the structure using the * notation. The alternative notation is

```
student1-> name = f;
student1-> marks = s1;
```

Points to Remember

1. You can access members of the structure using a pointer.
2. To access members of the structure, you have to first create a structure so that the address of the structure is assigned to the pointer.

16 Union

UNION

Introduction

Union is a composite type similar to structure. Even though it has members of different data types, it can hold data of only one member at a time.

Program

```
union marks           \\ A
{
   float perc;        \\ B
   char grade;        \\ C
}
main ( )
{
   union marks student1;    \\ E
   student1.perc = 98.5;    \\ F
   printf( "Marks are %f   address is  %16lu\n",
      student1.perc, &student1.perc);  \\ G
   student1.grade = 'A'';   \\ H
   printf( "Grade is  %c  address is  %16lu\n",
      student1.grade, &student1.grade);  \\ I
}
```

Explanation

1. Statement A declares a union of the type marks. It has two members: perc and grade. These two members are of different data types but they are allocated the same storage. The storage allocated by the union variable is equal to the maximum size of the members. In this case, the member grade

occupies 1 byte, while the member perc occupies 4 bytes, so the allocation is 4 bytes. The data is interpreted in bytes depending on which member you are accessing.

2. Statement E declares the variable `student1` of the type union.
3. Statement F assigns a value to a member of the union. In this case, the data is interpreted as the `float` data type.
4. Statement H assigns character 'A' to member `grade`. `student1.grade` interprets the data as character data.
5. When you print the value of the member `perc`, you have to use the placeholder `%type`. Note that the addresses printed by both `printf` statements are the same. This means that both members have the same memory location.

Points to Remember

1. In a union, the different members share the same memory location.
2. The total memory allocated to the union is equal to the maximum size of the member.
3. Since multiple members of different data types have the same location, the data is interpreted according to the type of the member.

17 Files

THE CONCEPT OF FILES

Introduction

A *file* is a data object whose lifetime may be greater than the lifetime of a program responsible for creating it, because it is created on secondary storage devices. It is used to store persistent data values and information. The files are used mainly for input and output of data to an external operating environment. The components of the file are called as records (this term has nothing to do with record data structure).

Types of Files

A file may be a *sequential file*, a *direct-access file*, or an *indexed sequential file*. A sequential file can be thought of as a linear sequence of components of the same type with no fixed maximum bound. The major operations on the sequential files are:

Open operation: When a file is to be used, it is first required to be opened. The open operation requires two operands: the name of the file and the access mode telling whether the file is to be opened for reading or writing. If the access mode is "read," then the file must exist. If the access mode is "write," then if the file already exists, that file is emptied and the file position pointer is set to the start of the file. If the file does not exist then the operating system is requested to create a new empty file with a given name. The open operation requests the information about the locations and other properties of the file from the operating system. The operating system allocates the storage for this information and for buffers, and sets the file-position pointer to the first component of the file. The runtime library of C provides an `fopen(name,mode)` function for it. This function returns a pointer to the internal structure called FILE (you get the definition of this structure in stdio.h). This pointer is called a *file descriptor*; it is used by the C program to refer to the file for reading or writing purposes.

Read operation: This operation transfers the current file component to the designated program variable. The runtime library of C provides a function fgetc(fp), where fp is a file descriptor, for fscanf(). fscanf() is similar to scanf() except that one extra parameter, fp, is required to be passed as the first parameter. The second and third parameters are the same as the first and second parameters of scanf().

Write operation: This operation transfers the contents of the designated program variable to the new component created at the current position. The runtime library of C provides a function fputc(c, fp), where fp is a file descriptor, and c is a character to be written in the file fprintf(). fprintf() is similar to printf() except that one extra parameter, fp, is required to be passed as the first parameter. The second and third parameters are the same as the first and second parameters of printf().

Close operation: This operation notifies the operating system that the file can be detached from the program and that it can deallocate the internal storage used for the file. The file generally gets closed implicitly when the program terminates without explicit action by the programmer. But when the access mode is required to be changed it is required to be closed explicitly and reopened in the new mode. The runtime library of C provides an fclose(fp) function for it.

Random Access

Each read and write operation takes place at a position in the file right after the previous one. But it is possible that you may need to read or write the file in any arbitrary order. The runtime library of C provides an fseek(fp, offset, from_where) function for this. This function forces the current position in the file, whose descriptor is fd, to move by offset bytes from either the beginning of the file, the current file pointer position, or from the end of the file, depending upon the value of from_where. The parameter from_where must have one of the values (0, 1, or 2) that represent three symbolic constants (defined in stdio.h) as shown in Table 17.1.

TABLE17.1 Random access

CONSTANT	WHERE	FILE LOCATION
SEEK_SET	0	File beginning
SEEK_CUR	1	Current file pointer position
SEEK_END	2	End-of-file

After fseek, the next operation on an update file can be either input or output.

Program

This program is designed to handle data such at rollno, name and marks of a student. In this program, the following operations are performed:

1. Information on a new student is entered and stored in the student.txt file.
2. The student.txt file is printed on screen on the operator's request.
3. The student.txt file is sorted on the basis of marks and stored in the file marks.txt.
4. Information on a student whose rollno is given is printed on screen.
5. The average marks of all students are calculated.

```
#include<stdio.h>

 int bubble(int*,int);
 void filewrite();
 void avgmarks();
 void fileprint();
 void filesort();
 void rollin();

/******************** SORTING FUNCTION ************************/
int bubble(int x[],int n)
{
    int hold,j,pass,i,switched = 1;
    for(pass = 0; pass < n-1 && switched == 1;pass++)
    {
        switched=0;
        for (j=0;j<n-pass-1;j++)
              if (x[j]>x[j+1])
```

```
                {
                        switched=1;
                        hold = x[j];
                        x[j] = x[j+1];
                        x[j+1]=hold;
                }
    }
    return(0);
}
/*************** FILE WRITING FUNCTION **********************/

void filewrite()
{
    int roll,ch,mark;
    char nam[50];
    FILE *fp;
    clrscr();
    fp = fopen("student.txt","a");
    printf("ENTER ROLL NUMBER, NAME , MARKS \n");
    ch =1;
    while(ch)
    {
        scanf("%d%s%d",&roll,&nam,&mark);
        fprintf(fp,"%d %s %d\n",roll,nam,mark);
        printf("\n\n press 1 to continue,0 to stop");
        scanf("%d",&ch);
    }
    fclose(fp) ;
}
/******************* OUTPUTTING DATA ON SCREEN***************/
void fileprint()
{
    int marks[100],rollno[100],x[100],i;
    char name[100][50];
    FILE *fp;

    clrscr();
    fp = fopen("student.txt","r");
    i=0;
    printf("ROLLNO      NAME        MARK\n");
    while(!feof(fp))
    {
```

```
            fscanf(fp,"%d %s %d\n",&rollno[i],&name[i],&marks[i]);
            printf(" %d          %s
%d\n",rollno[i],name[i],marks[i]);
            i=i+1;
        }
        fclose(fp);
        printf("\n\n\nPRESS ANY KEY");
        getch();

    }
    /****************** SORTING FILE **********************/
    void filesort()
    {
         int marks[100],rollno[100],x[100],n,i,j;
        char name[100][50];
        FILE *fp,*fm;

        fp = fopen("student.txt","r");
        fm = fopen("marks.txt","w");
        i=0;
        while(! feof(fp))
        {

            fscanf(fp,"%d %s %d\n",&rollno[i],&name[i],&marks[i]);
            x[i]= marks[i];
            i=i+1;
        }

        n=i;

        bubble(x,n);

        for(i=0;i<n;i++)
        {
            printf(" %d\t",x[i]);
        }

        for(i=0;i<n;i++)
        {
            for (j=0;j<n;j++)
            {
                    if(x[i]==marks[j])
```

```
                {
                        fprintf(fm,"%d %s
%d\n",rollno[j],name[j],marks[j]);
                }
            }
        }
        fclose(fm);
        fclose(fp);
        printf("\n\n\nPRESS ANY KEY");
        getch();

    }
    /******************** DATA USING ROLLNO************************/

    void rollin()
    {
        int i,roll,ch,mark,roll1;
        char nam[50];
        FILE *fm;

        ch=1;
        while(ch)
        {
            clrscr();
            fm = fopen("marks.txt","r");
            printf(" \n ENTER ROLL NUMBER - ");
            scanf("%d",&roll1);
            i=0;
            while(! feof(fm))
            {
                    fscanf(fm,"%d %s %d\n",&roll,&nam,&mark);
                    if(roll1==roll)
                    {
                        printf("\nROLLNO.    NAME        MARKS\n ");
                            printf(" %d           %s
%d\n",roll,nam,mark);
                            break;
                    }
                    else
                            i=i+1;
            }
            printf(
```

```
        "\n\npress 1 to see student info, 0 to return to main menu\n");
        scanf("%d",&ch);
        fclose(fm);
    }

}

void avgmarks()
{
    int marks[100],rollno[100],n,i;
    float avg,x;
    char name[100][50];
    FILE *fm;
    fm = fopen("marks.txt","r");
    i=0;
    while(! feof(fm))
    {

        fscanf(fm,"%d %s %d\n",&rollno[i],&name[i],&marks[i]);
        x = x + marks[i];
        i=i+1;
    }
    n = i;
    avg = x/n;
    printf("AVERAGE MARKS OF %d STUDENTS ARE -  %f ",n,avg);
    fclose(fm);
    printf("\n\n\nPRESS ANY KEY");
    getch();
}

/**************** FUNC. ENDS************************/
void main()
{
    int marks[100],rollno[100],x[100],n,i,j,roll,c,mark,roll1;
    char name[100][10],nam[50];

    while(c!=6)
    {
        clrscr();
        printf("GIVE CHOICE-\n");
        printf("   1 TO ENTER STUDENT INFO.\n");
        printf("   2 TO SEE STUDENT.TXT FILE\n");
```

```
        printf("   3 TO SORT FILE ON BASIS OF MARKS\n");
        printf("   4 TO PRINT STUDENT INFO. USING ROLL NO\n");
        printf("   5 TO FIND AVERAGE OF MARKS\n");
        printf("   6 TO EXIT\n\n-");
        scanf("%d",&c);
        clrscr();
        switch(c)
        {
        case 1:
                filewrite();
                break;
        case 2:
                fileprint();
                break;
        case 3:
                filesort();
                break;
        case 4:  rollin();
                break;
        case 5:  avgmarks();
                break;
        case 6:
                break;
        default:
                break;
        }
    }

}
```

Explanation

1. This program uses the following functions for its specified operation.

`int bubble(int*,int)` –	This bubble sorting technique is used for file sorting.
`void filewrite()` –	Used to write data of a new student in "student.txt" file
`void fileprint()` –	Used to print information on students.
`void filesort()` –	Used to sort the "student.txt" files on mark basis in "marks.txt"

void rollin() –	Used to find information on a student using his roll number.
void avgmarks() –	Used to find average marks of all students.

2. The filewrite() function opens the student.txt file in the append mode, and data entered is written in the same file. In the void fileprint() file, student.txt is opened in read mode and data is read from it. This data is printed on the screen.
3. filesort() opens the student.txt file in read mode and the file marks.txt in write mode. The data of all students is temporarily stored in one buffer consisting of three arrays: one for rollno, the second for name and the third for marks. At the same time, marks are stored in the x[ ] array for sorting purposes. The sorting is done by bubble sort. The result of sorting is available in x[ ]. At this stage, each x[i] is compared with marks [j]. If a match is found, the data on that student is stored in the marks.txt file. This process is done for all marks.
4. In this way, we get a marks.txt file that is sorted on the basis of marks. In the void rollin(), the file marks.txt is used to find the student whose roll number (rollno) is given. For every line in the file, the rollno in that file is compared with the rollno to be found. If a match exists in the file, the data on that student is printed on the screen. The avgmarks() function uses the file marks.txt. Marks of students are added to variable X, each time the file pointer is incremented. Then, the average marks (sre) is displayed on the screen.
5. In main function, the switch statement is used to invoke the function related to the option given by the user.
6. In this program, input consists of the rollno, marks, and name of each student.
7. Output depends on the user's choice. When information on students is to be printed, the program prints the content of the student.txt file on the screen. When information is sought on the basis of the roll number, the program prints the rollno, marks, and name of each student. When the average of marks is found, it prints that.

DIRECT ACCESS FILES

Introduction

A *direct access file* is a file in which any single component may be accessed at random. Every component has a key value associated with it. A write operation takes a component and its key value, writes the component into the file, and stores both the key and the location of the record in a file, an index. A read operation takes the key of the desired component, searches the index to find the location of the component, and retrieves the component from the file.

Program

A complete C program implementing a direct access file is given below:

```
#include <stdio.h>
#include <string.h>
#include <stdlib.h>
#define MAX 50

typedef struct
{
    char  name[10];
    int key;
}   file_record;
/* this function adds the relative address to the index for a key */
void create_index(long index[], int key, long rel_add )
{
    index[key] = rel_add;
}

/* this function writes a record to the file */
void write_rec(FILE *fp, file_record rec)
{
    fwrite(&rec,sizeof(rec),1,fp);
}

void main()
{
    long rel_add;
    int key;
```

```
        file_record frec;
        long index[MAX];/* an index list*/
        int n,i;

        FILE *recfile=NULL,*ifile=NULL;
        /* this initializes the index list to all -1 */
        for(i=0; i< MAX; i++)
           index[i]= (-1);

        recfile=fopen("mfile","w");
        if(recfile == NULL)
        {
           printf("Error in opening file mfile\n");
           exit(0);
        }
        rel_add = 0 ;
        do
        {
           printf(
    " Enter the data value and the key of the record to be added to file
mfile\n");
           scanf("%s %d",frec.name,&frec.key);
           while(index[frec.key] != (-1))
           {
                  printf(
    " A record with this key value already exist in a file enter record key
value\n");
                  scanf("%s %d",frec.name,&frec.key);
           }
           create_index(index,frec.key,rel_add);
           write_rec(recfile,frec);
           rel_add =  ftell(recfile);
           /* this sets the relative address for the next record to be
          the value of current file position pointer in bytes from
          the beginning of the file */
           printf("Enter 1 to continue adding records to the file\n");
           scanf("%d",&n);
        }while(n == 1);
        ifile=fopen("index_file","w");
        if(ifile == NULL)
        {
           printf("Error in opening file index_file\n");
```

```
            exit(0);
        }
        fwrite(index,sizeof(index),1,ifile); /*writes the complete index
into the index_file */
        fclose(recfile);
        fclose(ifile);
        printf("Enter 1 if you want to retrieve a record\n");
        scanf("%d",&n);
        if( n == 1)
        {
            ifile=fopen("index_file","r");
                   if(ifile == NULL)
                   {
                          printf("Error in opening file index_file\n");
                          exit(0);
                   }
                   fread(index,sizeof(index),1,ifile);
                   /* reads the complete index into the index list from the
            index_file*/
                   fclose(ifile);
              recfile=fopen("mfile","r");
              if(recfile == NULL)
                   {
                          printf("Error in opening file mfile\n");
                          exit(0);
                   }
        }
        printf("THE CONTENTS OF FILE IS \n");
        while( (fread(&frec,sizeof(frec),1,recfile)) != 0)
        printf("%s %d\n",frec.name,frec.key);
        do
        {
            printf("Enter the key of the record to be retrieved\n");
            scanf("%d",&key);
            rel_add = index[key]; /*gets the relative address of the record
from index list */
            if( (fseek(recfile,rel_add,SEEK_SET))!= 0)
            {
                   printf("Error\n");
                   exit(0);
            }
            fread(&frec,sizeof(frec),1,recfile);
```

```
        printf("The data value of the retrieved record is %s\n",
        frec.name);
        printf("Enter 1 if you want to retrieve a record\n");
        scanf("%d",&n);
    } while(n == 1);
    fclose(recfile);
}
```

Explanation

1. This program writes the names in the file. A unique integer value is assigned to every name as a key value.
2. The program takes the name to be stored in the file along with its key value, writes the name and key value in the file, obtains the relative address of that record, and stores it in a list called an index. The index is organized by key value.
3. When the addition process ends, it writes the complete index into an index file.
4. When retrieval of names is requested, the following occurs:
   A. The complete index is loaded into a list index from the index file.
   B. The index file is used to find the relative address of the record whose key value is given.
   C. The current position pointer is moved to that address.
   D. The record is read from the file.

Example

Input and Output

1.

Enter the data value and the key of the record to be added to file `mfile`

`logk 10`

Enter 1 to continue adding records to the file

1

Enter the data value and the key of the record to be added to file `mfile`

`psd 20`

```
Enter 1 to continue adding records to the file
1
Enter the data value and the key of the record to be added to file mfile
apg 3
Enter 1 to continue adding records to the file
1
Enter the data value and the key of the record to be added to file mfile
agk 5
Enter 1 to continue adding records to the file
1
Enter the data value and the key of the record to be added to file mfile
kdk 34
Enter 1 to continue adding records to the file
0
Enter 1 if you want to retrieve a record
1
THE CONTENTS OF FILE IS
logk 10
psd 20
apg 3
agk 5
kdk 34
Enter the key of the record to be retrieved
5
The data value of the retrieved record is agk
Enter 1 if you want to retrieve a record
1
Enter the key of the record to be retrieved
10
The data value of the retrieved record is logk
Enter 1 if you want to retrieve a record
1
```

Enter the key of the record to be retrieved
34
The data value of the retrieved record is kdk
Enter 1 if you want to retrieve a record
1
Enter the key of the record to be retrieved
20
The data value of the retrieved record is psd
Enter 1 if you want to retrieve a record
1
Enter the key of the record to be retrieved
10
The data value of the retrieved record is logk
Enter 1 if you want to retrieve a record
0

2.

Enter the data value and the key of the record to be added to file mfile
logk 10
Enter 1 to continue adding records to the file
1
Enter the data value and the key of the record to be added to file mfile
psd 20
Enter 1 to continue adding records to the file
1
Enter the data value and the key of the record to be added to file mfile
kdk 20
A record with this key value already exist in a file enter record key value
kdk 30
Enter 1 to continue adding records to the file
1
Enter the data value and the key of the record to be added to file mfile
asg 10

A record with this key value already exists in a file, enter record key value

asg 15

Enter 1 to continue adding records to the file

0

Enter 1 if you want to retrieve a record

1

THE CONTENTS OF FILE IS

```
logk 10
psd 20
kdk 30
asg 15
```

Enter the key of the record to be retrieved

20

The data value of the retrieved record is psd

Enter 1 if you want to retrieve a record

1

Enter the key of the record to be retrieved

15

The data value of the retrieved record is asg

Enter 1 if you want to retrieve a record

1

Enter the key of the record to be retrieved

10

The data value of the retrieved record is logk

Enter 1 if you want to retrieve a record

1

Enter the key of the record to be retrieved

30

The data value of the retrieved record is kdk

Enter 1 if you want to retrieve a record

0

Indexed Sequential Files

An indexed sequential file is like a direct access file with the additional facility of accessing the components in a sequential manner, beginning from the position of the component selected at random. This requires the index to be ordered by key values.

Exercises

1. Write a C program to implement a telephone directory. Your program should provide for retrieval of an arbitrary record, given the name of the telephone subscriber.
2. Write a C program to implement a database of people. Your program should provide for retrieval of information from any arbitrary record, given the code number of the person.

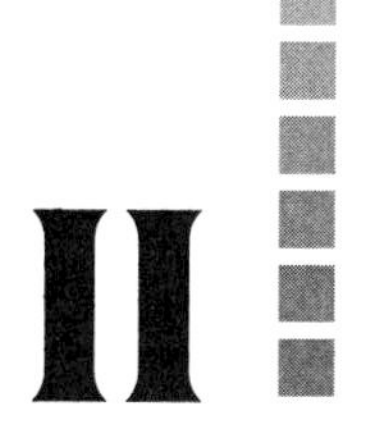

Data Structures

18 Arrays, Searching, and Sorting

ARRAYS

Introduction

An *array* is a fixed-sized, homogeneous, and widely-used data structure. By homogeneous, we mean that it consists of components which are all of the same type, called *element type* or *base type*. And by fixed sized, we mean that the number of components is constant, and so does not change during the lifetime of the structure. An array is also called a *random-access data structure*, because all components can be selected at random and are equally accessible. An array can be used to structure several data objects in the programming languages. A component of an array is selected by giving its *subscript*, which is an integer indicating the position of the component in the sequence. Therefore, an array is made of the pairs (value, index); it means that with every index, a value is associated. If every index is one single value then it is called a one-dimensional array, whereas if every index is a n-tuple $\{i_1,i_2,i_3,\ldots,in\}$, the array is called a n-dimensional array.

Memory Representation

An array is represented in memory by using a sequential mapping. The basic characteristic of the sequential mapping is that every element is at a fixed distance apart. Therefore, if the ith element is mapped into a location having an address a, then the $(i+1)^{th}$ element is mapped into the memory location having an address (a + 1), as shown in Figure 18.1.

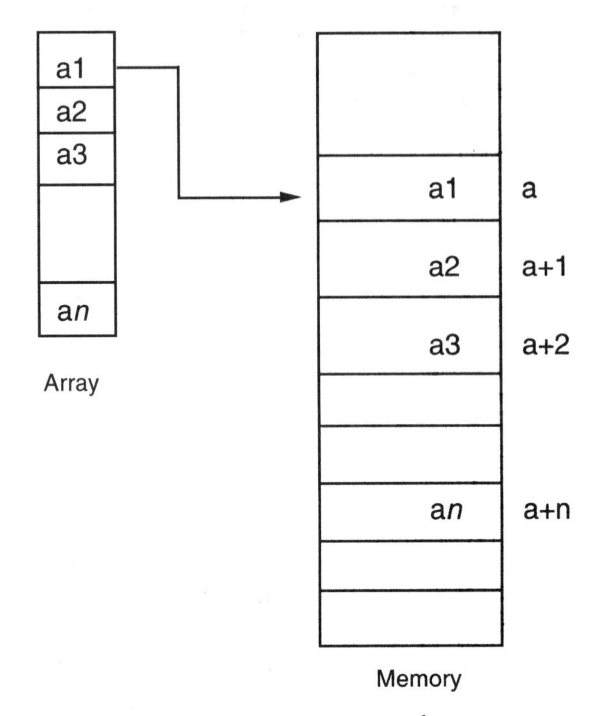

FIGURE 18.1 Representation of an array.

The address of the first element of an array is called the base address, so the address of the the ith element is

Base address + offset of the ith element from base address

where the offset is computed as:

offset of the ith element = number of elements before the ith element * size of each element.

If LB is the lower bound, then the offset computation becomes:

offset = (i – LB) * size.

Representation of Two-Dimensional Array

A two-dimensional array can be considered as a one-dimensional array whose elements are also one-dimensional arrays. So, we can view a two dimensional array as one single column of rows and map it sequentially as shown in Figure 18.2. Such a representation is called a *row-major representation*.

The address of the element of the ith row and the jth column therefore is:

addr(a[i,j]) = (number of rows placed before i^{th} row * size of a row) + (number of elements placed before the jth element in the ith row * size of element)

where

Number of rows placed before i^{th} row = (i – LB1), and LB1 is the lower bound of the first dimension.

Size of a row = number of elements in a row * a size of element.

Number of elements in a row = (UB2 – LB2+1) , where UB2 and LB2 are the upper and lower bounds of the second dimension, respectively.

Therefore:

addr(a[i,j]) =((i – LB1) * (UB2 – LB2+1) * size) + ((j – LB2)*size)

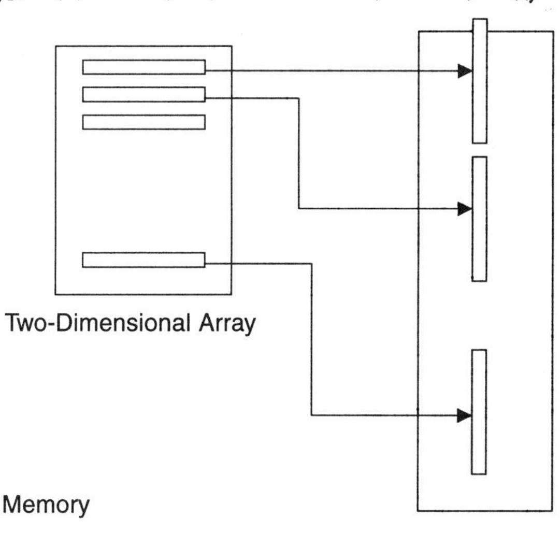

FIGURE 18.2 Row-major representation of a two-dimensional array.

It is also possible to view a two-dimensional array as one single row of columns and map it sequentially as shown in Figure 18.3. Such a representation is called a *column-major representation*.

The address of the element of the ith row and the jth column therefore is:

addr(a[i,j]) = (number of columns placed before j^{th} column * size of a column) + (number of elements placed before the ith element in the jth column * size of each element)

Number of columns placed before j^{th} column = (j – LB2) where LB2 is the lower bound of the second dimension.

Size of a column = number of elements in a column * size of element

Number of elements in a column = (UB1 – LB1 + 1), where UB1 and LB1 are the upper and lower bounds of the first dimension, respectively.

Therefore:

addr(a[i,j]) =((j – LB2) * (UB1 – LB1 + 1) * size) + ((i – LB1)*size)

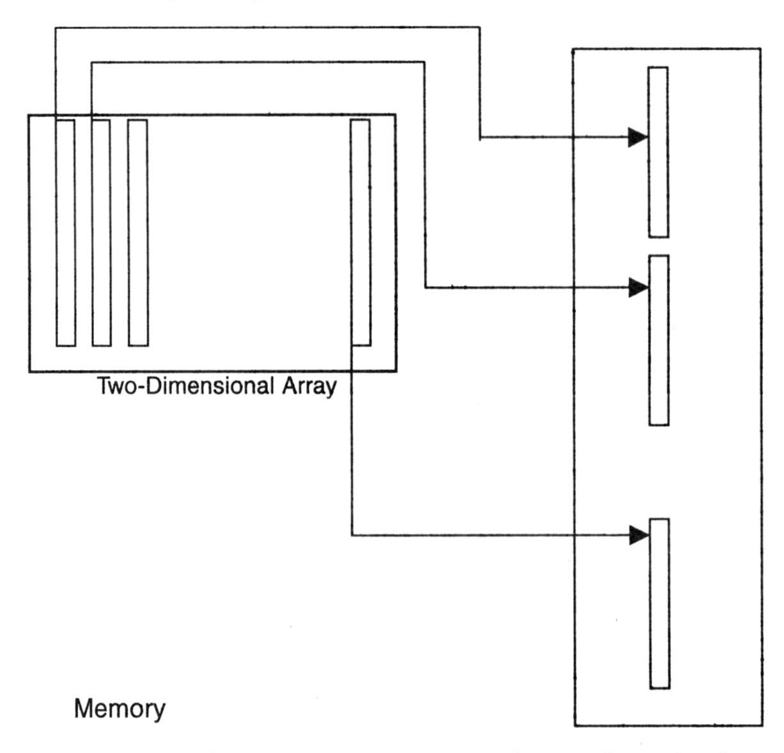

FIGURE 18.3 Column major representation of a two-dimensional array.

APPLICATION OF ARRAYS

Whenever we require a collection of data objects of the same type and want to process them as a single unit, an array can be used, provided the number of data items is constant or fixed. Arrays have a wide range of applications ranging from business data processing to scientific calculations to industrial projects.

Implementation of a Static Contiguous List

A *list* is a structure in which insertions, deletions, and retrieval may occur at any position in the list. Therefore, when the list is static, it can be implemented by using an array. When a list is implemented or realized by using an array, it is a *contiguous list*. By contiguous, we mean that the elements are placed consecutively one after another starting from some address, called the *base address*. The advantage of a list implemented using an array is that it is randomly accessible. The disadvantage of such a list is that insertions and deletions require moving of the entries, and so it is costlier. A static list can be implemented using an array by mapping the ith element of the list into the ith entry of the array, as shown in Figure 18.4.

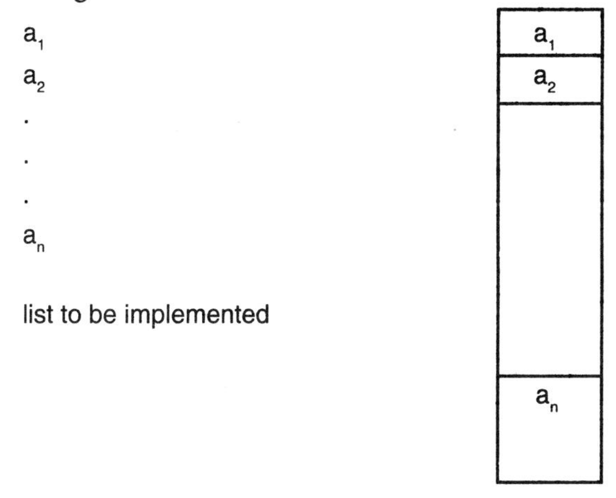

FIGURE 18.4 Implementation of a static contiguous list.

Program

A complete C program for implementing a list with operations for reading values of the elements of the list and displaying them is given here:

```
#include<stdio.h>
#include<conio.h>
void main()
{
    void read(int *,int);
    void dis(int *,int);
```

```
    int a[5],i,sum=0;

    clrscr();
    printf("Enter the elements of array \n");
    read(a,5);       /*read the array*/
    printf("The array elements are  \n");
    dis(a,5);
}

void read(int c[],int i)
{
    int j;
   for(j=0;j<i;j++)
    scanf("%d",&c[j]);
   fflush(stdin);
}

void dis(int d[],int i)
{
    int j;
    for(j=0;j<i;j++)
        printf("%d  ",d[j]);
    printf("\n");
}
```

Example

Input

Enter the elements of the first array

15

30

45

60

75

Output

The elements of the first array are

15 30 45 60 75

MANIPULATIONS ON THE LIST IMPLEMENTED USING AN ARRAY

Introduction

Shown next are C programs for carrying out manipulations such as finding the sum of elements of an array, adding two arrays, and reversing an array.

Program

ADDITION OF THE ELEMENTS OF THE LIST

```
#include<stdio.h>
#include<conio.h>
void main()
{
    void read(int *,int);
    void dis(int *,int);
    int a[5],i,sum=0;

    clrscr();
    printf("Enter the elements of list \n");
    read(a,5);      /*read the list*/
    printf("The list elements are  \n");
    dis(a,5);
    for(i=0;i<5;i++)
    {
        sum+=a[i];
   }
    printf("The sum of the elements of the list  is %d\n",sum);
    getch();
}

void read(int c[],int i)
{
    int j;
   for(j=0;j<i;j++)
    scanf("%d",&c[j]);
   fflush(stdin);
}

void dis(int d[],int i)
```

```
{
    int j;
   for(j=0;j<i;j++)
    printf("%d  ",d[j]);
    printf("\n");
}
```

Example

Input

Enter the elements of the first array

15

30

45

60

75

Output

The elements of the first array are

15 30 45 60 75

The sum of the elements of an array is 225.

Addition of the two lists

Suppose the first list is

1

2

3

4

5

and the second list is

5

6

8

9

10

The first element of first list is added to the first element of the second list, and the result of the addition is the first element of the third list.

In this example, 5 is added to 1, and the first element of third list is 6.

This step is repeated for all the elements of the lists and the resultant list after the addition is

6

8

11

13

15

```
#include<stdio.h>
   #include<conio.h>
   void main()
    {
     void read(int *,int);
     void dis(int *,int);
     void add(int * ,int *,int * ,int);
     int a[5],b[5],c[5],i;

     clrscr();
     printf("Enter the elements of first list \n");
     read(a,5);       /*read the  first list*/
     printf("The elements of first list are  \n");
     dis(a,5);  /*Display the first  list*/
     printf("Enter the elements of second list \n");
     read(b,5);       /*read the second list*/
     printf("The  elements of second list are  \n");
     dis(b,5);   /*Display the second  list*/
     add(a,b,c,i);
     printf("The resultant list  is \n");
     dis(c,5);
     getch();
    }

    void add(int a[],int b[],int c[],int i)
   {
    for(i=0;i<5;i++)
      {
       c[i]=a[i]+b[i];
      }
```

```
}
 void read(int c[],int i)
  {
   int j;
   for(j=0;j<i;j++)
    scanf("%d",&c[j]);
   fflush(stdin);
  }

  void dis(int d[],int i)
  {
   int j;
   for(j=0;j<i;j++)
    printf("%d  ",d[j]);
    printf("\n");
  }
```

Explanation

1. Repeat step (2) for `i=0,1,2,... (n-1)`, where `n` is the maximum number of elements in a list.
2. `c[i] = a[i]+b[i]`, where `a` is the first list, `b` is the second list, and `c` is the resultant list; `a[i]` denotes the ith element of list `a`.

Example

Input

Enter the elements of the first list

1

2

3

4

5

Output

The elements of the first list are

2 3 4 5

Input

Enter the elements of the second list

6

7

8

9

10

Output

The elements of the second list are

6 7 8 9 10

The resultant list is

7 9 11 13 15

Inverse of the list

The following program makes a reverse version of the list.

```
#include<stdio.h>
#include<conio.h>
void main()
{
    void read(int *,int);
    void  dis(int *,int);
    void inverse(int *,int);

    int a[5],i;
    clrscr();
    read(a,5);
    dis(a,5);
    inverse(a,5);
    dis(a,5);
    getch();
}

void read(int c[],int i)
{
    int j;
    printf("Enter the list \n");
   for(j=0;j<i;j++)
```

```
        scanf("%d",&c[j]);
    fflush(stdin);
}
void dis(int d[],int i)
{
    int j;
   printf("The list is \n");
   for(j=0;j<i;j++)
    printf("%d  ",d[j]);
    printf("\n");
}
void inverse(int inver_a[],int j)
{
    int i,temp;
    j--;
    for(i=0;i<(j/2);i++)
   {
        temp=inver_a[i];
        inver_a[i]=inver_a[j];
        inver_a[j]=temp;
        j--;
    }
}
```

Example

Input

Enter the list

10

20

30

40

50

Output

The list is

10 20 30 40 50

The inverse of the list is

50 40 30 20 10

This is another version of an inverse program, in which another list is used to hold the reversed list.

```
#include<stdio.h>
#include<conio.h>
void main()
{
    void read(int *,int);
    void  dis(int *,int);
    void  inverse(int *,int *,int);
    int a[5],b[5];
    clrscr();
    read(a,5);
    dis(a,5);
    inverse(a,b,5);
    dis(b,5);
    getch();
}

void read(int c[],int i)
{
    int j;
    printf("Enter the list \n");
   for(j=0;j<i;j++)
       scanf("%d",&c[j]);
   fflush(stdin);
}
void dis(int d[],int i)
{
    int j;
   printf("The list is \n");
   for(j=0;j<i;j++)
       printf("%d  ",d[j]);
    printf("\n");
}
void inverse(int a[],int inverse_b[],int j)
{
    int i,k;
    k=j-1;
    for(i=0;i<j;i++)
    {
       inverse_b[i]=a[k];
       k–;
    }
}
```

Example

Input

Enter the list

10

20

30

40

50

Output

The list is

10 20 30 40 50

The inverse of the list is

50 40 30 20 10

MERGING OF TWO SORTED LISTS

Introduction

Assume that two lists to be merged are sorted in descending order. Compare the first element of the first list with the first element of the second list. If the element of the first list is greater, then place it in the resultant list. Advance the index of the first list and the index of the resultant list so that they will point to the next term. If the element of the first list is smaller, place the element of the second list in the resultant list. Advance the index of the second list and the index of the resultant list so that they will point to the next term.

Repeat this process until all the elements of either the first list or the second list are compared. If some elements remain to be compared in the first list or in the second list, place those elements in the resultant list and advance the corresponding index of that list and the index of the resultant list.

Suppose the first list is 10 20 25 50 63, and the second list is 12 16 62 68 80. The sorted lists are 63 50 25 20 10 and 80 68 62 16 12.

The first element of the first list is 63, which is smaller than 80, so the first element of the resultant list is 80. Now, 63 is compared with 68; again it is smaller, so the second element in the resultant list is 68. Next, 63 is compared with 50. In this case it is greater, so the third element of the resultant list is 63.

Repeat this process for all the elements of the first list and the second list. The resultant list is 80 68 63 62 50 25 20 16 12 10.

Program

```
#include<stdio.h>
#include<conio.h>
void main()
{
    void read(int *,int);
    void dis(int *,int);
    void sort(int *,int);
    void merge(int *,int *,int *,int);
    int a[5],b[5],c[10];
    clrscr();
    printf("Enter the elements of first list \n");
    read(a,5);       /*read the list*/
    printf("The elements of first list are  \n");
    dis(a,5);  /*Display the first list*/
    printf("Enter the elements of second list \n");
    read(b,5);       /*read the list*/
    printf("The elements of second list are  \n");
    dis(b,5);  /*Display the second list*/
    sort(a,5);
    printf("The sorted list a is:\n");
    dis(a,5);
    sort(b,5);
    printf("The sorted list b is:\n");
    dis(b,5);

    merge(a,b,c,5);
    printf("The elements of merged list are  \n");
    dis(c,10);  /*Display the merged list*/
    getch();
}
void read(int c[],int i)
{
    int j;
   for(j=0;j<i;j++)
```

```
        scanf("%d",&c[j]);
    fflush(stdin);
}

void dis(int d[],int i)
{
    int j;
    for(j=0;j<i;j++)
        printf("%d  ",d[j]);
    printf("\n");
}
void sort(int arr[] ,int k)
{
    int temp;
    int i,j;
    for(i=0;i<k;i++)
    {
        for(j=0;j<k-i-1;j++)
        {
                if(arr[j]<arr[j+1])
                {
                        temp=arr[j];
                        arr[j]=arr[j+1];
                        arr[j+1]=temp;
                }
        }
    }
}
void merge(int a[],int b[],int c[],int k)
{
    int ptra=0,ptrb=0,ptrc=0;
    while(ptra<k && ptrb<k)
    {
        if(a[ptra] < b[ptrb])
        {
                c[ptrc]=b[ptrb];
            ptrb++;
        }
        else
        {
                c[ptrc]=a[ptra];
            ptra++;
```

```
        }
        ptrc++;
    }
    while(ptra<k)
    {
        c[ptrc]=a[ptra];
        ptra++;ptrc++;
    }
    while(ptrb<k)
    {
        c[ptrc]=b[ptrb];
        ptrb++;  ptrc++;

    }
}
```

Example

Input

Enter the elements of the first list

10 20 25 50 63

Output

The elements of first list are

20 25 50 63

Input

Enter the elements of the second list

16 62 68 80

Output

The elements of second list are

12 16 62 68 80

The sorted list a is

63 50 25 20 10

The sorted list b is

80 68 62 16 12

The elements of the merged list are

80 68 63 62 50 25 20 16 12 10

Explanation

(1) `ptra=0, ptrb=0, ptrc=0;`

(2) If the element in the first list pointed to by `ptra` is greater than the element in the second list pointed to by `ptrb`, place the element of the first list in the resultant list at the index equal to `ptrc`. Increment `ptra` or `ptrc` by one, or else place the element of the second list in the resultant list at the index equal to `ptrc`. Increment `ptrb` and `ptrc` by 1. Repeat this step until `ptra` is greater than the number of terms in the first list and `ptrb` is greater than the number of terms in the second list.

(3) If the first list has any elements, place one in the resultant list pointed to by `ptrc`, and increment `ptra` and `ptrc`. Repeat this step until `ptra` is greater than the number of terms in the first list.

(4) If the second list has any elements, place one in the resultant list pointed to by `ptrc`, and increment `ptrb` and `ptrc`. Repeat this step until `ptrb` is greater than the number of terms in the first list.

TRANSPOSE OF A MATRIX

Introduction

The *transpose* of a matrix is obtained by interchanging the rows with the corresponding columns. Let matrix a be

12 13 14
15 16 17
18 19 11

The diagonal elements are the same both in matrix a and in the matrix obtained by transposing a. In this example, in the 0th row, interchange 13 with 15 and 14 with 18. After interchanging, the matrix becomes

12 15 18
13 16 17
14 19 11

In the first row, interchange the element that has not yet been interchanged in the 0th row; 17 with 19. After interchanging the elements, the matrix becomes:

12 15 18

13 16 19

11 17 11

In the next iteration, search for the nondiagonal un-swapped element. In this example, no such element is there, so the result of transposing matrix a is

12 15 18

13 16 19

14 19 11

Program

```
#include<stdio.h>
#include<conio.h>
#define ROW 3
#define COL 3

void main()
{
    void read(int a[][COL],int,int);
    void dis(int a[][COL],int,int);
    void trans(int a[][COL],int,int);
    int a[3][3];
    clrscr();
    read(a,ROW,COL);
    printf("\nThe matrix is \n");
    dis(a,ROW,COL);
    trans(a,ROW,COL);
    printf("The tranpose of the matrix is\n");
    dis(a,ROW,COL);
    getch();
}
void read(int c[3][3] ,int i ,int k)
{
    int j,l;
   printf("Enter the array \n");
   for(j=0;j<i;j++)
       for(l=0;l<k;l++)
              scanf("%d",&c[j][l]);
   fflush(stdin);
}
```

```
void dis(int d[3][3 ],int i,int k)
{
    int j,l;
   for(j=0;j<i;j++)
    {
       for(l=0;l<k;l++)
              printf("%d  ",d[j][l]);
       printf("\n");
   }
}
void trans(int mat[][3],int k ,int l)
{
    int i,j,temp;
    for(i=0;i<k;i++)
       for(j=i+1;j<l;j++)
       {
              temp=mat[i][j];
              mat[i][j]=mat[j][i];
              mat[j][i]=temp;
       }
}
```

Explanation

Basic steps:

1. Repeat step (2) for `i=0,1,2, ............(k-1 )` where `k` is the number of rows in the matrix.
2. Repeat step (3–5) for `j=(i+1),(i+2).....{l-1)` where `l` is the number of columns in the matrix .
3. `temp = mat[i][j]`
4. `mat[i][j] = mat[j][i]`
5. `mat[j][i] = temp`

Example

Input

Enter the array

12

13

14

15

16

17

18

19

11

Output

The matrix is

12 13 14

15 16 17

18 19 11

The transpose of the matrix is

12 15 18

13 16 19

14 17 11

Alternative Version of the Program

This is another version of the transpose program. Here a separate matrix is used to hold the result of transposition.

```
#include<stdio.h>
#include<conio.h>
#define ROW 3
#define COL 3

void main()
{
    void read(int a[][COL],int,int);
    void dis(int a[][COL],int,int);
    void trans(int a[][COL],int b[][COL],int,int);
    int a[3][3],b[3][3],i,j;
    clrscr();
    read(a,ROW,COL);
    printf("\nThe matrix is \n");
    dis(a,ROW,COL);
```

```
    trans(a,b,ROW,COL);
    printf("The tranpose of the matrix is\n");
    dis(b,ROW,COL);
    getch();
}
void read(int c[3][3] ,int i ,int k)
{
    int j,l;
   printf("Enter the array \n");
   for(j=0;j<i;j++)
       for(l=0;l<k;l++)
              scanf("%d",&c[j][l]);
   fflush(stdin);
}
void dis(int d[3][3 ],int i,int k)
{
    int j,l;
   for(j=0;j<i;j++)
    {
       for(l=0;l<k;l++)
              printf("%d  ",d[j][l]);
       printf("\n");
    }
}
void trans(int mat[][3],int tr_mat[][3], int k ,int l)
{
    int i,j;
    for(i=0;i<k;i++)
       for(j=0;j<l;j++)
       {
              tr_mat[i][j]=mat[j][i];
       }
}
```

Example

Input

Enter the array

1

2

3

4

5

6

7

8

9

Output

The matrix is

1 2 3

4 5 6

7 8 9

The transpose of the matrix is

1 4 7

2 5 8

3 6 9

FINDING THE SADDLE POINT OF A MATRIX

Introduction

A matrix a is said to have a *saddle point* if some entry a[I][j] is the smallest value in the ith row and the largest value in the jth column. A matrix may have more than one saddle point.

Program

```
#include<stdio.h>
#include<conio.h>
#define ROW 3
#define COL 3

void main()
{
   void read(int a[][COL],int,int);
```

```
        void dis(int a[][COL],int,int);
        int sadd_pt(int a[][COL],int,int,int *,int*);
        int i,a[3][3],m=0,n=0;
        clrscr();
        read(a,ROW,COL);
        printf("\nThe matrix is \n");
        dis(a,ROW,COL);
        i=sadd_pt(a,3,3,&m,&n);
        printf("The saddle point is %d &its position is row : %d col :
%d\n",
            i,m+1,n+1);
        getch();
    }
    void read(int c[][3] ,int i ,int k)
    {
        int j,l;
        printf("Enter the array \n");
        for(j=0;j<i;j++)
            for(l=0;l<k;l++)
                scanf("%d",&c[j][l]);
        fflush(stdin);
    }
    void dis(int d[][3],int i,int k)
    {
        int j,l;
        for(j=0;j<i;j++)
        {
            for(l=0;l<k;l++)
                printf("\t%d",d[j][l]);
            printf("\n");
        }
    }
    int sadd_pt(int mat[][3],int k ,int l,int *row,int *col)
    {
        int min=32767,i=0,j,m,n,p=0;
        while(i<k)
        {
            min=32767;
            m=i;
            p=0;
            for(j=0;j<l;j++)
            {
```

```
                if(mat[i][j]<min)
                {
                        min=mat[i][j];
                        n=j;
                }
        }
        for(j=0;j<k;j++)
                if(min>=mat[j][n])
                        p++;

        if(p==3)
        {
                *row=m;
                *col=n;
                return(min);
        }
        i++;
    }
    printf("No saddle point exists\n");
    getch();
    exit(0);
}
```

Example

Input

Enter the array

20 30 40

56 78 45

1 2 3

Output

The matrix is

20 30 40

56 78 45

1 2 3

The saddle point is 45 and its position is row 2 , column 3.

IMPLEMENTATION OF HEAPS

A *heap* is a list with the following attributes:

- Each entry contains a key.
- For all positions k in the list, the key at position k is least as large as the keys in positions 2k and 2k+1, provided these positions exist in the list. Therefore, an array can be used to implement a heap as shown in Figure 18.5.

1	97
2	58
3	75
4	53
5	42
6	53
7	48
8	32
9	20

FIGURE 18.5 A heap.

A heap is definitely not an ordered list because the first entry in the heap has the largest key, and there is no necessary ordering between the keys in locations k and k+1, if k > 1.

A heap is used in sorting a continuous list of length n in $O(n \log_2(n))$ comparisons and movements of entries, even in the worst case. The corresponding sorting method is called *heapsort*.

SORTING AND SEARCHING

We encounter several applications that require an ordered list. So it is required to order the elements of a given list either in ascending/increasing order or decending/decreasing order, as per the requirement. This process is called sorting. There are many techniques available for sorting an array-based list.

These techniques differ in their time and space complexities. Some of the important sorting techniques are discussed here.

BUBBLE SORT

Introduction

Bubble sorting is a simple sorting technique in which we arrange the elements of the list by forming pairs of adjacent elements. That means we form the pair of the ith and $(i+1)^{th}$ element. If the order is ascending, we interchange the elements of the pair if the first element of the pair is greater than the second element. That means for every pair (list[i],list[i+1]) for i :=1 to (n–1) if list[i] > list[i+1], we need to interchange list[i] and list[i+1]. Carrying this out once will move the element with the highest value to the last or n^{th} position. Therefore, we repeat this process the next time with the elements from the first to $(n-1)th$ positions. This will bring the highest value from among the remaining $(n-1)$ values to the $(n-1)^{th}$ position. We repeat the process with the remaining $(n-2)$ values and so on. Finally, we arrange the elements in ascending order. This requires to perform $(n-1)$ passes. In the first pass we have $(n-1)$ pairs, in the second pass we have $(n-2)$ pairs, and in the last (or $(n-1)^{th}$) pass, we have only one pair. Therefore, the number of probes or comparisons that are required to be carried out is

$$(n-1) + (n-2) + (n-3) + \ldots + 1$$

$$= n(n-1)/2,$$

and the order of the algorithm is $O(n^2)$.

Program

```
#include <stdio.h>
#define MAX 10
void swap(int *x,int *y)
{
    int temp;
    temp = *x;
    *x = *y;
    *y = temp;
}
void bsort(int list[], int n)
```

```
{
    int i,j;
    for(i=0;i<(n-1);i++)
        for(j=0;j<(n-(i+1));j++)
                if(list[j] > list[j+1])
                        swap(&list[j],&list[j+1]);
}

void readlist(int list[],int n)
{
    int i;
    printf("Enter the elements\n");
    for(i=0;i<n;i++)
        scanf("%d",&list[i]);
}

void printlist(int list[],int n)
{
    int i;
    printf("The elements of the list are: \n");
    for(i=0;i<n;i++)
        printf("%d\t",list[i]);
}

void main()
{
    int list[MAX], n;
    printf("Enter the number of elements in the list max = 10\n");
    scanf("%d",&n);
    readlist(list,n);
    printf("The list before sorting is:\n");
    printlist(list,n);
    bsort(list,n);
    printf("The list after sorting is:\n");
    printlist(list,n);
}
```

Example

Input

Enter the number of elements in the list, max = 10

5

Enter the elements

23

5

4

9

1

Output

The list before sorting is:

The elements of the list are:

23 5 4 9 1

The list after sorting is:

The elements of the list are:

1 4 5 9 23

QUICK SORT

Introduction

In the *quick sort* method, an array a[1],.....,a[n] is sorted by selecting some value in the array as a key element. We then swap the first element of the list with the key element so that the key will be in the first position. We then determine the key's proper place in the list. The proper place for the key is one in which all elements to the left of the key are smaller than the key, and all elements to the right are larger.

To obtain the key's proper position, we traverse the list in both directions using the indices `i` and `j`, respectively. We initialize `i` to that index that is one more than the index of the key element. That is, if the list to be sorted has the indices running from *m* to *n*, the key element is at index m, hence we initialize i to (m+1). The index i is incremented until we get an element at the ith position that is greater than the key value. Similarly, we initialize j to n and go on decrementing j until we get an element with a value less than the key's value.

We then check to see whether the values of i and j have crossed each other. If not, we interchange the elements at the key (m^{th}) position with the elements at the jth position. This brings the key element to the jth position, and we find that the elements to its left are less than it, and the elements to its right are greater than it. Therefore we can split the list into two sublists. The first sublist is composed of elements from the mth position to the $(j-1)^{th}$ position, and the second sublist consists of elements from the $(j+1)^{th}$ position to the nth position. We then repeat the same procedure on each of the sublists separately.

Choice of the key

We can choose any entry in the list as the key. The choice of the first entry is often a poor choice for the key, since if the list has already been sorted, there will be no element less than the first element selected as the key. So, one of the sublists will be empty. So we choose a key near the center of the list in the hope that our choice will partition the list in such a manner that about half of the elements will end up on one side of the key, and half will end up on the other.

Therefore the function `getkeyposition` is

```
int getkeyposition(int i,j)
{
   return(( i+j )/ 2);
}
```

The choice of the key near the center is also arbitrary, so it is not necessary to always divide the list exactly in half. It may also happen that one sublist is much larger than the other. So some other method of selecting a key should be used. A good way to choose a key is to use a random number generator to choose the position of the next key in each activation of quick sort. Therefore, the function `getkeyposition` is:

```
int getkeyposition(int i,j)
{
   return(random number in the range of i to j);
}
```

Program

```
#include <stdio.h>
#define MAX 10
void swap(int *x,int *y)
{
   int temp;
   temp = *x;
   *x = *y;
   *y = temp;
}
int getkeyposition(int i,int j )
{
   return((i+j) /2);
}

void qsort(int list[],int m,int n)
{
   int key,i,j,k;
   if( m < n)
   {
      k = getkeyposition(m,n);
      swap(&list[m],&list[k]);
      key = list[m];
      i = m+1;
      j = n;
      while(i <= j)
      {
            while((i <= n) && (list[i] <= key))
                  i++;
            while((j >= m) && (list[j] > key))
                  j–;
            if( i < j)
                  swap(&list[i],&list[j]);
      }
      swap(&list[m],&list[j]);
      qsort(list,m,j-1);
      qsort(list,j+1,n);
   }
}

void readlist(int list[],int n)
{
```

```
    int i;
    printf("Enter the elements\n");
    for(i=0;i<n;i++)
        scanf("%d",&list[i]);
}

void printlist(int list[],int n)
{
    int i;
    printf("The elements of the list are: \n");
    for(i=0;i<n;i++)
        printf("%d\t",list[i]);
}

void main()
{
    int list[MAX], n;
    printf("Enter the number of elements in the list max = 10\n");
    scanf("%d",&n);
    readlist(list,n);
    printf("The list before sorting is:\n");
    printlist(list,n);
    qsort(list,0,n-1);
    printf("\nThe list after sorting is:\n");
    printlist(list,n);
}
```

Example

Input

Enter the number of elements in the list, max = 10

10

Enter the elements

7

99

23

11

65

43

23
21
21
77

Output

The list before sorting is:
The elements of the list are:
7 99 23 11 65 43 23 21 21 77
The list after sorting is:
The elements of the list are:
7 11 21 21 23 23 43 65 77 99

Explanation

Consider the following list:

0	1	2	3	4	5	6 ← *indices*
10	5	23	67	20	30	60

1. When `qsort` is activated the first time, key = 67, `i` =1, and `j` =6. `i` is incremented until it becomes 7, because there is no element greater than the key. `j` is not decremented, because at position 6, the value that we have is less than the key. Since `i > j`, we interchange the key element (the element at position 0) with the element at position 6, and call `qsort` recursively, with the left sublist made of elements from positions 0 to 5, and the right sublist empty as shown here:

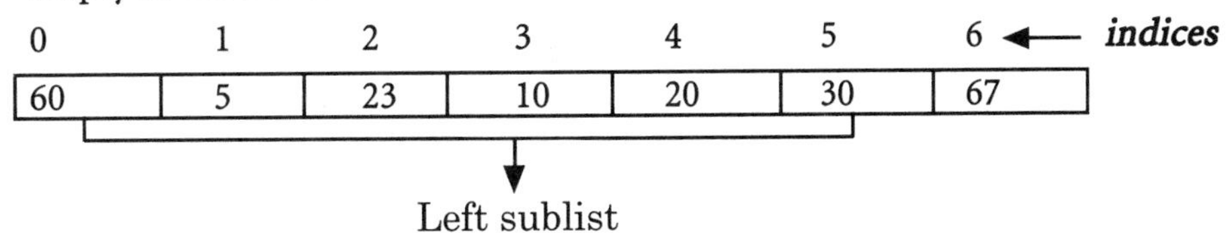

2. When `qsort` is activated the second time on the left sublist as shown, key = 23, `i` =1, and `j` =5. `i` is incremented until it reaches 2. Because the element at position 2 is greater than the key, `j` is decremented to 4 because the value at position 4 is less than the key. Since `i < j`, the elements at positions 2 and 4 are swapped. `i` is then incremeneted to 4 and `j` is decremented to 3. Since `i > j`, we interchange the key element (the element at position 0), with the element at position 3, and call `qsort` recursively with the left sublist made

of elements from position 0 to 2, and the right sublist made of elements from position 4 to 5, as shown here:

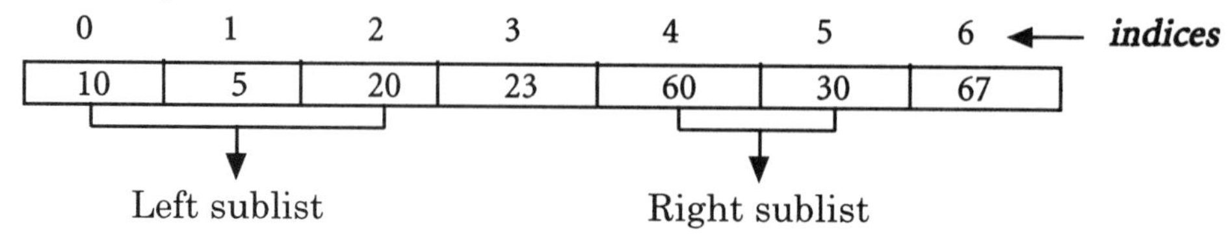

3. By continuing in this fashion, we eventually get the list sorted.
4. The average case-time complexity of the quick sort algorithm can be determined as follows:

   We assume that every time this is done, the list gets split into two approximately equal-sized sublists. If the size of a given list is n, it gets split into two sublists of size approximately $n/2$. Each of these sublists gets further split into two sublists of size n/4, and this is continued until the size equals 1. When the quick sort works with a list of size n, it places the key element (which takes the first element of the list under consideration) in its proper position in the list. This requires no more than n iterations. After placing the key element in its proper position in the list of size n, quick sort activates itself twice to work with the left and right sublists, each assumed to be of size $n/2$. Therefore T(n) is the time required to sort a list of size n. Since the time required to sort the list of size n is equal to the sum of the time required to place the key element in its proper position in the list of size n, and the time required to sort the left and right sublists, each assumed to be of size $n/2$. $T(n)$ turns out to be:

   $\therefore \; T(n) = c*n + 2*T(n/2)$

   where c is a constant and $T(n/2)$ is the time required to sort the list of size n/2.

5. Similarly, the time required to sort a list of size n/2 is equal to the sum of the time required to place the key element in its proper position in the list of size n/2 and the time required to sort the left and right sublists each assumed to be of size n/4. T(n/2) turns out to be:

   $T(n/2) = c*n/2 + 2*T(n/4)$

   where $T(n/4)$ is the time required to sort the list of size $n/4$.

   $\therefore \; T(n/4) = c*n/4 + 2*T(n/8)$, and so on. We eventually we get $T(1) = 1$.

   $\therefore \; T(n) = c*n + 2(c*n(n/2) + 2T(n/4))$

   $\therefore \; T(n) = c*n + c*n + 4T(n/4)) = 2*c*n + 4T(n/4) = 2*c*n + 4(c*(n/4) + 2T(n/8))$

   $\therefore \; T(n) = 2*c*n + c*n + 8T(n/8) = 3*c*n + 8T(n/8)$

   $\therefore \; T(n) = (\log n)*c*\mathrm{n} + nT(n/n) = (\log n)*c*n + nT(1) = n + n*(\log n)*c$

$\therefore$ T(n) µ $n$log(n)

6. Therefore, we conclude that the average complexity of the quick sort algorithm is O(nlog n). But the worst-case time complexity is of the O(n^2). The reason for this is, in the worst case, one of the two sublists will always be empty and the other will be of size (n–1), where n is the size of the original list. Therefore, in the worst case, T(n) turns out to be

   $T(n) = c{*}n + T(n-1)$

   $= c{*}n + c{*}(n-1) + T(n-2)$

   $= 2{*}c{*}n - c + T(n-2)$

   $= 2{*}c{*}n - c + c{*}(n-2) + T(n-3)$

   $= 3{*}c{*}n - 3{*}c + T(n-3)$

   ...

   ...

   $= n{*}c{*}n - n{*}c + T(1)$

   $= n^2c - nc + 1$

   Therefore T(n) µ n^2, so the order is O(n^2).

7. Space complexity: The average-case space complexity is $\log_2 n$, because the space complexity depends on the maximum number of activations that can exist. We find that if we assume that every time the list gets split into approximately two equal-sized lists, the maximum number of activations that will exist simultaneously will be $\log_2 n$.

   In the worst case, there exist n activations, because the depth of the recursion is n. So the worst-case space complexity is O(n).

MERGE SORT

Introduction

This is another sorting technique having the same average-case and worst-case time complexities, but requiring an additional list of size n.

The technique that we use is the merging of the two sorted lists of size m and n to form a single sorted list of size (m + n). Given a list of size n to be sorted, instead of viewing it to be one single list of size n, we start by viewing it to be n lists each of size 1, and merge the first list with the second list to form a single sorted list of size 2.

Similarly, we merge the third and the fourth lists to form a second single sorted list of size 2, and so on. This completes one pass. We then consider the first sorted list of size 2 and the second sorted list of size 2, and merge them to form a single sorted list of size 4.

Similarly, we merge the third and the fourth sorted lists, each of size 2, to form the second single sorted list of size 4, and so on. This completes the second pass.

In the third pass, we merge these adjacent sorted lists, each of size 4, to form sorted lists of size 8. We continue this process until we finally obtain a single sorted list of size n as shown next.

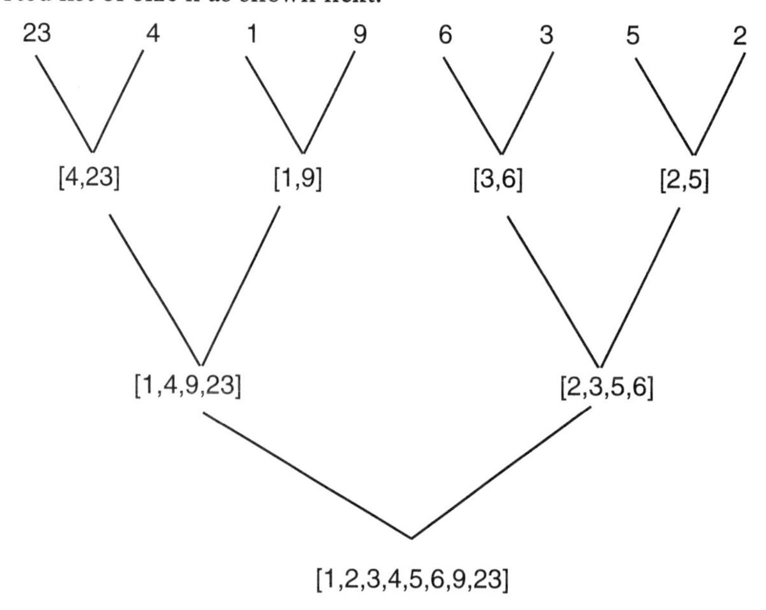

To carry out this task, we require a function to merge the two sorted lists of size m and n to form a single sorted list of size (m + n). We also require a function to carry out one pass of the list to merge the adjacent sorted lists of the specified size. This is because we have to carry out repeated passes of the given list.

In the first pass, we merge the adjacent lists of size 1. In the second pass, we merge the adjacent lists of size 2, and so on. Therefore, we will call this function by varying the size of the lists to be merged.

Program

```
#include <stdio.h>
#define MAX 10
void merge(int list[],int list1[],int k,int m,int n)
{
    int i,j;
   i=k;
   j = m+1;
   while( i <= m && j <= n)
   {
       if(list[i] <= list[j])
      {
              list1[k] = list[i];
        i++;
        k++;
      }
      else
      {
              list1[k] = list[j];
        j++;
        k++;
      }
    }
    while(i <= m)
   {
       list1[k] = list[i];
      i++;
      k++;
    }
   while( j <= n)
   {
       list1[k] = list[j];
      j++;
      k++;
    }
}

void mpass( int list[],int list1[],int l,int n)
{
    int i;
   i = 0;
   while( i <= (n-2*l+1))
```

```
        {
            merge(list,list1,i,(i+l-1),(i+2*l-1));
          i = i + 2*l;
        }
        if((i+l-1) < n)
            merge(list,list1,i,(i+l-1),n);
        else
            while (i <= n )
            {
                    list1[i] = list[i];
                    i++;
            }
}

void msort(int list[], int n )
{
    int l;
    int list1[MAX];
    l =1;
   while (l <= n )
    {
        mpass(list,list1,l,n);
        l = l*2;
        mpass(list1,list,l,n);
        l = l*2;
    }
}
void readlist(int list[],int n)
{
    int i;
   printf("Enter the elements\n");
   for(i=0;i<n;i++)
        scanf("%d",&list[i]);
}

void printlist(int list[],int n)
{
    int i;
   printf("The elements of the list are: \n");
   for(i=0;i<n;i++)
        printf("%d\t",list[i]);
}
```

```
void main()
{
   int list[MAX], n;
   printf("Enter the number of elements in the list max = 10\n");
   scanf("%d",&n);
   readlist(list,n);
   printf("The list before sorting is:\n");
   printlist(list,n);
   msort(list,n-1);
   printf("The list after sorting is:\n");
   printlist(list,n);
}
```

Example

Input

Enter the number of elements in the list, max = 10

10

Enter the elements

11

2

45

67

33

22

11

0

34

23

Output

The list before sorting has the following elements:

11 2 45 67 33 22 11 0 34 23

The list after sorting has the following elements:

0 2 11 11 22 23 33 34 45 67

Explanation

1. The merging of two sublists, the first running from the index 0 to m, and the second running from the index $(m + 1)$ to $(n - 1)$ requires no more than $(n-l + 1)$ iterations.
2. So if $l = 1$, then no more than n iterations are required, where n is the size of the list to be sorted.
3. Therefore, if n is the size of the list to be sorted, every pass that a merge routine performs requires a time proportional to $O(n)$, since the number of passes required to be performed is log2n.
4. The time complexity of the algorithm is $O(n \log_2(n))$, for both average-case and worst-case. The merge sort requires an additional list of size n.

HEAPSORT

Introduction

Heapsort is a sorting technique that sorts a contiguous list of length n with $O(n \log_2(n))$ comparisons and movement of entries, even in the worst case. Hence it achieves the worst-case bounds better than those of quick sort, and for the contiguous list, it is better than merge sort, since it needs only a small and constant amount of space apart from the list being sorted.

Heapsort proceeds in two phases. First, all the entries in the list are arranged to satisfy the heap property, and then the top of the heap is removed and another entry is promoted to take its place repeatedly. Therefore, we need a function that builds an initial heap to arrange all the entries in the list to satisfy the heap property. The function that builds an initial heap uses a function that adjusts the ith entry in the list, whose entries at 2i and 2i + 1 positions already satisfy the heap property in such a manner that the entry at the ith position in the list will also satisfy the heap property.

Program

```
#include <stdio.h>
#define MAX 10
void swap(int *x,int *y)
{
```

```
    int temp;
    temp = *x;
    *x = *y;
    *y = temp;
}

void adjust( int list[],int i, int n)
{
    int j,k,flag;
    k = list[i];
    flag = 1;
    j = 2 * i;
    while(j <= n && flag)
    {
        if(j < n && list[j] < list[j+1])
        j++;
        if( k >= list[j])
                flag =0;
        else
        {
                list[j/2] = list[j];
                j = j *2;
        }
    }
   list [j/2] = k;
}

void build_initial_heap( int list[], int n)
{
    int i;
    for(i=(n/2);i>=0;i–)
        adjust(list,i,n-1);
}

void heapsort(int list[],int n)
{
    int i;
    build_initial_heap(list,n);
    for(i=(n-2); i>=0;i–)
    {
        swap(&list[0],&list[i+1]);
        adjust(list,0,i);
```

```
    }
}

void readlist(int list[],int n)
{
    int i;
   printf("Enter the elements\n");
   for(i=0;i<n;i++)
       scanf("%d",&list[i]);
}

void printlist(int list[],int n)
{
    int i;
   printf("The elements of the list are: \n");
   for(i=0;i<n;i++)
       printf("%d\t",list[i]);
}

void main()
{
    int list[MAX], n;
   printf("Enter the number of elements in the list max = 10\n");
   scanf("%d",&n);
   readlist(list,n);
   printf("The list before sorting is:\n");
   printlist(list,n);
   heapsort(list,n);
   printf("The list after sorting is:\n");
   printlist(list,n);
}
```

Example

Input

Enter the number of elements in the list, max = 10

10

Enter the elements

56

1

34

42

90

66

87

12

21

11

Output

The list before sorting is:

The elements of the list are:

56 1 34 42 90 66 87 12 21 11

The list after sorting is:

The elements of the list are:

1 11 12 21 34 42 56 66 87 90

Explanation

In each pass of the `while` loop in the function `adjust(x,i,n)`, the position `i` is doubled, so the number of passes cannot exceed log(*n*/i). Therefore, the computation time of adjust is O(log*n*/i).

The function `build_initial_heap` calls the adjust procedure n/2 for values ranging from n1/2 to 0. Hence the total number of iterations will be:

$$\log(n) + \log(n/2) + \ldots + \log(n/n/2)$$

$$= \sum_{i=1}^{n/2} \log(n/i)$$

$$= n/2\log(n) - \log(!n/2)$$

This turns out to be some constant time n. So the computation time of build_initial_heap is O(n). The heapsort function calls the adjust (x,1, i) ($n-1$) times. So the total number of iterations made in the heapsort will be

$$\log(i/1)$$

$$= n-1$$

log (i)

i=1

$= \log(1) + \log(2) + \ldots + \log(n-1)$

which turns out to be approximately n log(n). So the computing time of heapsort is O(n log(n)) + O(n). The only additional space needed by heapsort is the space for one record to carry out the exchange.

SEARCHING TECHNIQUES: LINEAR OR SEQUENTIAL SEARCH

Introduction

There are many applications requiring a search for a particular element. Searching refers to finding out whether a particular element is present in the list. The method that we use for this depends on how the elements of the list are organized. If the list is an unordered list, then we use linear or sequential search, whereas if the list is an ordered list, then we use binary search.

The search proceeds by sequentially comparing the key with elements in the list, and continues until either we find a match or the end of the list is encountered. If we find a match, the search terminates successfully by returning the index of the element in the list which has matched. If the end of the list is encountered without a match, the search terminates unsuccessfully.

Program

```
#include <stdio.h>
#define MAX 10

void lsearch(int list[],int n,int element)
{
    int i, flag = 0;
   for(i=0;i<n;i++)
    if(  list[i] == element)
   {
        printf(" The element whose value is %d is present at position %d
in list\n",
               element,i);
        flag =1;
```

```
            break;
         }
        if( flag == 0)
            printf("The element whose value is %d is not present in the
list\n",
                    element);
}

void readlist(int list[],int n)
{
    int i;
    printf("Enter the elements\n");
    for(i=0;i<n;i++)
        scanf("%d",&list[i]);
}

void printlist(int list[],int n)
{
    int i;
    printf("The elements of the list are: \n");
    for(i=0;i<n;i++)
        printf("%d\t",list[i]);
}

void main()
{
    int list[MAX], n, element;
    printf("Enter the number of elements in the list max = 10\n");
    scanf("%d",&n);
    readlist(list,n);
    printf("\nThe list before sorting is:\n");
    printlist(list,n);
    printf("\nEnter the element to be searched\n");
    scanf("%d",&element);
    lsearch(list,n,element);
}
```

Example

Input

Enter the number of elements in the list, max = 10

10

Enter the elements

23

1

45

67

90

100

432

15

77

55

Output

The list before sorting is:

The elements of the list are:

23 1 45 67 90 100 432 15 77 55

Enter the element to be searched

100

The element whose value is 100 is present at position 5 in list

Input

Enter the number of elements in the list max = 10

10

Enter the elements

23

1

45

67

90

101

23

56

44

22

Output

The list before sorting is:

The elements of the list are:

23 1 45 67 90 101 23 56 44 22

Enter the element to be searched

100

The element whose value is 100 is not present in the list

Explanation

1. In the best case, the search procedure terminates after one comparison only, whereas in the worst case, it will do n comparisons.
2. On average, it will do approximately $n/2$ comparisons, since the search time is proportional to the number of comparisons required to be performed.
3. The linear search requires an average time proportional to $O(n)$ to search one element. Therefore to search n elements, it requires a time proportional to $O(n^2)$.
4. We conclude that this searching technique is preferable when the value of n is small. The reason for this is the difference between n and n^2 is small for smaller values of n.

BINARY SEARCH

Introduction

The prerequisite for using binary search is that the list must be a sorted one. We compare the element to be searched with the element placed approximately in the middle of the list.

If a match is found, the search terminates successfully. Otherwise, we continue the search for the key in a similar manner either in the upper half or the lower half. If the elements of the list are arranged in ascending order, and the key is less than the element in the middle of the list, the search is continued in the lower half. If the elements of the list are arranged in descending order, and the key is greater than the element in the middle of the list, the search is continued in the upper half of the list. The procedure for the binary search is given in the following program.

Program

```
#include <stdio.h>
#define MAX 10

void bsearch(int list[],int n,int element)
{
     int l,u,m, flag = 0;
    l = 0;
    u = n-1;
    while(l <= u)
     {
         m = (l+u)/2;
        if( list[m] == element)
         {
                printf(" The element whose value is %d is present at
position %d in list\n",
                         element,m);
                flag =1;
                break;
         }
        else
                if(list[m] < element)
                         l = m+1;
                else
                         u = m-1;
     }

    if( flag == 0)
    printf("The element whose value is %d is not present in the list\n",
         element);
}

void readlist(int list[],int n)
{
     int i;
    printf("Enter the elements\n");
    for(i=0;i<n;i++)
         scanf("%d",&list[i]);
}
```

```
void printlist(int list[],int n)
{
    int i;
   printf("The elements of the list are: \n");
   for(i=0;i<n;i++)
       printf("%d\t",list[i]);
}
void main()
{
   int list[MAX], n, element;
   printf("Enter the number of elements in the list max = 10\n");
   scanf("%d",&n);
   readlist(list,n);
   printf("\nThe list before sorting is:\n");
   printlist(list,n);
   printf("\nEnter the element to be searched\n");
   scanf("%d",&element);
   bsearch(list,n,element);
}
```

Example

Input

Enter the number of elements in the list, max = 10

10

Enter the elements

34

2

1

789

99

45

66

33

22

11

Output

The elements of the list before sorting are:

34 2 1 789 99 45 66 33 22 11

1 2 3 4 5 6 7 8 9 10

Enter the element to be searched

99

The element whose value is 99 is present at position 5 in the list

Input

Enter the number of elements in the list max = 10

10

Enter the elements

54

89

09

43

66

88

77

11

22

33

Output

The elements of the list before sorting are:

54 89 9 43 66 88 77 11 22 33

Enter the element to be searched

100

The element whose value is 100 is not present in the list.

In the binary search, the number of comparisons required to search one element in the list is no more than log2*n*, where n is the size of the list. Therefore, the binary search algorithm has a time complexity of O(n *(log 2*n*.).)

HASHING

Introduction

A data object called a *symbol table* is required to be defined and implemented in many applications, such as compiler/assembler writing. A symbol table is nothing but a set of pairs (name, value), where *value* represents a collection of attributes associated with the name, and the collection of attributes depends on the program element identified by the name.

For example, if a name x is used to identify an array in a program, then the attributes associated with x are the number of dimensions, lower bound and upper bound of each dimension, and element type. Therefore, a symbol table can be thought of as a linear list of pairs (name, value), and we can use a list data object for realizing a symbol table.

A symbol table is referred to or accessed frequently for adding a name, or for storing or retrieving the attributes of a name.

Therefore, accessing efficiency is a prime concern when designing a symbol table. The most common method of implementing a symbol table is to use a hash table.

Hashing is a method of directly computing the index of the table by using a suitable mathematical function called a hash function.

The hash function operates on the name to be stored in the symbol table, or whose attributes are to be retrieved from the symbol table.

If h is a hash function and x is a name, then h(x) gives the index of the table where x, along with its attributes, can be stored. If x is already stored in the table, then h(x) gives the index of the table where it is stored, in order to retrieve the attributes of x from the table.

There are various methods of defining a hash function. One is the division method. In this method, we take the sum of the values of the characters, divide it by the size of the table, and take the remainder. This gives us an integer value lying in the range of 0 to (n–1), if the size of the table is n.

Another method is the *mid-square method*. In this method, the identifier is first squared and then the appropriate number of bits from the middle of the square is used as the hash value. Since the middle bits of the square usually depend on all the characters in the identifier, it is expected that different identifiers will result in different values. The number of middle bits that we select depends on the table size. Therefore, if r is the number of middle bits that we are using to form the hash value, then the table size will be 2^r. So when we use this method, the table size is required to be a power of 2.

A third method is *folding*, in which the identifier is partitioned into several parts, all but the last part being of the same length. These parts are then added together to obtain the hash value.

To store the name or to add attributes of the name, we compute the hash value of the name, and place the name or attributes, as the case may be, at that place in the table whose index is the hash value of the name.

To retrieve the attribute values of the name kept in the symbol table, we apply the hash function of the name to that index of the table where we get the attributes of the name. So we find that no comparisons are required to be done; the time required for the retrieval is independent of the table size. The retrieval is possible in a constant amount of time, which will be the time taken for computing the hash function.

Therefore a hash table seems to be the best for realization of the symbol table, but there is one problem associated with the hashing, and that is collision.

Hash collision occurs when the two identifiers are mapped into the same hash value. This happens because a hash function defines a mapping from a set of valid identifiers to the set of those integers that are used as indices of the table.

Therefore we see that the domain of the mapping defined by the hash function is much larger than the range of the mapping, and hence the mapping is of a many-to-one nature. Therefore, when we implement a hash table, a suitable collision-handling mechanism is to be provided, which will be activated when there is a collision.

Collision handling involves finding an alternative location for one of the two colliding symbols. For example, if x and y are the different identifiers and h(x = h(y), x and y are the colliding symbols. If x is encountered before y, then the ith entry of the table will be used for accommodating the symbol x, but later on when y comes, there is a hash collision. Therefore we have to find a suitable alternative location either for x or y. This means we can either accommodate y in that location, or we can move x to that location and place y in the ith location of the table.

Various methods are available to obtain an alternative location to handle the collision. They differ from each other in the way in which a search is made for an alternative location. The following are commonly used collision-handling techniques:

Linear Probing or Linear Open Addressing

In this method, if for an identifier x, h(x) = i, and if the ith location is already occupied, we search for a location close to the ith location by doing a linear search, starting from the $(i+1)^{th}$ location to accommodate x. This means we start from the $(i+1)^{th}$ location and do the linear search until we get an empty location; once we get an empty location we accommodate x there.

Rehashing

In *rehashing* we find an alternative empty location by modifying the hash function and applying the modified hash function to the colliding symbol. For example, if x is the symbol and h(x) = i, and if the ith location is already occupied, then we modify the hash function h to h_1, and find out $h_1(x)$, if $h_1(x) = j$. If the jth location is empty, then we accommodate x in the jth location. Otherwise, we once again modify h_1 to some h_2 and repeat the process until the collision is handled. Once the collision is handled, we revert to the original hash function before considering the next symbol.

Overflow chaining

Overflow chaining is a method of implementing a hash table in which the collisions are handled automatically. In this method, we use two tables: a symbol table to accommodate identifiers and their attributes, and a *hash table*, which is an array of pointers pointing to symbol table entries. Each symbol table entry is made of three fields: the first for holding the identifier, the second for holding the attributes, and the third for holding the link or pointer that can be made to point to any symbol table entry. The insertions into the symbol table are done as follows:

If x is the symbol to be inserted, it will be added to the next available entry of the symbol table. The hash value of x is then computed. If h(x) = i, then the ith hash table pointer is made to point to the symbol table entry in which x is stored, if the ith hash table pointer does not point to any symbol table entry. If the ith hash table pointer is already pointing to some symbol table entry, then the link field of the symbol table entry containing x is made to point to that symbol table entry to which the ith hash table pointer is pointing to, and the ith hash table pointer is made to point to the symbol entry containing x. This is equivalent to building a linked list on the ith index of the hash table. The retrieval of attributes is done as follows:

If x is a symbol, then we obtain h(x), use this value as the index of the hash table, and traverse the list built on this index to get that entry which contains x. A typical hash table implemented using this technique is shown here.

The symbols to b stored are x_1,y_1,z_1,x_2,y_2,z_2. The hash function that we use is

h(symbol) = (value of first letter of the symbol) mod n,

where n is the size of table.

if $h(x_1) = i$

$h(y_1) = j$

$h(z_1) = k$

then

$h(x_2) = i$

$h(y_2) = j$

$h(z_2) = k$

Therefore, the contents of the symbol table will be the one shown in Figure 18.6.

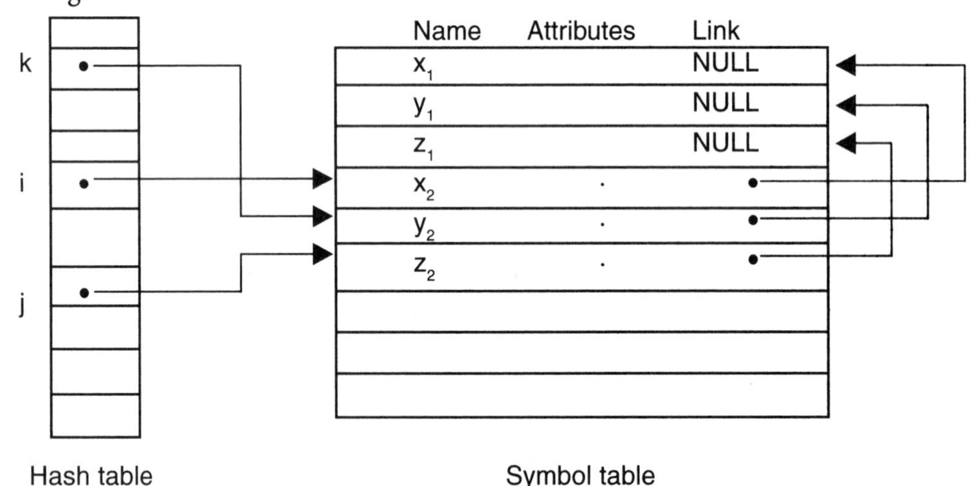

FIGURE 18.6 Hash table implementation using overflow chaining for collision handling.

HASHING FUNCTIONS

Some of the methods of defining a hash function are discussed in the following paragraphs.

Modular Arithmetic

In *modular arithmetic*, first the key is converted to an integer, then it is divided by the size of the index range, and the remainder is taken to be the hash value. The spread achieved depends very much on the modulus. If the modulus is the power of small integers such as 2 or 10, then many keys tend to map into the same index, while other indices remain unused. The best choice for the modulus is often, but not always, a prime number, which usually has the effect of spreading the keys quite uniformly.

Truncation

Truncation ignores part of the key, and uses the remainder directly as the hash value (using numeric code to represent non-numeric field data). If the keys, for example, are eight-digit numbers and the hash table has 1000 entries, then the first, second, and fifth digits from the right side might make the hash value. So, 62538194 maps to 394. It is a fast method, but it often fails to distribute keys evenly.

Folding

In *folding,* the identifier is partitioned into several parts, all but the last part being of the same length. These parts are then added together to obtain the hash value. For example, an eight-digit integer can be divided into groups of three, three, and two digits. The groups are then added together, and truncated, if necessary, to be in the proper range of indices. So 62538149 maps to 625 + 381 + 94 = 1100, truncated to 100. Since all information in the key can affect the value of the function, folding often achieves a better spread of indices than truncation.

Mid-square method

In this method, the identifier is squared (using numeric code to represent non-numeric field data), and then the appropriate number of bits from the middle of the square are used to get the hash value. Since the middle bits of the square usually depend on all the characters in the identifier, it is expected that different identifiers will result in different values. The number of middle bits that we select depends on table size. Therefore, if r is the number of middle bits used to form the hash value, then the table size will be 2^r. So when we use the mid-square method, the table size should be a power of 2.

Program

A complete C program for implementation of a hash table is given here:

```
#include <stdio.h>
#include <string.h>
#include <stdlib.h>
#define SIZE 50
#define MAX 10
 typedef struct node
    {
    char symbol[MAX];
       int value;
       struct node *next;
    } entry;
    typedef entry *entry_ptr;

     int hash_value(char * name)
     {
          int sum=0;
          while( *name != '\0')
          {
            sum += *name;
            name++;
          }
          return(sum % SIZE);
        }
      void initialize( entry_ptr table[])
       {
       int i=0;
        for(i=0; i<SIZE; i++)
             table[i] = NULL;
       }
       void insert( entry_ptr table[], char *name, int val)
     {
          int h, flag = 1;
          entry_ptr temp;
          h = hash_value(name);
          temp = table[h];
          while(  temp != NULL && flag )
     {
              if( strcmp(temp->symbol,name) == 0)
                {
              printf("The symbol %s is already present in the
table\n",name);
                   flag =0;;
                }
```

```
                temp=temp->next;
              }
              if(flag)
              {
              temp = (entry_ptr) malloc(sizeof( entry));
               if(temp == NULL)
        {
                    printf("ERRRR .......\n");
                    exit(0);
                  }
               strcpy(temp->symbol,name);
               temp->value = val;
               temp->next = table[h];
               table[h]=temp;
               }
          }
          void retrieve( entry_ptr table[],char *name)
        {
            int h,flag =1;
            entry_ptr temp;
            h = hash_value(name);
            temp = table[h];
            while(  temp != NULL && flag)
        {
              if( strcmp(temp->symbol,name) == 0)
          {
        printf("The symbol %s is  present in the table and having value =
%d\n",
    name,temp->value);
                     flag =0;
                  }
              temp=temp->next;
            }
            if(flag == 1)
               printf("The symbol %s is not present in the table
\n",name);
          }
          void main()
          {
            entry_ptr table[SIZE];
            char name[MAX];
            int value,n;
            initialize(table);
            do
            {
```

```
        do
    {
          printf("Enter the symbol and value pair to be inserted\n");
          scanf("%s %d",name,&value);
          insert(table,name,value);
          printf("Enter 1 to continue\n");
          scanf("%d",&n);
    }
    while(n == 1);
    do
    {
          printf("Enter the symbol whose value is to be retrieved\n");
          scanf("%s",name);
          retrieve(table,name);
          printf("Enter 1 to continue\n");
          scanf("%d",&n);
    }while( n == 1);
        printf("Eneter 1 to continue\n");
        scanf("%d",&n);
        }while(n == 1);
      }
```

Example

Input and Output

Enter the symbol and value pair to be inserted

    ogk10

Enter 1 to continue

    1

Enter the symbol and value pair to be inserted

    psd20

Enter 1 to continue

    0

Enter the symbol whose value is to be retrieved

    ogk

The symbol ogk is present in the table with the value = 10

Enter 1 to continue

    1

Enter the symbol whose value is to be retrieved

    psd

The symbol psd is present in the table with the value = 20

Enter 1 to continue

    1

Enter the symbol whose value is to be retrieved

    asg

The symbol asg is not present in the table

Enter 1 to continue

    0

Eneter 1 to continue

    1

Enter the symbol and value pair to be inserted

    asg30

Enter 1 to continue

    0

Enter the symbol whose value is to be retrieved

    asg

The symbol asg is present in the table with the value = 30

Enter 1 to continue

    0

Eneter 1 to continue

    0

Exercises

1. Consider an unsorted array A[n] of integer elements that may have many elements present more than once. It is required to store only the distinct elements of the array A in a separate array B. The information about the number of times each element is replicated is maintained in a third array C. For example, C[0] would indicate the number of times the element B[0] occurs in array A. Write a C program to generate the arrays B and C, given an array A.
2. Write a C program that finds the largest and the second largest elements in an unsorted array A. The program should make just a single scan of the array.

3. Sort the following list by applying the bubble sort method.

   10

   01

   11

   100

   23

   21

   11

   99

   78

4. Sort the following list by applying the heapsort method.

   44

   23

   67

   88

   22

   43

   90

   04

5. Consider the following list whose elements are arranged in ascending order. Assume that a binary search technique is used. Determine the number of probes required to find each entry in the list.

   11

   22

   43

   56

   67

   71

   89

19 Stacks and Queues

THE CONCEPT OF STACKS AND QUEUES

There are many applications requiring the use of the data structures *stacks* and *queues*. The most striking use of a data structure stack is the runtime stack that a programming language uses to implement a function call and return. Similarly, one of the important uses of a data structure queue is the process queue maintained by the scheduler. Both these data structures are modified versions of the list data structure, so they can be implemented using arrays or linked representation.

STACKS

A stack is simply a list of elements with insertions and deletions permitted at one end—called the stack top. That means that it is possible to remove elements from a stack in reverse order from the insertion of elements into the stack. Thus, a stack data structure exhibits the LIFO (last in first out) property. Push and pop are the operations that are provided for insertion of an element into the stack and the removal of an element from the stack, respectively. Shown in Figure 19.1 are the effects of push and pop operations on the stack.

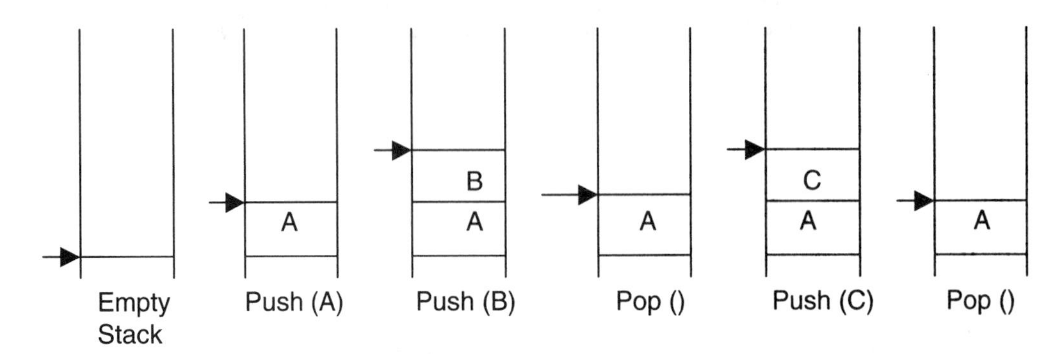

FIGURE 19.1 Stack operations.

Since a stack is basically a list, it can be implemented by using an array or by using a linked representation.

Array Implementation of a Stack

When an array is used to implement a stack, the push and pop operations are realized by using the operations available on an array. The limitation of an array implementation is that the stack cannot grow and shrink dynamically as per the requirement.

Program

A complete C program to implement a stack using an array appears here:

```
#include <stdio.h>
#define MAX 10 /* The maximum size of the stack */
#include <stdlib.h>

void push(int stack[], int *top, int value)
{
    if(*top < MAX )
   {
       *top = *top + 1;
      stack[*top] = value;
   }
   else
   {
```

```
        printf("The stack is full can not push a  value\n");
        exit(0);
    }
}
void pop(int stack[], int *top, int * value)
{
    if(*top >= 0 )
    {
        *value = stack[*top];
       *top = *top - 1;
    }
    else
    {
        printf("The stack is empty can not pop a  value\n");
        exit(0);
    }
}

void main()
{
    int stack[MAX];
   int top = -1;
   int n,value;
   do
   {
        do
      {
               printf("Enter the element to be pushed\n");
         scanf("%d",&value);
         push(stack,&top,value);
               printf("Enter 1 to continue\n");
         scanf("%d",&n);
        } while(n == 1);

      printf("Enter 1 to pop an element\n");
      scanf("%d",&n);
      while( n == 1)
      {
               pop(stack,&top,&value);
         printf("The value poped is %d\n",value);
               printf("Enter 1 to pop an element\n");
               scanf("%d",&n);
```

```
        }
        printf("Enter 1 to continue\n");
        scanf("%d",&n);
    } while(n == 1);
}
```

Example

Enter the element to be pushed

10

Enter 1 to continue

1

Enter the element to be pushed

20

Enter 1 to continue

0

Enter 1 to pop an element

1

The value popped is 20

Enter 1 to pop an element

0

Enter 1 to continue

1

Enter the element to be pushed

40

Enter 1 to continue

1

Enter the element to be pushed

50

Enter 1 to continue

0

Enter 1 to pop an element

1

The value popped is 50

```
Enter 1 to pop an element
    1
The value popped is 40
Enter 1 to pop an element
    1
The value popped is 10
Enter 1 to pop an element
    0
Enter 1 to continue
    0
```

Implementation of a Stack Using Linked Representation

Initially the list is empty, so the top pointer is NULL. The push function takes a pointer to an existing list as the first parameter and a data value to be pushed as the second parameter, creates a new node by using the data value, and adds it to the top of the existing list. A pop function takes a pointer to an existing list as the first parameter, and a pointer to a data object in which the popped value is to be returned as a second parameter. Thus it retrieves the value of the node pointed to by the top pointer, takes the top point to the next node, and destroys the node that was pointed to by the top.

If this strategy is used for creating a stack with the previously used four data values: 10, 20, 30, and 40, then the stack is created as shown in Figure 19.2.

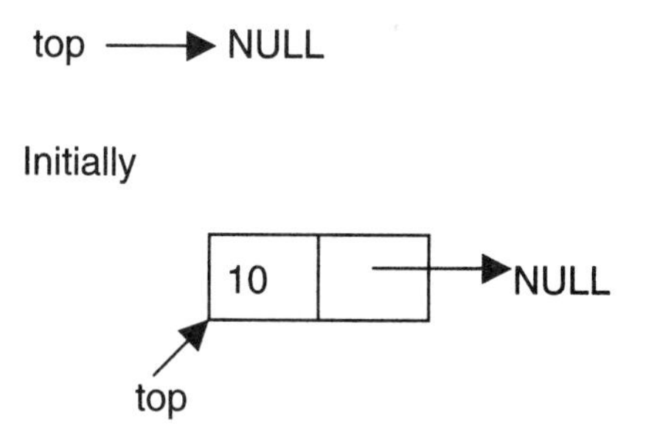

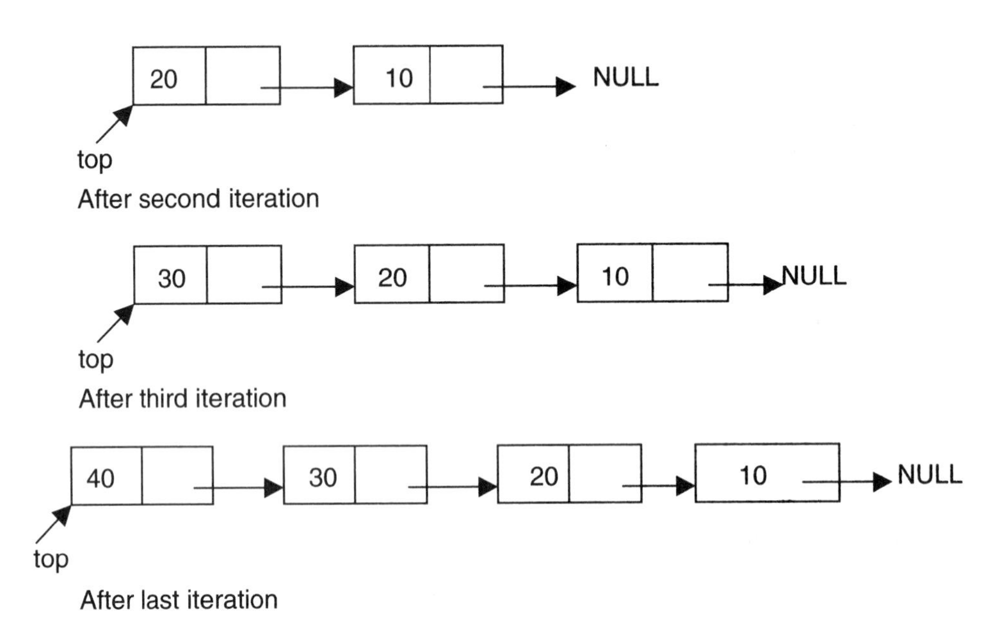

FIGURE 19.2 Linked stack.

Program

A complete C program for implementation of a stack using the linked list is given here:

```
# include <stdio.h>
# include <stdlib.h>
struct node
{
    int data;
    struct node *link;
};
struct node *push(struct node *p , int value)
{
    struct node *temp;
    temp=(struct node *)malloc(sizeof(struct node));
        /* creates new node
        using data value
        passed as  parameter */
    if(temp==NULL)
   {
        printf("No Memory available Error\n");
```

```
        exit(0);
     }
    temp->data = value;
    temp->link = p;
    p = temp;
    return(p);
}

struct node *pop(struct node *p , int *value)
{
     struct node *temp;
    if(p==NULL)
    {
        printf(" The stack is empty can not pop Error\n");
        exit(0);
     }
     *value = p->data;
    temp = p;
    p = p->link;
    free(temp);
    return(p);
}

void main()
{
     struct node *top = NULL;
    int n,value;
    do
    {
        do
       {
              printf("Enter the element to be pushed\n");
          scanf("%d",&value);
          top = push(top,value);
          printf("Enter 1 to continue\n");
          scanf("%d",&n);
        } while(n == 1);

       printf("Enter 1 to pop an element\n");
       scanf("%d",&n);
       while( n == 1)
       {
```

```
            top = pop(top,&value);
        printf("The value poped is %d\n",value);
            printf("Enter 1 to pop an element\n");
            scanf("%d",&n);
      }
      printf("Enter 1 to continue\n");
      scanf("%d",&n);
   } while(n == 1);
}
```

Example

Input and Output

Enter the element to be pushed

10

Enter 1 to continue

1

Enter the element to be pushed

20

Enter 1 to continue

0

Enter 1 to pop an element

1

The value popped is 20

Enter 1 to pop an element

1

The value poped is 10

Enter 1 to pop an element

0

Enter 1 to continue

1

Enter the element to be pushed

30

Enter 1 to continue

1

```
Enter the element to be pushed
    40
Enter 1 to continue
    0
Enter 1 to pop an element
    1
The value popped is 40
Enter 1 to pop an element
    0
Enter 1 to continue
    1
Enter the element to be pushed
    50
Enter 1 to continue
    0
Enter 1 to pop an element
    1
The value popped is 50
Enter 1 to pop an element
    1
The value popped is 30
Enter 1 to pop an element
    0
Enter 1 to continue
    0
```

APPLICATIONS OF STACKS

Introduction

One of the applications of the stack is in expression evaluation. A complex assignment statement such as a = b + c*d/e–f may be interpreted in many different ways. Therefore, to give a unique meaning, the precedence and associativity rules are used. But still it is difficult to evaluate an expression by computer in its present form, called the infix notation. In *infix notation*, the binary operator comes in between the operands. A unary operator comes before the operand. To get it evaluated, it is first converted to the postfix form, where the operator comes after the operands. For example, the postfix form for the expression a*(b–c)/d is abc–*d/. A good thing about postfix expressions is that they do not require any precedence rules or parentheses for unique definition. So, evaluation of a postfix expression is possible using a stack-based algorithm.

Program

Convert an infix expression to prefix form.

```
#include <stdio.h>
#include <string.h>
#include <ctype.h>

#define N 80

typedef enum {FALSE, TRUE} bool;

#include "stack.h"
#include "queue.h"

#define NOPS 7

char operators [] = "()^/*+-";
int  priorities[] = {4,4,3,2,2,1,1};
char associates[] = "  RLLLL";

char t[N]; char *tptr = t;  // this is where prefix will be saved.
int getIndex( char op ) {
    /*
     * returns index of op in operators.
     */
    int i;
    for( i=0; i<NOPS; ++i )
       if( operators[i] == op )
              return i;
    return -1;
}
```

```
int getPriority( char op ) {
    /*
     * returns priority of op.
     */
    return priorities[ getIndex(op) ];
}

char getAssociativity( char op ) {
    /*
     * returns associativity of op.
     */
    return associates[ getIndex(op) ];
}

void processOp( char op, queue *q, stack *s ) {
    /*
     * performs processing of op.
     */
    switch(op) {
        case ')':
                printf( "\t S pushing )...\n" );
                sPush( s, op );
                break;
        case '(':
                while( !qEmpty(q) ) {
                        *tptr++ = qPop(q);
                        printf( "\tQ popping %c...\n", *(tptr-1) );
                }
                while( !sEmpty(s) ) {
                        char popop = sPop(s);
                        printf( "\tS popping %c...\n", popop );
                        if( popop == ')' )
                                break;
                        *tptr++ = popop;
                }
                break;
        default: {
                int priop;    // priority of op.
                char topop;   // operator on stack top.
                int pritop;   // priority of topop.
                char asstop;  // associativity of topop.
```

```
            while( !sEmpty(s) ) {
                  priop  = getPriority(op);
                  topop  = sTop(s);
                  pritop = getPriority(topop);
                  asstop = getAssociativity(topop);

    if( pritop < priop || (pritop == priop && asstop == 'L')
       || topop == ')' )     // IMP.
                        break;
                  while( !qEmpty(q) ) {
                        *tptr++ = qPop(q);
                  printf( "\tQ popping %c...\n", *(tptr-1) );
                  }
                  *tptr++ = sPop(s);
                  printf( "\tS popping %c...\n", *(tptr-1) );
            }
            printf( "\tS pushing %c...\n", op );
            sPush( s, op );
            break;
      }
   }
}

bool isop( char op ) {
   /*
    * is op an operator?
    */
   return (getIndex(op) != -1);
}

char *in2pre( char *str ) {
   /*
    * returns valid infix expr in str to prefix.
    */
   char *sptr;
   queue q = {NULL};
   stack s = NULL;
   char *res = (char *)malloc( N*sizeof(char) );
   char *resptr = res;
   tptr = t;
   for( sptr=str+strlen(str)-1; sptr!=str-1; -sptr ) {
      printf( "processing %c tptr-t=%d...\n", *sptr, tptr-t );
```

```
        if( isalpha(*sptr) ) // if operand.
                qPush( &q, *sptr );
        else if( isop(*sptr) )      // if valid operator.
                processOp( *sptr, &q, &s );
        else if( isspace(*sptr) )   // if whitespace.
                ;
        else {
                fprintf( stderr, "ERROR:invalid char %c.\n", *sptr );
                return "";
        }
    }
    while( !qEmpty(&q) ) {
        *tptr++ = qPop(&q);
        printf( "\tQ popping %c...\n", *(tptr-1) );
    }
    while( !sEmpty(&s) ) {
        *tptr++ = sPop(&s);
        printf( "\tS popping %c...\n", *(tptr-1) );
    }
    *tptr = 0;
    printf( "t=%s.\n", t );
    for( –tptr; tptr!=t-1; –tptr ) {
        *resptr++ = *tptr;
    }
    *resptr = 0;

    return res;
}

int main() {
    char s[N];

    puts( "enter infix freespaces max 80." );
    gets(s);
    while(*s) {
        puts( in2pre(s) );
        gets(s);
    }

    return 0;
}
```

Explanation

1. In an infix expression, a binary operator separates its operands (a unary operator precedes its operand). In a postfix expression, the operands of an operator precede the operator. In a prefix expression, the operator precedes its operands. Like postfix, a prefix expression is parenthesis-free, that is, any infix expression can be unambiguously written in its prefix equivalent without the need for parentheses.
2. To convert an infix expression to reverse-prefix, it is scanned from right to left. A queue of operands is maintained noting that the order of operands in infix and prefix remains the same. Thus, while scanning the infix expression, whenever an operand is encountered, it is pushed in a queue. If the scanned element is a right parenthesis (')'), it is pushed in a stack of operators. If the scanned element is a left parenthesis ('('), the queue of operands is emptied to the prefix output, followed by the popping of all the operators up to, but excluding, a right parenthesis in the operator stack.
3. If the scanned element is an arbitrary operator o, then the stack of operators is checked for operators with a greater priority then o. Such operators are popped and written to the prefix output after emptying the operand queue. The operator o is finally pushed to the stack.
4. When the scanning of the infix expression is complete, first the operand queue, and then the operator stack, are emptied to the prefix output. Any whitespace in the infix input is ignored. Thus the prefix output can be reversed to get the required prefix expression of the infix input.

Example

If the infix expression is a*b + c/d, then different snapshots of the algorithm, while scanning the expression from right to left, are shown in Table 19.1.

TABLE 19.1 Scanning the infex expression a*b+c/d from right to left

STEP	REMAINING EXPRESSION	SCANNED ELEMENT	QUEUE OF OPERANDS	STACK OF OPERATORS	PREFIX OUTPUT
0	a*b+c/d	nil	empty	empty	nil
1	a*b+c/	d	d	empty	nil
2	a*b+c	/	d	/	nil
3	a*b+	c	d c	/	nil
4	a*b	+	empty	+	dc/
5	a*	b	b	+	dc/
6	a	*	b	* +	dc/
7	nil	a	b a	* +	dc/
8	nil	nil	empty	empty	dc/ba*+

The final prefix output that we get is dc/ba*+ whose reverse is +*ab/cd, which is the prefix equivalent of the input infix expression a*b+c*d. Note that all the operands are simply pushed to the queue in steps 1, 3, 5, and 7. In step 2, the operator / is pushed to the empty stack of operators.

In step 4, the operator + is checked against the elements in the stack. Since / (division) has higher priority than + (addition), the queue is emptied to the prefix output (thus we get 'dc' as the output) and then the operator / is written (thus we get 'dc/' as the output). The operator + is then pushed to the stack. In step 6, the operator * is checked against the stack elements. Since * (multiplication) has a higher priority than + (addition), * is pushed to the stack. Step 8 signifies that all of the infix expression is scanned. Thus, the queue of operands is emptied to the prefix output (to get 'dc/ba'), followed by the emptying of the stack of operators (to get 'dc/ba*+').

Points to remember

1. A prefix expression is parenthesis-free.
2. To convert an infix expression to the postfix equivalent, it is scanned from right to left. The prefix expression we get is the reverse of the required prefix equivalent.

3. Conversion of infix to prefix requires a queue of operands and a stack, as in the conversion of an infix to postfix.
4. The order of operands in a prefix expression is the same as that in its infix equivalent.
5. If the scanned operator o1 and the operator o2 at the stack top have the same priority, then the associativity of o2 is checked. If o2 is right-associative, it is popped from the stack.

QUEUES

Introduction

A *queue* is also a list of elements with insertions permitted at one end—called the rear, and deletions permitted from the other end—called the front. This means that the removal of elements from a queue is possible in the same order in which the insertion of elements is made into the queue. Thus, a queue data structure exhibits the *FIFO* (*first in first out*) property. `insert` and `delete` are the operations that are provided for insertion of elements into the queue and the removal of elements from the queue, respectively. Shown in Figure 19.3 are the effects of `insert` and `delete` operations on the queue.

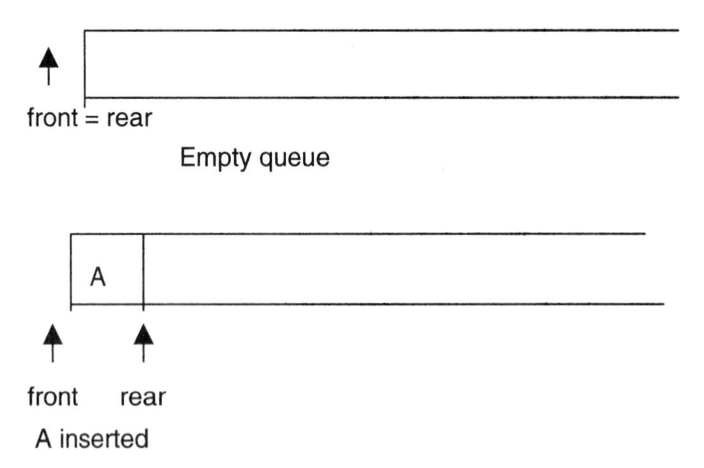

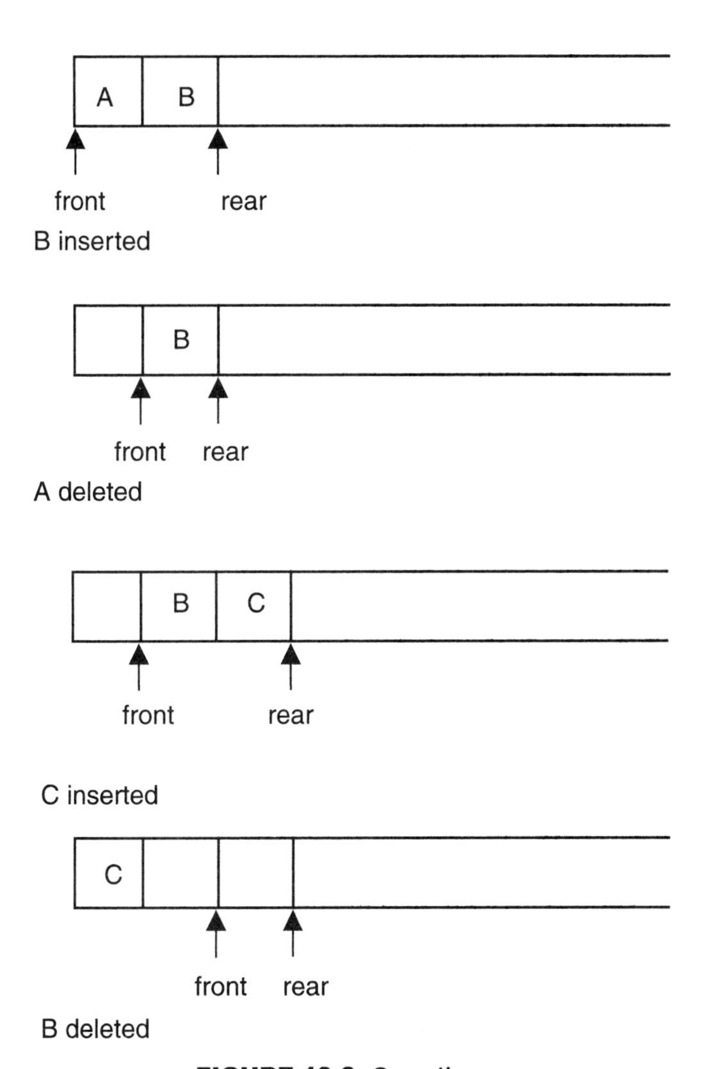

FIGURE 19.3 Operations on a queue.

IMPLEMENTATION OF QUEUES

Introduction

Since a queue is also a list, it can be implemented using an array or it can be implemented using a linked representation.

Array Implementation of a Stack

When an array is used to implement a queue, then the insert and delete operations are realized using the operations available on an array. The limitation of an array implementation is that the queue cannot grow and shrink dynamically as per the requirement.

Program

A complete C program to implement a queue by using an array is shown here:

```
#include <stdio.h>
#define MAX 10 /* The maximum size of the queue */
#include <stdlib.h>

void insert(int queue[], int *rear, int value)
{
    if(*rear < MAX-1)
   {
       *rear= *rear +1;
      queue[*rear] = value;
    }
   else
    {
       printf("The queue is full can not insert a  value\n");
       exit(0);
    }
}

void delete(int queue[], int *front, int rear, int * value)
{
    if(*front == rear)
   {
       printf("The queue is empty can not delete a  value\n");
       exit(0);
    }
    *front = *front + 1;
   *value = queue[*front];
}

void main()
{
```

```
    int queue[MAX];
   int front,rear;
   int n,value;
   front=rear=(-1);
   do
   {
      do
     {
           printf("Enter the element to be inserted\n");
       scanf("%d",&value);
       insert(queue,&rear,value);
       printf("Enter 1 to continue\n");
           scanf("%d",&n);
      } while(n == 1);

     printf("Enter 1 to delete an element\n");
     scanf("%d",&n);
     while( n == 1)
     {
           delete(queue,&front,rear,&value);
       printf("The value deleted is %d\n",value);
           printf("Enter 1 to delete an element\n");
       scanf("%d",&n);
      }
      printf("Enter 1 to continue\n");
      scanf("%d",&n);
   } while(n == 1);
}
```

Example

Input and Output

Enter the element to be inserted

10

Enter 1 to continue

1

Enter the element to be inserted

20

Enter 1 to continue

1

```
Enter the element to be inserted
    30
Enter 1 to continue
    0
Enter 1 to delete an element
    1
The value deleted is 10
Enter 1 to delete an element
    1
The value deleted is 20
Enter 1 to delete an element
    0
Enter 1 to continue
    1
Enter the element to be inserted
    40
Enter 1 to continue
    1
Enter the element to be inserted
    50
Enter 1 to continue
    0
Enter 1 to delete an element
    1
The value deleted is 30
Enter 1 to delete an element
    1
The value deleted is 40
Enter 1 to delete an element
    0
Enter 1 to continue
    0
```

CIRCULAR QUEUES

Introduction

The problem with the previous implementation is that the insert function gives a queue-full signal even if a considerable portion is free. This happens because the queue has a tendency to move to the right unless the 'front' catches up with the 'rear' and both are reset to 0 again (in the delete procedure). To overcome this problem, the elements of the array are required to shift one position left whenever a deletion is made. But this will make the deletion process inefficient. Therefore, an efficient way of overcoming this problem is to consider the array to be circular, as shown in Figure 19.4.

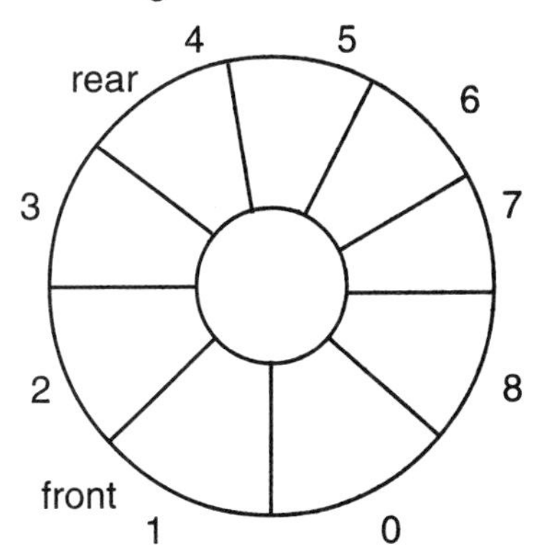

FIGURE 19.4 Circular queue.

Program

```
#include <stdio.h>
#define MAX 10 /* The maximum size of the queue */
#include <stdlib.h>

void insert(int queue[], int *rear, int front, int value)
{
    *rear= (*rear +1) % MAX;
   if(*rear == front)
   {
      printf("The queue is full can not insert a  value\n");
      exit(0);
   }
   queue[*rear] = value;
```

```
}

void delete(int queue[], int *front, int rear, int * value)
{
    if(*front == rear)
    {
        printf("The queue is empty can not delete a  value\n");
        exit(0);
    }
    *front = (*front + 1) % MAX;
    *value = queue[*front];
}

void main()
{
    int queue[MAX];
    int front,rear;
    int n,value;
    front=0; rear=0;
    do
    {
        do
        {
                printf("Enter the element to be inserted\n");
            scanf("%d",&value);
            insert(queue,&rear,front,value);
            printf("Enter 1 to continue\n");
            scanf("%d",&n);
        } while(n == 1);

    printf("Enter 1 to delete an element\n");
    scanf("%d",&n);
    while( n == 1)
    {
        delete(queue,&front,rear,&value);
        printf("The value deleted is %d\n",value);
        printf("Enter 1 to delete an element\n");
        scanf("%d",&n);
    }
    printf("Enter 1 to continue\n");
    scanf("%d",&n);
    } while(n == 1);
}
```

Example

Input and Output

Enter the element to be inserted

10

Enter 1 to continue

1

Enter the element to be inserted

20

Enter 1 to continue

1

Enter the element to be inserted

30

Enter 1 to continue

1

Enter the element to be inserted

40

Enter 1 to continue

1

Enter the element to be inserted

50

Enter 1 to continue

1

Enter the element to be inserted

60

Enter 1 to continue

1

Enter the element to be inserted

70

Enter 1 to continue

1

Enter the element to be inserted

80

```
Enter 1 to continue
    1
Enter the element to be inserted
    90
Enter 1 to continue
    0
Enter 1 to delete an element
    1
The value deleted is 10
Enter 1 to delete an element
    1
The value deleted is 20
Enter 1 to delete an element
    0
Enter 1 to continue
    1
Enter the element to be inserted
    100
Enter 1 to continue
    1
Enter the element to be inserted
    110
Enter 1 to continue
    0
Enter 1 to delete an element
    1
The value deleted is 30
Enter 1 to delete an element
    1
The value deleted is 40
```

```
Enter 1 to delete an element
    0
Enter 1 to continue
    1
Enter the element to be inserted
    120
Enter 1 to continue
    1
Enter the element to be inserted
    130
Enter 1 to continue
    0
Enter 1 to delete an element
    0
Enter 1 to continue
    0
```

IMPLEMENTATION OF A QUEUE USING LINKED REPRESENTATION

Introduction

Initially, the list is empty, so both the front and rear pointers are `NULL`. The `insert` function creates a new node, puts the new data value in it, appends it to an existing list, and makes the rear pointer point to it. A `delete` function checks whether the queue is empty, and if not, retrieves the data value of the node pointed to by the front, advances the front, and frees the storage of the node whose data value has been retrieved.

If the above strategy is used for creating a queue with four data values —10, 20, 30, and 40, the queue gets created as shown in Figure 19.5.

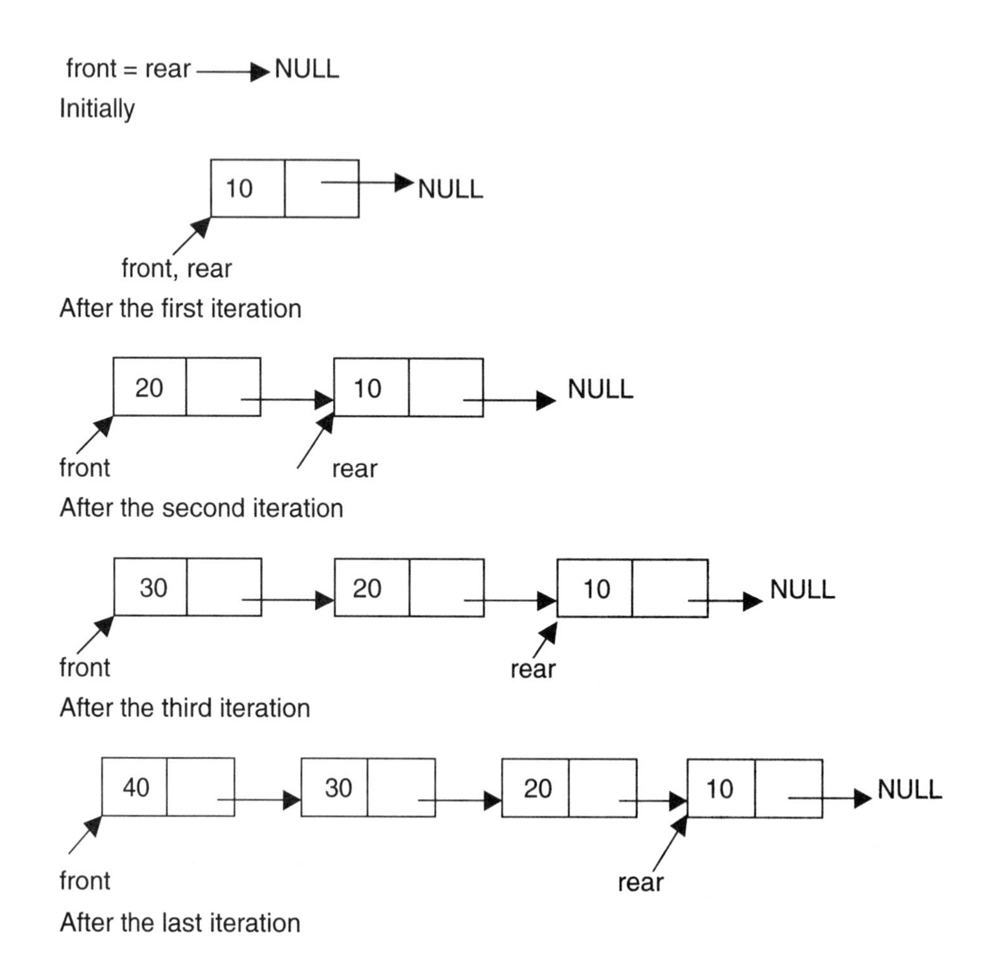

FIGURE 19.5 Linked queue.

Program

A complete C program for implementation of a stack using the linked list is shown here:

```
# include <stdio.h>
# include <stdlib.h>
struct node
{
    int data;
    struct node *link;
```

```
};
void insert(struct node **front, struct node **rear, int value)
{
    struct node *temp;
    temp=(struct node *)malloc(sizeof(struct node));
        /* creates new node
        using data value
        passed as  parameter */
    if(temp==NULL)
   {
        printf("No Memory available Error\n");
      exit(0);
    }
   temp->data = value;
   temp->link=NULL;
    if(*rear == NULL)
   {
        *rear = temp;
        *front = *rear;
    }
   else
    {
        (*rear)->link = temp;
        *rear = temp;
    }

}

void delete(struct node **front, struct node **rear, int *value)
{
    struct node *temp;
   if((*front == *rear)  && (*rear == NULL))
   {
        printf(" The queue is empty cannot delete Error\n");
        exit(0);
    }
    *value = (*front)->data;
   temp = *front;
   *front = (*front)->link;
   if(*rear == temp)
    *rear = (*rear)->link;
   free(temp);
}
```

```
void main()
{
    struct node *front=NULL,*rear = NULL;
   int n,value;
   do
   {
      do
     {
            printf("Enter the element to be inserted\n");
       scanf("%d",&value);
       insert(&front,&rear,value);
       printf("Enter 1 to continue\n");
       scanf("%d",&n);
      } while(n == 1);

      printf("Enter 1 to delete an element\n");
     scanf("%d",&n);
      while( n == 1)
     {
            delete(&front,&rear,&value);
            printf("The value deleted is %d\n",value);
            printf("Enter 1 to delete an element\n");
            scanf("%d",&n);
      }
     printf("Enter 1 to continue\n");
     scanf("%d",&n);
   } while(n == 1);
}
```

Example

Input and Output

Enter the element to be inserted

10

Enter 1 to continue

1

Enter the element to be inserted

20

```
Enter 1 to continue
    1
Enter the element to be inserted
    30
Enter 1 to continue
    0
Enter 1 to delete an element
    1
The value deleted is 10
Enter 1 to delete an element
    1
The value deleted is 20
Enter 1 to delete an element
    0
Enter 1 to continue
    1
Enter the element to be inserted
    40
Enter 1 to continue
    1
Enter the element to be inserted
    50
Enter 1 to continue
    0
Enter 1 to delete an element
    1
The value deleted  is 30
Enter 1 to pop an element
    1
The value deleted is 40
Enter 1 to delete an element
    1
```

The value deleted is 50

Enter 1 to delete an element

1

The queue is empty, cannot delete Error

APPLICATIONS OF QUEUES

Introduction

One application of the queue data structure is in the implementation of priority queues required to be maintained by the scheduler of an operating system. It is a queue in which each element has a priority value and the elements are required to be inserted in the queue in decreasing order of priority. This requires a change in the function that is used for insertion of an element into the queue. No change is required in the `delete` function.

Program

A complete C program implementing a priority queue is shown here:

```
# include <stdio.h>
# include <stdlib.h>
struct node
{
    int data;
   int priority;
    struct node *link;
};
void insert(struct node **front, struct node **rear, int value, int
priority)
{
    struct node *temp,*temp1;
    temp=(struct node *)malloc(sizeof(struct node));
        /* creates new node using data value
    passed as  parameter */
   if(temp==NULL)
    {
        printf("No Memory available Error\n");
      exit(0);
```

```
      }
      temp->data = value;
      temp->priority = priority;
      temp->link=NULL;
      if(*rear == NULL)   /* This is the first node */
      {
          *rear = temp;
          *front = *rear;
      }
      else
          if((*front)->priority < priority)
                 /* the element to be inserted has
          highest priority hence should
                 be the first element*/
          {
                 temp->link = *front;
                 *front = temp;
          }
        else
                 if( (*rear)->priority > priority)
                        /* the element to be inserted has
                        lowest priority hence should
                        be the last element*/
                 {
                        (*rear)->link = temp;
                        *rear = temp;
                 }
           else
           {
                        temp1 = *front;
                        while((temp1->link)->priority >= priority)
          /* find the position and insert the new element */
                               temp1=temp1->link;
                  temp->link = temp1->link;
                  temp1->link = temp;
                 }
   }
   void delete(struct node **front, struct node **rear, int *value, int
*priority)
   {
       struct node *temp;
      if((*front == *rear)  && (*rear == NULL))
```

```
        {
            printf(" The queue is empty cannot delete Error\n");
          exit(0);
        }
        *value = (*front)->data;
        *priority = (*front)->priority;
        temp = *front;
        *front = (*front)->link;
        if(*rear == temp)
            *rear = (*rear)->link;
         free(temp);
    }
    void main()
    {
         struct node *front=NULL,*rear = NULL;
        int n,value, priority;
        do
        {
            do
            {
                   printf("Enter the element to be inserted and its
priority\n");
              scanf("%d %d",&value,&priority);
              insert(&front,&rear,value,priority);
              printf("Enter 1 to continue\n");
              scanf("%d",&n);
            } while(n == 1);

            printf("Enter 1 to delete an element\n");
          scanf("%d",&n);
          while( n == 1)
          {
                   delete(&front,&rear,&value,&priority);
                   printf("The value deleted is %d\ and its priority is %d
\n",
                          value,priority);
                   printf("Enter 1 to delete an element\n");
              scanf("%d",&n);
          }
           printf("Enter 1 to continue\n");
          scanf("%d",&n);
        } while(n == 1);
    }
```

Example

Input and Output

Enter the element to be inserted and its priority

10 90

Enter 1 to continue

1

Enter the element to be inserted and its priority

5 8

Enter 1 to continue

1

Enter the element to be inserted and its priority

11 60

Enter 1 to continue

1

Enter the element to be inserted and its priority

12 75

Enter 1 to continue

1

Enter the element to be inserted and its priority

13 10

Enter 1 to continue

1

Enter the element to be inserted and its priority

14 6

Enter 1 to continue

0

Enter 1 to delete an element

1

The value deleted is 10 and its priority is 90

Enter 1 to delete an element

1

```
The value deleted is 12 and its priority is 75
Enter 1 to delete an element
    1
The value deleted is 11 and its priority is 60
Enter 1 to delete an element
    1
The value deleted is 13 and its priority is 10
Enter 1 to delete an element
    1
The value deleted is 5 and its priority is 8
Enter 1 to delete an element
    1
The value deleted is 14 and its priority is 6
Enter 1 to delete an element
    1
The queue is empty cannot delete Error
```

Points to Remember

1. A stack is basically a list with insertions and deletions permitted from only one end, called the stack-top, so it is a data structure that exhibits the LIFO property.
2. The operations that are permitted to manipulate a stack are push and pop.
3. One of the important applications of a stack is in the implementation of recursion in the programming language.
4. A queue is also a list with insertions permitted from one end, called rear, and deletions permitted from the other end, called front. So it is a data structure that exhibits the FIFO property.
5. The operations that are permitted on a queue are `insert` and `delete`.
6. A circular queue is a queue in which the element next to the last element is the first element.
7. When the size of the stack/queue is known beforehand, the array implementation can be used and is more efficient.

8. When the size of the stack/queue is not known beforehand, then the linked representation is used. It provides more flexibility.

Exercises

1. Write a C program to implement a stack of characters.
2. Write a C program to implement a double-ended queue, which is a queue in which insertions and deletions may be performed at either end. Use a linked representation.

20 Linked Lists

THE CONCEPT OF THE LINKED LIST

Introduction

When dealing with many problems we need a dynamic list, dynamic in the sense that the size requirement need not be known at compile time. Thus, the list may grow or shrink during runtime. A *linked list* is a data structure that is used to model such a dynamic list of data items, so the study of the linked lists as one of the data structures is important.

Concept

An array is represented in memory using sequential mapping, which has the property that elements are fixed distance apart. But this has the following disadvantage: It makes insertion or deletion at any arbitrary position in an array a costly operation, because this involves the movement of some of the existing elements.

When we want to represent several lists by using arrays of varying size, either we have to represent each list using a separate array of maximum size or we have to represent each of the lists using one single array. The first one will lead to wastage of storage, and the second will involve a lot of data movement.

So we have to use an alternative representation to overcome these disadvantages. One alternative is a linked representation. In a linked representation, it is not necessary that the elements be at a fixed distance apart. Instead, we can place elements anywhere in memory, but to make it a part of the same list, an element is required to be linked with a previous element of the list. This can be done by storing the address of the next element in the previous element itself. This requires that every element be capable of holding the data as well as the address of the next element. Thus every element must be a structure with a minimum of two fields, one for holding the data value, which we call a data field, and the other for holding the address of the next element, which we call link field.

Therefore, a linked list is a list of elements in which the elements of the list can be placed anywhere in memory, and these elements are linked with each other using an explicit link field, that is, by storing the address of the next element in the link field of the previous element.

Program

Here is a program for building and printing the elements of the linked list:

```
# include <stdio.h>
# include <stdlib.h>
struct node
{
int data;
struct node *link;
};
struct node *insert(struct node *p , int n)
{
struct node *temp;
 /* if the existing list is empty then insert a new node as the
starting node */
 if(p==NULL)
   {
     p=(struct node *)malloc(sizeof(struct node)); /* creates new node
data value passes
 as  parameter */

     if(p==NULL)
     {
   printf("Error\n");
         exit(0);
     }
      p-> data = n;
      p-> link = p; /* makes the pointer pointing to itself because it
is a circular list*/
   }
   else
   {
     temp = p;
 /* traverses the existing list to get the pointer to the last node of
```

```
it */
     while (temp-> link != p)
         temp = temp-> link;
          temp-> link =  (struct node *)malloc(sizeof(struct node)); /*
creates new node using
           data value passes as
             parameter  and puts its
            address in the link field
            of last node of the
            existing list*/
          if(temp -> link == NULL)
          {
        printf("Error\n");
              exit(0);
          }
          temp = temp-> link;
          temp-> data = n;
          temp-> link = p;
         }
         return (p);
    }
    void printlist ( struct node *p   )
    {
     struct node *temp;
      temp = p;
     printf("The data values in the list are\n");
        if(p!= NULL)
        {
           do
                 {
                 printf("%d\t",temp->data);
                 temp=temp->link;
                 } while (temp!= p);
          }
          else
             printf("The list is empty\n");
     }

    void main()
    {
          int n;
          int x;
          struct node *start = NULL ;
```

```
        printf("Enter the nodes to be created \n");
        scanf("%d",&n);
        while ( n-- > 0 )
        {
    printf( "Enter the data values to be placed in a node\n");
            scanf("%d",&x);
            start = insert ( start , x );
        }
        printf("The created list is\n");
        printlist ( start );
}
```

Explanation

1. This program uses a strategy of inserting a node in an existing list to get the list created. An `insert` function is used for this.
2. The `insert` function takes a pointer to an existing list as the first parameter, and a data value with which the new node is to be created as a second parameter, creates a new node by using the data value, appends it to the end of the list, and returns a pointer to the first node of the list.
3. Initially the list is empty, so the pointer to the starting node is `NULL`. Therefore, when `insert` is called first time, the new node created by the `insert` becomes the start node.
4. Subsequently, the insert traverses the list to get the pointer to the last node of the existing list, and puts the address of the newly created node in the link field of the last node, thereby appending the new node to the existing list.
5. The main function reads the value of the number of nodes in the list. Calls iterate that many times by going in a `while` loop to create the links with the specified number of nodes.

Points to Remember

1. Linked lists are used when the quantity of data is not known prior to execution.
2. In linked lists, data is stored in the form of nodes and at runtime, memory is allocated for creating nodes.
3. Due to overhead in memory allocation and deallocation, the speed of the program is lower.

4. The data is accessed using the starting pointer of the list.

INSERTING A NODE BY USING RECURSIVE PROGRAMS

Introduction

A linked list is a recursive data structure. A *recursive data structure* is a data structure that has the same form regardless of the size of the data. You can easily write recursive programs for such data structures.

Program

```
# include <stdio.h>
# include <stdlib.h>
struct node
{
int data;
struct node *link;
};
struct node *insert(struct node *p , int n)
{
  struct node *temp;
   if(p==NULL)
   {
      p=(struct node *)malloc(sizeof(struct node));
      if(p==NULL)
      {
    printf("Error\n");
            exit(0);
      }
       p-> data = n;
       p-> link = NULL;
    }
    else
            p->link = insert(p->link,n);/* the while loop replaced by
recursive call */

   return (p);
}
```

```
void printlist ( struct node *p  )
{
        printf("The data values in the list are\n");
        while (p!= NULL)
       {
    printf("%d\t",p-> data);
          p = p-> link;
       }
}
void main()
{
       int n;
       int x;
       struct node *start = NULL ;
       printf("Enter the nodes to be created \n");
       scanf("%d",&n);
       while ( n- > 0 )
       {
    printf( "Enter the data values to be placed in a node\n");
          scanf("%d",&x);
          start = insert ( start , x );
       }
       printf("The created list is\n");
       printlist ( start );

}
```

Explanation

1. This recursive version also uses a strategy of inserting a node in an existing list to create the list.
2. An `insert` function is used to create the list. The `insert` function takes a pointer to an existing list as the first parameter, and a data value with which the new node is to be created as the second parameter. It creates the new node by using the data value, then appends it to the end of the list. It then returns a pointer to the first node of the list.
3. Initially, the list is empty, so the pointer to the starting node is `NULL`. Therefore, when `insert` is called the first time, the new node created by the `insert` function becomes the start node.
4. Subsequently, the `insert` function traverses the list by recursively calling itself.

5. The recursion terminates when it creates a new node with the supplied data value and appends it to the end of the list.

Points to Remember

1. A linked list has a recursive data structure.
2. Writing recursive programs for such structures is programmatically convenient.

SORTING AND REVERSING A LINKED LIST

Introduction

To sort a linked list, first we traverse the list searching for the node with a minimum data value. Then we remove that node and append it to another list which is initially empty. We repeat this process with the remaining list until the list becomes empty, and at the end, we return a pointer to the beginning of the list to which all the nodes are moved, as shown in Figure 20.1.

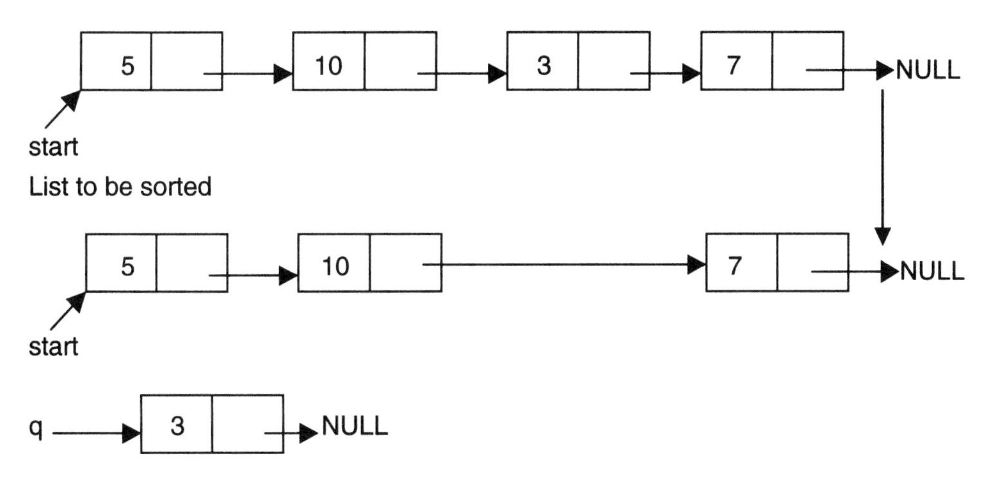

FIGURE 20.1 Sorting of a linked list.

To reverse a list, we maintain a pointer each to the previous and the next node, then we make the link field of the current node point to the previous, make the previous equal to the current, and the current equal to the next, as shown in Figure 20.2.

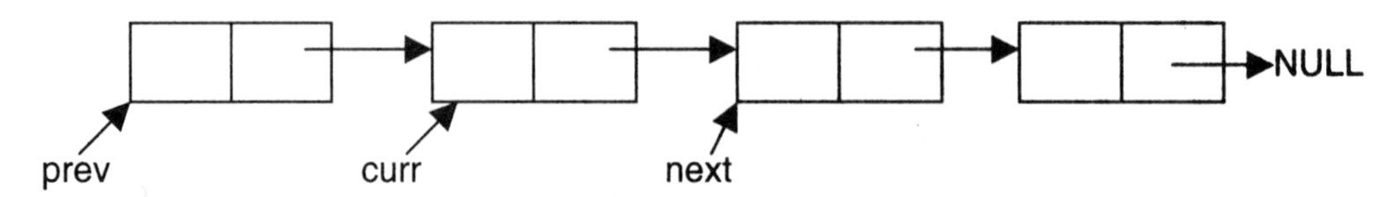

FIGURE 20.2 A linked list showing the previous, current, and next nodes at some point during reversal process.

Therefore, the code needed to reverse the list is

```
Prev = NULL;
While (curr != NULL)
{
    Next = curr->link;
    Curr -> link = prev;
    Prev = curr;
    Curr = next;
}
```

Program

```
# include <stdio.h>
# include <stdlib.h>
struct node
{
    int data;
    struct node *link;
};
struct node *insert(struct node *p , int n)
{
    struct node *temp;
   if(p==NULL)
   {
      p=(struct node *)malloc(sizeof(struct node));
      if(p==NULL)
      {
            printf("Error\n");
         exit(0);
      }
       p-> data = n;
```

```
          p-> link = NULL;
      }
      else
      {
        temp = p;
        while (temp-> link!= NULL)
          temp = temp-> link;
        temp-> link =  (struct node *)malloc(sizeof(struct node));
        if(temp -> link == NULL)
        {
                printf("Error\n");
           exit(0);
        }
          temp = temp-> link;
         temp-> data = n;
        temp-> link = NULL;
       }
       return (p);
}

void printlist ( struct node *p  )
{
        printf("The data values in the list are\n");
        while (p!= NULL)
       {
                printf("%d\t",p-> data);
           p = p-> link;
       }
}

/* a function to reverse a list */
struct node *reverse( struct node *p )
{
   struct node *prev, *curr;
   prev = NULL;
   curr = p;
   while ( curr!= NULL)
   {
      p = p-> link;
      curr-> link = prev;
      prev = curr;
      curr = p;
   }
```

```
    return(prev);
}

/* a function to sort a list */
struct node *sortlist(struct node *p)
{
     struct node *temp1,*temp2,*min,*prev,*q;
    q = NULL;
    while(p != NULL)
    {
        prev = NULL;
        min = temp1 = p;
        temp2 = p -> link;
        while ( temp2 != NULL )
        {
                if(min -> data > temp2 -> data)
            {
                    min = temp2;
                    prev = temp1;
            }
            temp1 = temp2;
            temp2 = temp2-> link;
        }
         if(prev == NULL)
                p = min -> link;
         else
                prev -> link = min -> link;
        min -> link = NULL;
        if( q == NULL)
       q = min; /* moves the node with lowest data value in the list
pointed to by p to the list
    pointed to by q as a first node*/
        else
        {
            temp1 = q;
            /* traverses the list pointed to by q to get pointer to its
last node */
            while( temp1 -> link != NULL)
                    temp1 = temp1 -> link;
            temp1 -> link = min; /* moves the node with lowest data value
in the list pointed to
    by p to the list pointed to by q at the end of list pointed by
    q*/
```

```
            }
        }
        return (q);
    }

    void main()
    {
            int n;
            int x;
            struct node *start = NULL ;
            printf("Enter the nodes to be created \n");
            scanf("%d",&n);
            while ( n- > 0 )
            {
                    printf( "Enter the data values to be placed in a
node\n");
              scanf("%d",&x);
              start = insert ( start ,x);
            }
            printf("The created list is\n");
            printlist ( start );
            start = sortlist(start);
            printf("The sorted list is\n");
            printlist ( start );
            start = reverse(start);
            printf("The reversed list is\n");
            printlist ( start );
    }
```

Explanation

The working of the sorting function on an example list is shown in Figure 20.3.

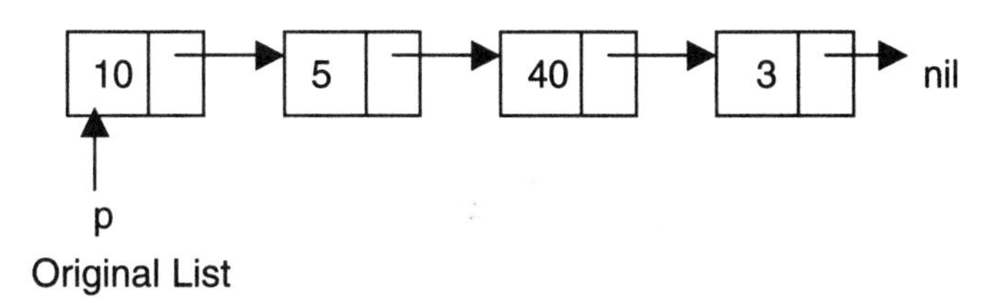

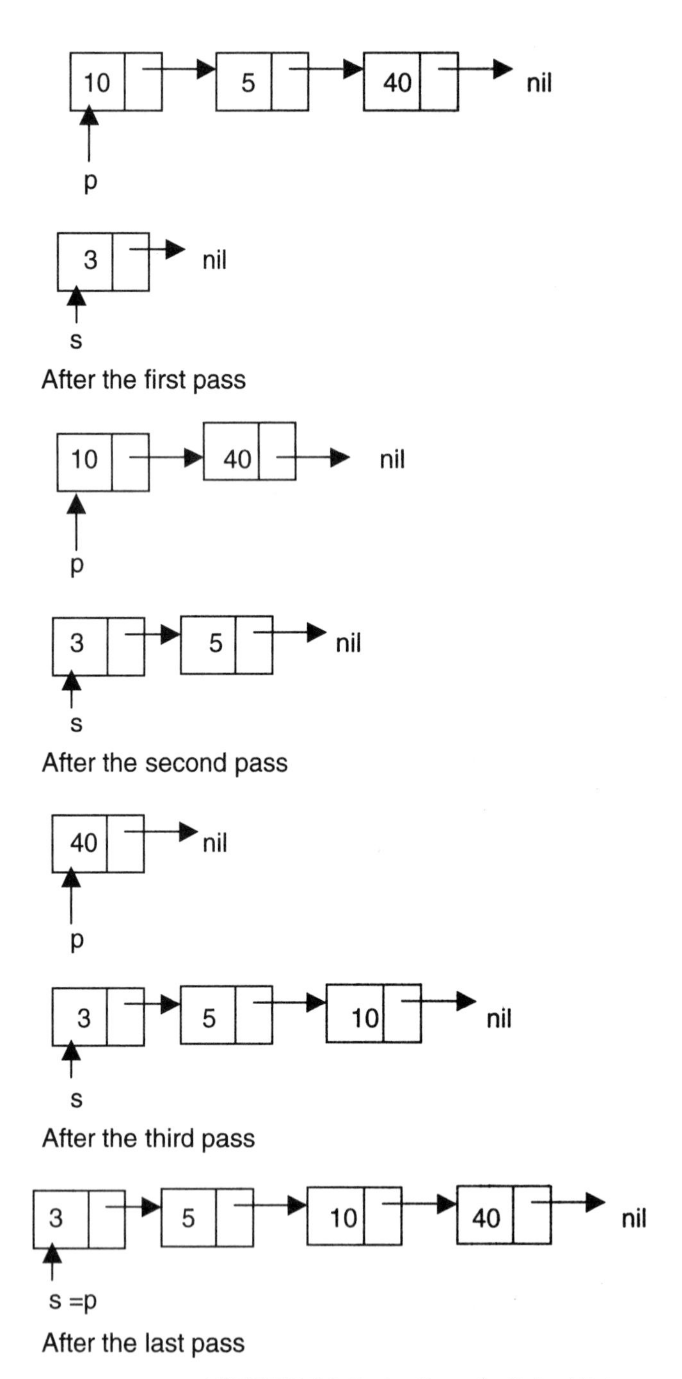

FIGURE 20.3 Sorting of a linked list.

The working of a reverse function is shown in Figure 20.4.

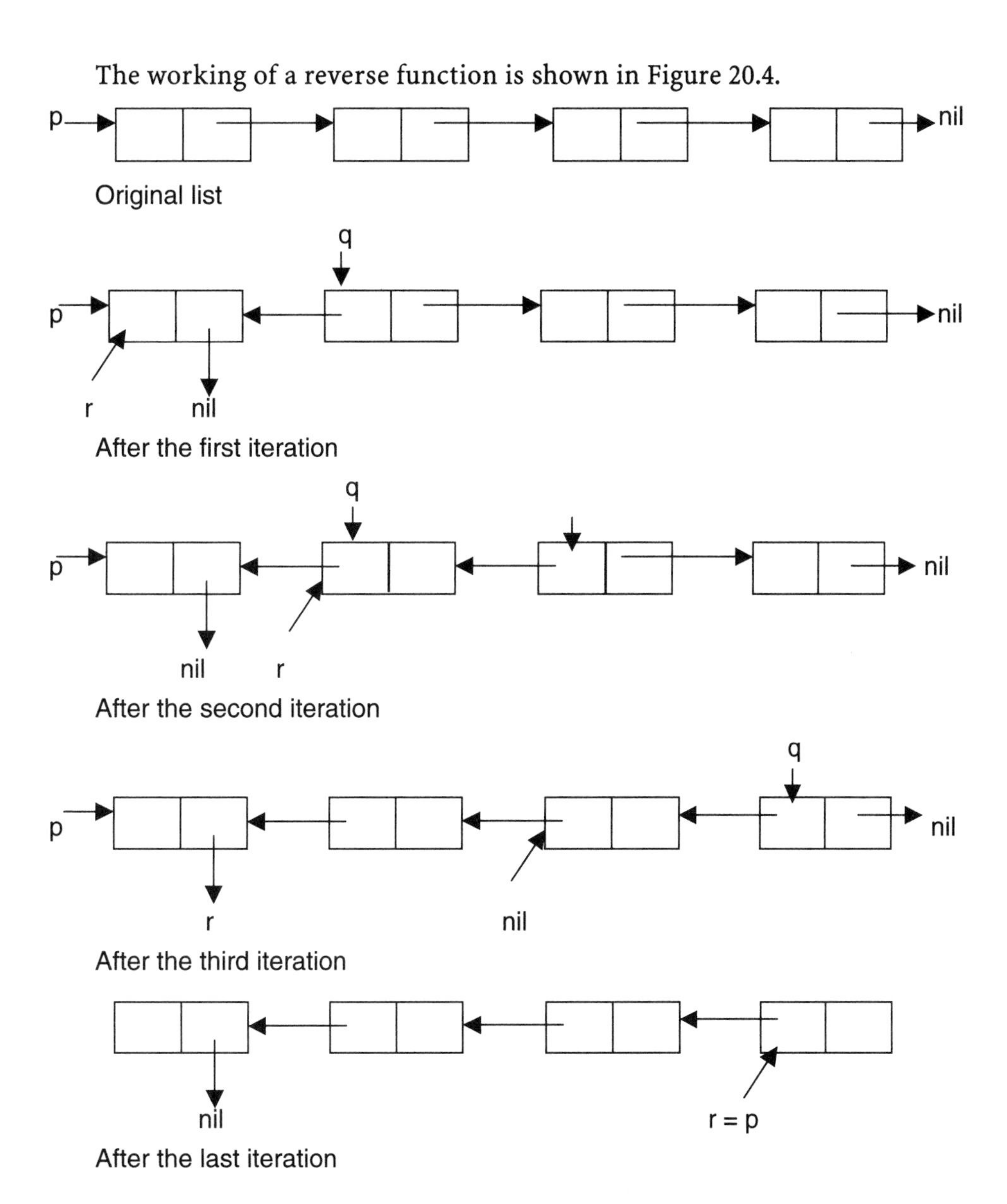

FIGURE 20.4 Reversal of a list.

DELETING THE SPECIFIED NODE IN A SINGLY LINKED LIST

Introduction

To delete a node, first we determine the node number to be deleted (this is based on the assumption that the nodes of the list are numbered serially from 1 to n). The list is then traversed to get a pointer to the node whose number is given, as well as a pointer to a node that appears before the node to be deleted. Then the link field of the node that appears before the node to be deleted is made to point to the node that appears after the node to be deleted, and the node to be deleted is freed. Figures 20.5 and 20.6 show the list before and after deletion, respectively.

Program

```
# include <stdio.h>
# include <stdlib.h>
struct node *delet ( struct node *, int );
int length ( struct node * );
struct node
{
    int data;
    struct node *link;
};
struct node *insert(struct node *p , int n)
{
    struct node *temp;
   if(p==NULL)
   {
      p=(struct node *)malloc(sizeof(struct node));
      if(p==NULL)
      {
              printf("Error\n");
         exit(0);
      }
      p-> data = n;
      p-> link = NULL;
    }
    else
    {
      temp = p;
```

```
      while (temp-> link != NULL)
       temp = temp-> link;
       temp-> link  =  (struct node *)malloc(sizeof(struct node));
      if(temp -> link == NULL)
      {
              printf("Error\n");
         exit(0);
      }
       temp = temp-> link;
      temp-> data = n;
      temp-> link= NULL;
     }
     return (p);
}

void printlist ( struct node *p  )
{
   printf("The data values in the list are\n");
   while (p!= NULL)
    {
       printf("%d\t",p-> data);
      p = p-> link;
    }
}
void main()
{
    int n;
   int x;
   struct node *start = NULL;
   printf("Enter the nodes to be created \n");
   scanf("%d",&n);
   while ( n-- > 0 )
   {
       printf( "Enter the data values to be placed in a node\n");
      scanf("%d",&x);
      start = insert ( start, x );
   }
    printf(" The list before deletion is\n");
   printlist ( start );
   printf(" \n Enter the node no \n");
   scanf ( " %d",&n);
   start = delet (start , n );
```

```
    printf(" The list after deletion is\n");
    printlist ( start );
}

  /* a function to delete the specified node*/
struct node *delet ( struct node *p , int node_no )
{
    struct node *prev , *curr ;
    int i;

    if (p == NULL )
    {
        printf("There is no node to be deleted \n");
    }
    else
    {
        if ( node_no > length (p))
        {
                printf("Error\n");
        }
        else
        {
                prev = NULL;
                curr = p;
                i = 1 ;
                while ( i < node_no )
                {
                        prev = curr;
                        curr = curr-> link;
                        i = i+1;
                }
                if ( prev == NULL )
                {
                        p = curr -> link;
                        free ( curr );
                }
                else
                {
                        prev -> link  = curr -> link ;
                        free ( curr );
                }
        }
```

```
    }
    return(p);
}
/* a function to compute the length of a linked list */
int length ( struct node *p )
{
    int count = 0 ;
    while ( p != NULL )
    {
        count++;
        p = p->link;
    }
    return ( count ) ;

}
```

Explanation

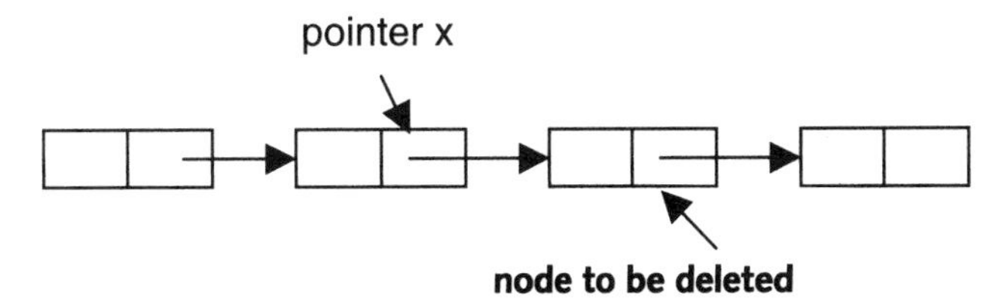

FIGURE 20.5 Before deletion.

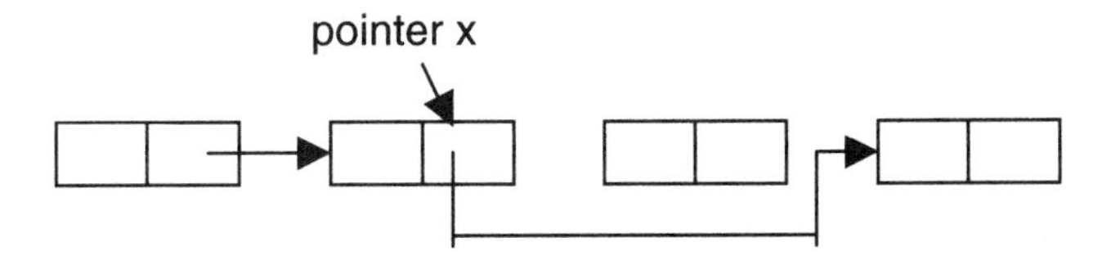

FIGURE 20.6 After deletion.

INSERTING A NODE AFTER THE SPECIFIED NODE IN A SINGLY LINKED LIST

Introduction

To insert a new node after the specified node, first we get the number of the node in an existing list after which the new node is to be inserted. This is based on the assumption that the nodes of the list are numbered serially from 1 to n. The list is then traversed to get a pointer to the node, whose number is given. If this pointer is x, then the link field of the new node is made to point to the node pointed to by x, and the link field of the node pointed to by x is made to point to the new node. Figures 20.7 and 20.8 show the list before and after the insertion of the node, respectively.

Program

```
# include <stdio.h>
# include <stdlib.h>
int length ( struct node * );
struct node
{
    int data;
    struct node *link;
};

/* a function which appends a new node to an existing list used for
building a list */
struct node *insert(struct node *p , int n)
{
    struct node *temp;
   if(p==NULL)
   {
       p=(struct node *)malloc(sizeof(struct node));
      if(p==NULL)
      {
              printf("Error\n");
         exit(0);
       }
      p-> data = n;
      p-> link = NULL;
```

```
        }
       else
       {
           temp = p;
           while (temp-> link != NULL)
                  temp = temp-> link;
           temp-> link  =  (struct node *)malloc(sizeof(struct node));
          if(temp -> link == NULL)
          {
                  printf("Error\n");
            exit(0);
           }
           temp = temp-> link;
           temp-> data = n;
          temp-> link= NULL;
        }
       return (p);
    }

    /* a function which inserts a newly created node after the specified
node */
    struct node * newinsert ( struct node *p, int node_no, int value )
    {
        struct node *temp , * temp1;
       int i;
       if ( node_no <= 0 || node_no > length (p))
       {
           printf("Error! the specified node does not exist\n");
          exit(0);
        }
       if ( node_no == 0)
       {
           temp = ( struct node * )malloc ( sizeof ( struct node ));
           if ( temp == NULL )
          {
                  printf( " Cannot allocate \n");
                  exit (0);
           }
          temp -> data = value;
          temp -> link = p;
          p = temp ;
        }
```

```
    else
    {
        temp = p ;
       i = 1;
       while ( i < node_no )
       {
              i = i+1;
              temp = temp-> link ;
       }
       temp1 = ( struct node * )malloc ( sizeof(struct node));
       if ( temp == NULL )
       {
              printf("Cannot allocate \n");
              exit(0);
       }
       temp1 -> data = value ;
       temp1 -> link = temp -> link;
       temp -> link = temp1;
    }
     return (p);
}

void printlist ( struct node *p  )
{
     printf("The data values in the list are\n");
     while (p!= NULL)
     {
        printf("%d\t",p-> data);
        p = p-> link;
     }
}
void main ()
{
     int n;
     int x;
     struct node *start = NULL;
     printf("Enter the nodes to be created \n");
     scanf("%d",&n);
     while ( n– > 0 )
     {
```

```
        printf( "Enter the data values to be placed in a node\n");
        scanf("%d",&x);
        start = insert ( start, x );
    }
    printf(" The list before deletion is\n");
    printlist ( start );
    printf(" \n Enter the node no after which the insertion is to be
done\n");
    scanf ( " %d",&n);
    printf("Enter the value of the node\n");
    scanf("%d",&x);
    start = newinsert(start,n,x);
    printf("The list after insertion is \n");
    printlist(start);
}
```

Explanation

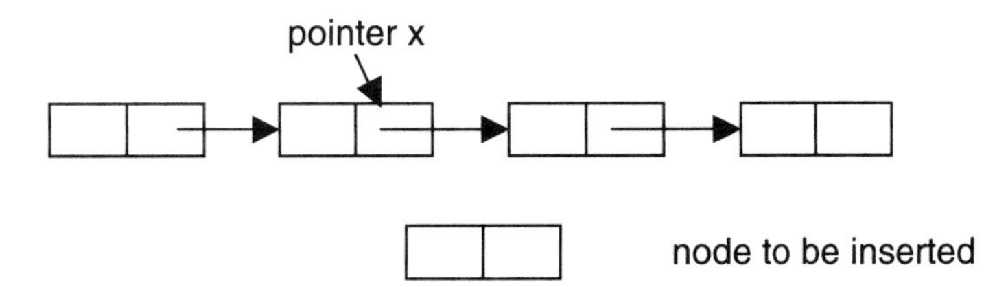

FIGURE 20.7 Before insertion.

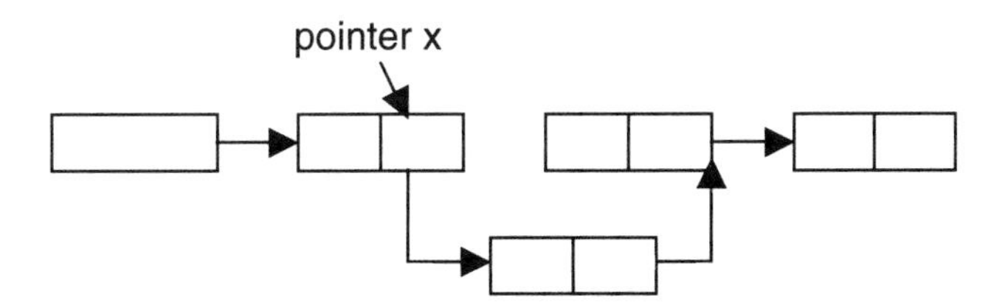

FIGURE 20.8 After insertion.

INSERTING A NEW NODE IN A SORTED LIST

Introduction

To insert a new node into an already sorted list, we compare the data value of the node to be inserted with the data values of the nodes in the list starting from the first node. This is continued until we get a pointer to the node that appears immediately before the node in the list whose data value is greater than the data value of the node to be inserted.

Program

Here is a complete program to insert an element in a sorted list of elements using the linked list representation so that after insertion, it will remain a sorted list.

```
# include <stdio.h>
# include <stdlib.h>
struct node
{
int data;
struct node *link;
};
struct node *insert(struct node * , int);
struct node *sinsert(struct node*  , int );
void printlist ( struct node *  );
struct node *sortlist(struct node *);

struct node *insert(struct node *p , int n)
{
    struct node *temp;
   if(p==NULL)
   {
      p=(struct node *)malloc(sizeof(struct node));
      if(p==NULL)
      {
    printf("Error\n");
            exit(0);
      }
       p-> data = n;
       p-> link = NULL;
    }
```

```
    else
    {
      temp = p;
      while (temp-> link!= NULL)
     temp = temp-> link;
      temp-> link =  (struct node *)malloc(sizeof(struct node));
      if(temp -> link == NULL)
      {
    printf("Error\n");
         exit(0);
      }
    temp = temp-> link;
      temp-> data = n;
      temp-> link = NULL;
     }
     return (p);
}


void printlist ( struct node *p  )
{
       printf("The data values in the list are\n");
       while (p!= NULL)
      {
    printf("%d\t",p-> data);
         p = p-> link;
      }
}


/* a function to sort a list */
struct node *sortlist(struct node *p)
{
    struct node *temp1,*temp2,*min,*prev,*q;
   q = NULL;
   while(p != NULL)
   {
    prev = NULL;
      min = temp1 = p;
      temp2 = p -> link;
      while ( temp2 != NULL )
      {
```

```
        if(min -> data > temp2 -> data)
            {
        min = temp2;
                prev = temp1;
            }
            temp1 = temp2;
            temp2 = temp2-> link;
          }
    if(prev == NULL)
          p = min -> link;
          else
          prev -> link = min -> link;
          min -> link = NULL;
          if( q == NULL)
          q = min; /* moves the node with lowest data value in the list
pointed to by p to the list
    pointed to by q as a first node*/
          else
          {
            temp1 = q;
            /* traverses the list pointed to by q to get pointer to its
last node */
            while( temp1 -> link != NULL)
            temp1 = temp1 -> link;
            temp1 -> link = min; /* moves the node with lowest data value
in the list pointed to
    by p to the list pointed to by q at the end of list pointed by
    q*/

          }
        }
        return (q);
    }

    /* a function to insert a node with data value n in a sorted list
pointed to by p*/
    struct node *sinsert(struct node *p , int n)
    {
         struct node *curr, *prev;
         curr =p;
         prev = NULL;
        while(curr ->data < n)
```

```
            {
                                         prev = curr;
                              curr = curr->link;
            }
      if ( prev == NULL) /* the element is to be inserted at the start of
the list because
                        it is less than the data value of the first node*/
      {
            curr = (struct node *) malloc(sizeof(struct node));
            if( curr == NULL)
              {
                printf("error cannot allocate\n");
                exit(0);
              }
           curr->data = n;
           curr->link = p;
           p = curr;
       }
      else
       {
           curr->data = n;
           curr->link = prev->link;
           prev->link = curr;
       }
   return(p);

   }

   void main()
   {
        int n;
        int x;
        struct node *start = NULL ;
        printf("Enter the nodes to be created \n");
        scanf("%d",&n);
        while ( n-- > 0 )
        {
      printf( "Enter the data values to be placed in a node\n");
            scanf("%d",&x);
            start = insert ( start ,x);
        }
```

```
        printf("The created list is\n");
        printlist ( start );
        start = sortlist(start);
        printf("The sorted list is\n");
        printlist ( start );
        printf("Enter the value to be inserted\n");
        scanf("%d",&n);
        start = sinsert(start,n);
        printf("The list after insertion is\n");
        printlist ( start );
}
```

Explanation

1. If this pointer is prev, then prev is checked for a NULL value.
2. If prev is NULL, then the new node is created and inserted as the first node in the list.
3. When prev is not NULL, then a new node is created and inserted after the node pointed by prev, as shown in Figure 20.9.

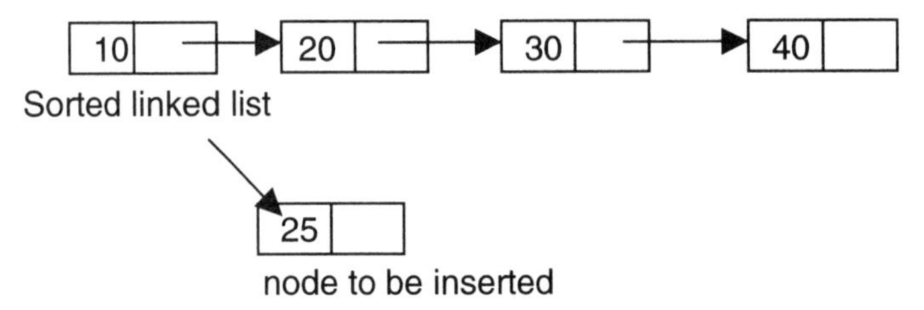

Before insertion

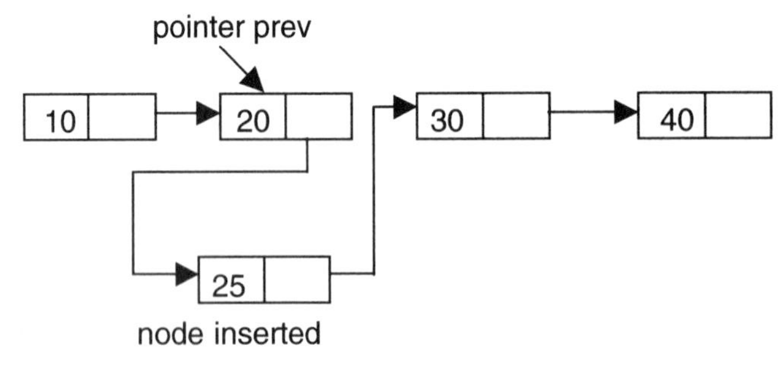

After insertion

FIGURE 20.9 Insertion in a sorted list.

COUNTING THE NUMBER OF NODES OF A LINKED LIST

Introduction

Counting the number of nodes of a singly linked list requires maintaining a counter that is initialized to 0 and incremented by 1 each time a node is encountered in the process of traversing a list from the start.

Here is a complete program that counts the number of nodes in a singly linked chain p, where p is a pointer to the first node in the list.

Program

```
# include <stdio.h>
# include <stdlib.h>
struct node
{
int data;
struct node *link;
};
struct node *insert(struct node * , int);
int nodecount(struct node*);
void printlist ( struct node *  );

struct node *insert(struct node *p , int n)
{
    struct node *temp;
   if(p==NULL)
   {
      p=(struct node *)malloc(sizeof(struct node));
      if(p==NULL)
      {
    printf("Error\n");
            exit(0);
      }
       p-> data = n;
       p-> link = NULL;
    }
    else
    {
      temp = p;
      while (temp-> link!= NULL)
```

```
     temp = temp-> link;
       temp-> link =  (struct node *)malloc(sizeof(struct node));
       if(temp -> link == NULL)
       {
    printf("Error\n");
           exit(0);
       }
     temp = temp-> link;
       temp-> data = n;
       temp-> link = NULL;
      }
      return (p);
}


void printlist ( struct node *p  )
{
        printf("The data values in the list are\n");
        while (p!= NULL)
       {
    printf("%d\t",p-> data);
           p = p-> link;
       }
}
/* A function to count the number of nodes in a singly linked list */
int  nodecount (struct node *p )
{
    int count=0;
    while (p != NULL)
    {
        count ++;
        p = p->link;
    }
      return(count);
}


void main()
{
      int n;
      int x;
```

```
    struct node *start = NULL ;
    printf("Enter the nodes to be created \n");
    scanf("%d",&n);
    while ( n-- > 0 )
    {
  printf( "Enter the data values to be placed in a node\n");
        scanf("%d",&x);
        start = insert ( start ,x);
    }
    printf("The created list is\n");
    printlist ( start );
    n = nodecount(start);
    printf("The number of nodes in a list are: %d\n",n);
}
```

MERGING OF TWO SORTED LISTS

Introduction

Merging of two sorted lists involves traversing the given lists and comparing the data values stored in the nodes in the process of traversing.

If p and q are the pointers to the sorted lists to be merged, then we compare the data value stored in the first node of the list pointed to by p with the data value stored in the first node of the list pointed to by q. And, if the data value in the first node of the list pointed to by p is less than the data value in the first node of the list pointed to by q, make the first node of the resultant/merged list to be the first node of the list pointed to by p, and advance the pointer p to make it point to the next node in the same list.

If the data value in the first node of the list pointed to by p is greater than the data value in the first node of the list pointed to by q, make the first node of the resultant/merged list to be the first node of the list pointed to by q, and advance the pointer q to make it point to the next node in the same list.

Repeat this procedure until either p or q becomes NULL. When one of the two lists becomes empty, append the remaining nodes in the non-empty list to the resultant list.

Program

```
# include <stdio.h>
# include <stdlib.h>
struct node
{
int data;
struct node *link;
};

struct node *merge (struct node *, struct node *);
struct node *insert(struct node *p , int n)
{
    struct node *temp;
   if(p==NULL)
   {
      p=(struct node *)malloc(sizeof(struct node));
      if(p==NULL)
      {
    printf("Error\n");
           exit(0);
      }
       p-> data = n;
       p-> link = NULL;
    }
    else
    {
      temp = p;
      while (temp-> link!= NULL)
     temp = temp-> link;
      temp-> link =  (struct node *)malloc(sizeof(struct node));
      if(temp -> link == NULL)
      {
    printf("Error\n");
           exit(0);
      }
    temp = temp-> link;
      temp-> data = n;
      temp-> link = NULL;
     }
     return (p);
 }
```

```
void printlist ( struct node *p  )
{
   printf("The data values in the list are\n");
   while (p!= NULL)
      {
    printf("%d\t",p-> data);
         p = p-> link;
      }
}

/* a function to sort a list */
struct node *sortlist(struct node *p)
{
    struct node *temp1,*temp2,*min,*prev,*q;
   q = NULL;
   while(p != NULL)
   {
    prev = NULL;
      min = temp1 = p;
      temp2 = p -> link;
      while ( temp2 != NULL )
      {
    if(min -> data > temp2 -> data)
         {
    min = temp2;
            prev = temp1;
         }
         temp1 = temp2;
         temp2 = temp2-> link;
      }
if(prev == NULL)
      p = min -> link;
      else
      prev -> link = min -> link;
      min -> link = NULL;
      if( q == NULL)
      q = min; /* moves the node with lowest data value in the list
pointed to by p to the list
   pointed to by q as a first node*/
      else
      {
         temp1 = q;
```

```
            /* traverses the list pointed to by q to get pointer to its
last node */
            while( temp1 -> link != NULL)
            temp1 = temp1 -> link;
            temp1 -> link = min; /* moves the node with lowest data value
in the list pointed to
   by p to the list pointed to by q at the end of list pointed by
   q*/

         }
      }
      return (q);
   }

   void main()
   {
         int n;
         int x;
         struct node *start1 = NULL ;
         struct node *start2 = NULL;
         struct node *start3 = NULL;
        /* The following code creates and sorts the first list */
         printf("Enter the number of nodes in the first list \n");
         scanf("%d",&n);
         while ( n-- > 0 )
         {
       printf( "Enter the data value to be placed in a node\n");
            scanf("%d",&x);
            start1 = insert ( start1 ,x);
         }
         printf("The first list is\n");
         printlist ( start1);
         start1 = sortlist(start1);
         printf("The sorted list1 is\n");
         printlist ( start1 );
        /* the following creates and sorts the second list*/
        printf("Enter the number of nodes in the second list \n");
         scanf("%d",&n);
         while ( n-- > 0 )
         {
```

```
            printf( "Enter the data value to be placed in a node\n");
            scanf("%d",&x);
            start2 = insert ( start2 ,x);
        }
        printf("The second list is\n");
        printlist ( start2);
        start2 = sortlist(start2);
        printf("The sorted list2 is\n");
        printlist ( start2 );
        start3 = merge(start1,start2);
        printf("The merged list is\n");
        printlist ( start3);
}

/* A function to merge two sorted lists */
struct node *merge (struct node *p, struct node *q)
  {
    struct node *r=NULL,*temp;
        if (p == NULL)
             r =  q;
        else
        if(q == NULL)
             r = p;
        else
            {
                if (p->data < q->data )
                {
    r = p;
    temp = p;
    p = p->link;
    temp->link = NULL;
                }
               else
                {
    r = q;
    temp =q;
    q =q->link;
    temp->link = NULL;
              }
             while((p!= NULL) &&  (q != NULL))
             {
```

```
                  if (p->data < q->data)
             {
             temp->link =p;
             p = p->link;
             temp =temp->link;
             temp->link =NULL;
             }
                       else
            {
             temp->link =q;
             q = q->link;
             temp =temp->link;
             temp->link =NULL;
             }

    }
    if (p!= NULL)
            temp->link = p;
    if (q != NULL)
             temp->link  = q;
      }
return( r) ;
}
```

Explanation

If the following lists are given as input, then what would be the output of the program after each pass? This is shown here:

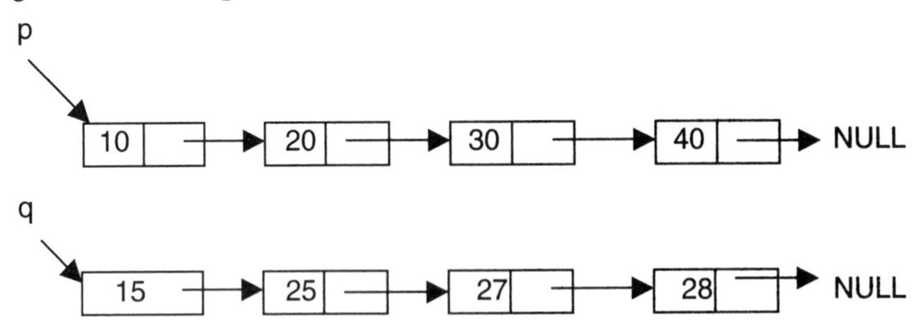

Two sorted lists before merging

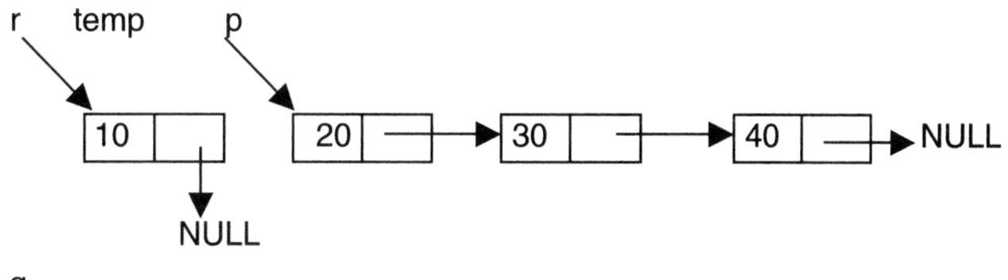

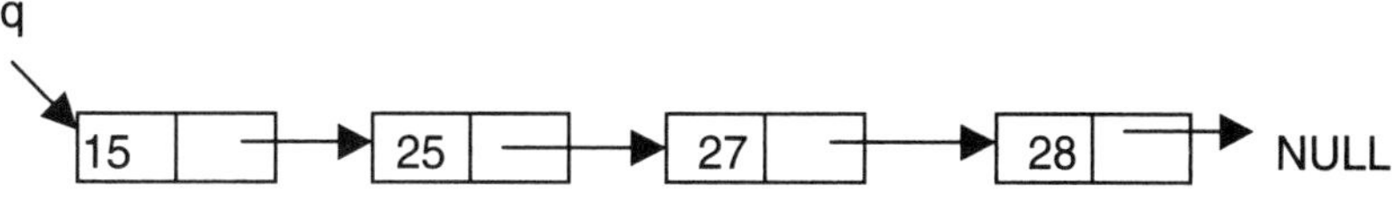

After the first pass

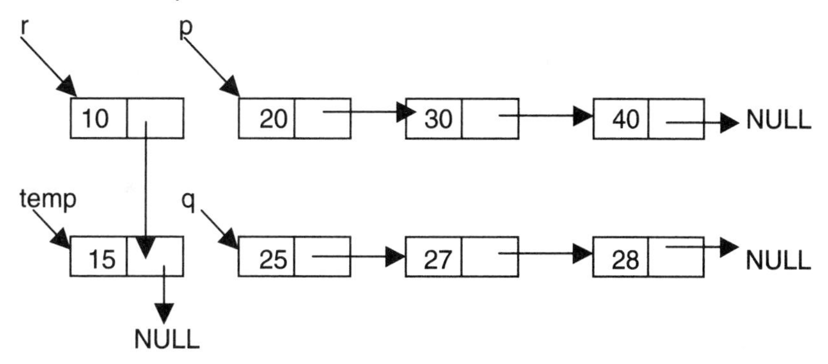

After the second pass

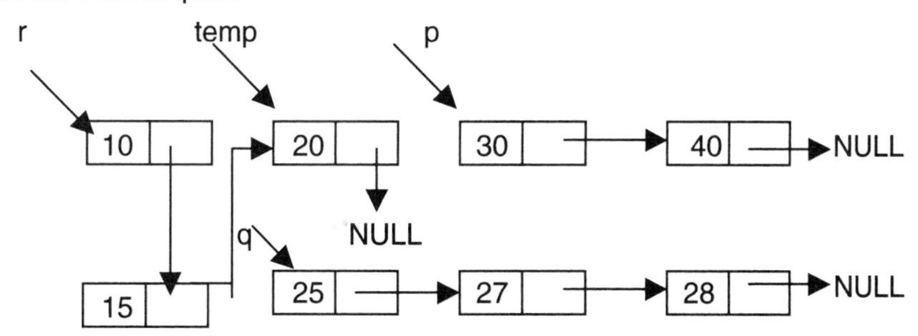

After the third pass

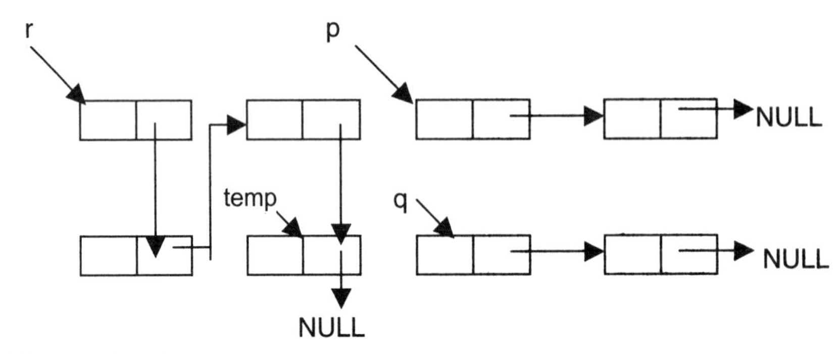

After the fourth pass

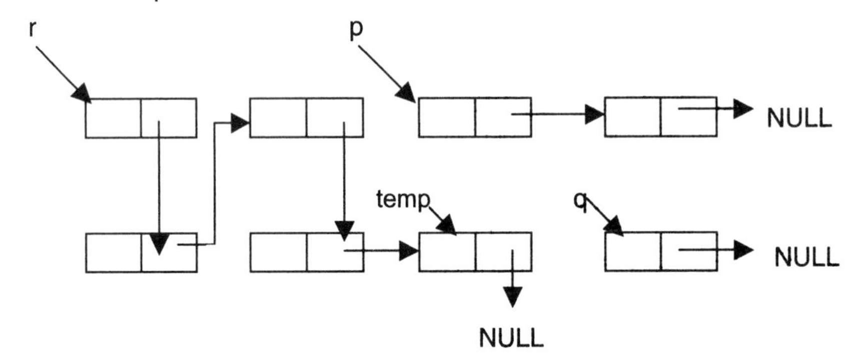

After the fifth pass

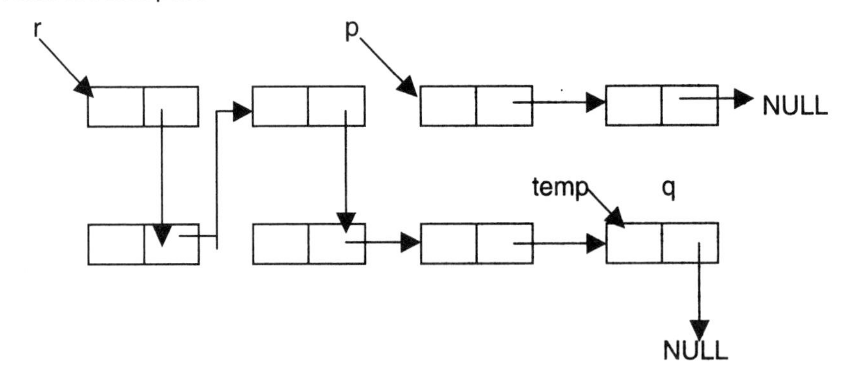

After the sixth pass

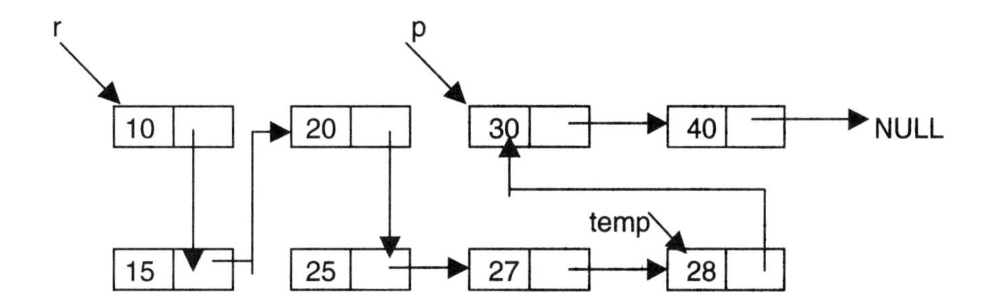

Final merged list

ERASING A LINKED LIST

Introduction

Erasing a linked list involves traversing the list starting from the first node, freeing the storage allocated to the nodes, and then setting the pointer to the list to NULL. If p is a pointer to the start of the list, the actions specified through the following code will erase the list:

```
while(p != NULL)
   {
          temp  = p;
          p =  p->link;
         free(t);
   }
```

But a better strategy of erasing a list is to mark all the nodes of the list to be erased as free nodes without actually freeing the storage of these nodes. That means to maintain this list, a list of free nodes, so that if a new node is required it can be obtained from this list of free nodes.

Program

Here is a complete program that erases a list pointed to by p by adding the nodes of a list pointed by p to the free list.

```
# include <stdio.h>
# include <stdlib.h>
struct node
```

```
{
int data;
struct node *link;
};
struct node *insert(struct node * , int);
void erase(struct node **,struct node **);
void printlist ( struct node *  );

void erase (struct node **p, struct node **free)
{
     struct node *temp;
    temp = *p;
    while (temp->link != NULL)
          temp  = temp ->link;
    temp->link = (*free);
    *free = *p;
    *p = NULL;
}

struct node *insert(struct node *p , int n)
{
    struct node *temp;
   if(p==NULL)
   {
      p=(struct node *)malloc(sizeof(struct node));
      if(p==NULL)
      {
    printf("Error\n");
           exit(0);
      }
       p-> data = n;
       p-> link = NULL;
    }
    else
    {
      temp = p;
      while (temp-> link!= NULL)
     temp = temp-> link;
      temp-> link =  (struct node *)malloc(sizeof(struct node));
      if(temp -> link == NULL)
      {
    printf("Error\n");
```

```
            exit(0);
         }
       temp = temp-> link;
         temp-> data = n;
         temp-> link = NULL;
        }
        return (p);
   }

   void printlist ( struct node *p  )
   {
          printf("The data values in the list are\n");
          while (p!= NULL)
         {
       printf("%d\t",p-> data);
            p = p-> link;
         }
   }

   void main()
   {
         int n;
         int x;
         struct node *start = NULL ;
         struct node *free=NULL;

         /* this code will create a free list for the test purpose*/
        printf("Enter the number of nodes in the initial free list \n");
         scanf("%d",&n);
         while ( n-- > 0 )
        {
          printf( "Enter the data values to be placed in a node\n");
          scanf("%d",&x);
          free = insert ( free ,x);
       }

         /* this code will create a list to be erased*/
        printf("Enter the number of nodes in the list to be created for
erasing \n");
         scanf("%d",&n);
         while ( n-- > 0 )
        {
```

```
            printf( "Enter the data values to be placed in a node\n");
            scanf("%d",&x);
            start = insert ( start ,x);
        }

           printf("The free list islist is:\n");
           printlist ( free );
           printf("The list to be erased is:\n");
           printlist ( start);
           erase(&start,&free);
           printf("The free list after adding all the nodes from the list to
be erased is:\n");
           printlist ( free );
     }
```

Explanation

The method of erasing a list requires adding all the nodes of the list to be erased to the list of free nodes, as shown here.

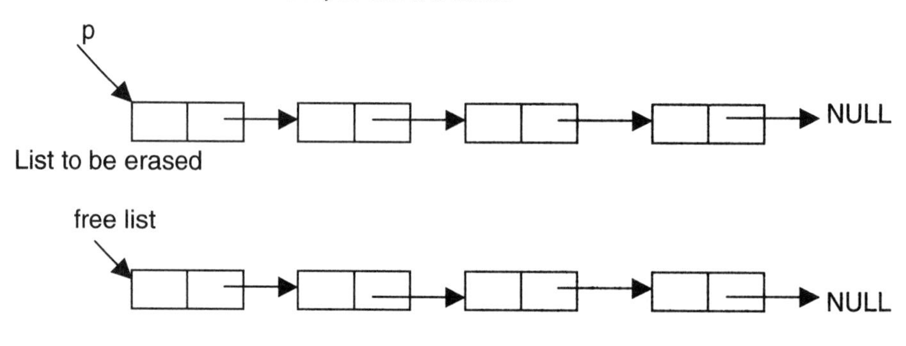

Before erasing the list pointed to by p

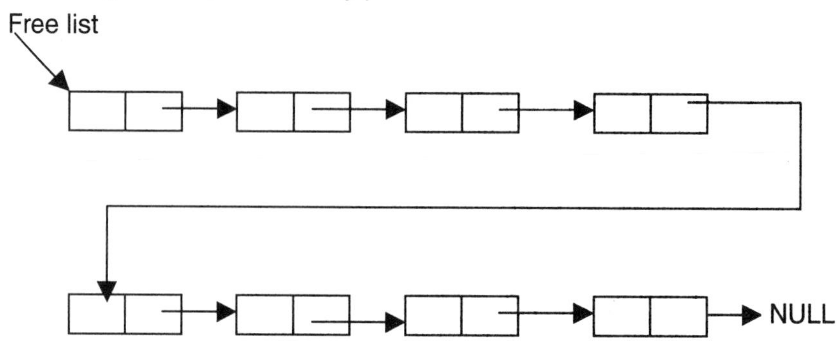

After erasing the list pointed to by p

POLYNOMIAL REPRESENTATION

Introduction

One of the problems that a linked list can deal with is manipulation of symbolic polynomials. By symbolic, we mean that a polynomial is viewed as a list of coefficients and exponents. For example, the polynomial

$3x^2+2x+4$,

can be viewed as list of the following pairs

(3,2),(2,1),(4,0)

Therefore, we can use a linked list in which each node will have three fields, as shown in Figure 20.10.

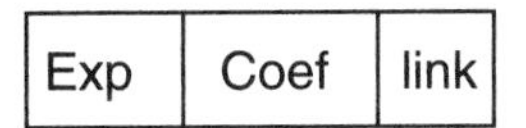

A polynomial $10x^4 + 5x^2 + 1$ can be represented as shown here:

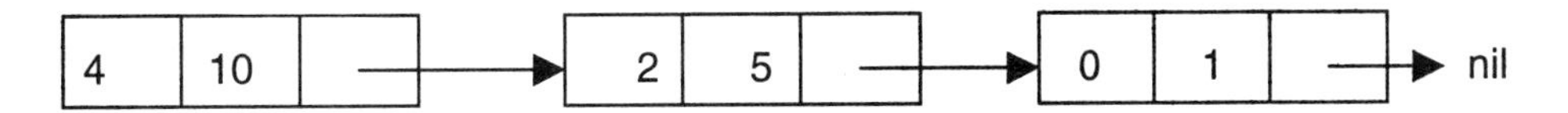

FIGURE 20.10 Polynomial representation.

The procedure to add these two polynomials using the linked list is in the following program.

Program

```
# include <stdio.h>
# include <stdlib.h>
struct  pnode
    {
       int exp;
       double coeff;
       struct pnode *link;
    };

struct pnode *insert(struct pnode * , int,double);
void printlist ( struct pnode *  );
```

```
struct pnode *polyadd(struct pnode *, struct pnode *);
struct pnode *sortlist(struct pnode *);


struct pnode *insert(struct pnode *p , int e,double c)
{
  struct pnode *temp;
   if(p==NULL)
   {
      p=(struct pnode *)malloc(sizeof(struct pnode));
      if(p==NULL)
      {
    printf("Error\n");
           exit(0);
      }
       p-> exp = e;
       p->coeff = c;
       p-> link = NULL;
    }
    else
    {
      temp = p;
      while (temp-> link!= NULL)
     temp = temp-> link;
      temp-> link =  (struct pnode *)malloc(sizeof(struct pnode));
      if(temp -> link == NULL)
      {
    printf("Error\n");
         exit(0);
      }
    temp = temp-> link;
      temp-> exp = e;
      temp->coeff = c;
      temp-> link = NULL;
     }
     return (p);
}

/* a function to sort a list */
struct pnode *sortlist(struct pnode *p)
{
```

```
    struct pnode *temp1,*temp2,*max,*prev,*q;
    q = NULL;
    while(p != NULL)
    {
     prev = NULL;
       max = temp1 = p;
       temp2 = p -> link;
       while ( temp2 != NULL )
       {
     if(max -> exp < temp2 -> exp)
          {
     max = temp2;
            prev = temp1;
          }
          temp1 = temp2;
          temp2 = temp2-> link;
       }
 if(prev == NULL)
       p = max -> link;
       else
       prev -> link = max -> link;
       max -> link = NULL;
       if( q == NULL)
       q = max; /* moves the node with highest data value in the list
pointed to by p to the list
 pointed to by q as a first node*/
       else
       {
          temp1 = q;
          /* traverses the list pointed to by q to get pointer to its
last node */
          while( temp1 -> link != NULL)
          temp1 = temp1 -> link;
          temp1 -> link = max; /* moves the node with highest data value
in the list pointed to
 by p to the list pointed to by q at the end of list pointed by
 q*/

       }
    }
    return (q);
 }
```

```
/* A function to add two polynomials */
struct pnode *polyadd(struct pnode *p, struct pnode *q)
{
    struct pnode *r = NULL;
    int e;
    double c;
    while((p!=NULL) && (q != NULL))
            {
               if(p->exp > q->exp)
                {
                     r = insert(r,p->exp,p->coeff);
                     p = p->link;
                 }
               else
                  if(p->exp < q->exp)
                {
                     r = insert(r,q->exp,q->coeff);
                     q = q->link;
                 }
                 else
    {
        c = p->coeff  +  q->coeff;
        e = q->exp;
        r = insert( r , e ,c);
        p = p->link;
        q = q->link;
                      }
             }
while(p != NULL)
        {
               r = insert( r , p->exp ,p->coeff);
               p = p->link;
        }
        while(q!=NULL)
        {
                r = insert( r , q->exp ,q->coeff);
                q = q->link;
        }
return(r);
}

void printlist ( struct pnode *p  )
```

```
{
    printf("The polynomial is\n");
    while (p!= NULL)
        {
     printf("%d  %lf\t",p-> exp,p->coeff);
                p = p-> link;
        }
}
void main()
{
        int e;
        int n,i;
        double c;
        struct  pnode *poly1 = NULL ;
        struct  pnode *poly2=NULL;
        struct pnode *result;
        printf("Enter the terms in the polynomial1 \n");
        scanf("%d",&n);
        i=1;
        while ( n-- > 0 )
        {
     printf( "Enter the exponent and coefficient of the term number
%d\n",i);
               scanf("%d %lf",&e,&c);
               poly1 = insert ( poly1,e,c);
        }
      printf("Enter the terms in the polynomial2 \n");
        scanf("%d",&n);
        i=1;
        while ( n-- > 0 )
        {
     printf( "Enter the exponent and coefficient of the term number
%d\n",i);
               scanf("%d %lf",&e,&c);
               poly2 = insert ( poly2,e,c);
        }
        poly1 = sortlist(poly1);
        poly2 = sortlist(poly2);
        printf("The polynomial 1 is\n");
        printlist ( poly1 );
        printf("The polynomial 2 is\n");
        printlist ( poly2 );
```

```
        result = polyadd(poly1,poly2);
         printf("The result of addition is\n");
         printlist ( result );

}
```

Explanation

1. If the polynomials to be added have n and m terms, respectively, then the linked list representation of these polynomials contains m and n terms, respectively.
2. Since `polyadd` traverses each of these lists, sequentially, the maximum number of iterations that `polyadd` will make will not be more than $m + n$. So the computation time of `polyadd` is $O(m + n)$.

REPRESENTATION OF SPARSE MATRICES

Introduction

A matrix is a two-dimensional data object made of m rows and n columns, therefore having $m \, ' \, n$ values. When $m=n$, we call it a square matrix.

The most natural representation is to use two-dimensional array A[m][n] and access the element of ith row and j^{th} column as A[i][j]. If a large number of elements of the matrix are zero elements, then it is called a sparse matrix.

Representing a sparse matrix by using a two-dimensional array leads to the wastage of a substantial amount of space. Therefore, an alternative representation must be used for sparse matrices. One such representation is to store only non-zero elements along with their row positions and column positions. That means representing every non-zero element by using triples (i,j,value), where i is a row position and j is a column position, and store these triples in a linear list. It is possible to arrange these triples in the increasing order of row indices, and for the same row index in the increasing order of column indices. Each triple (i,j,value) can be represented by using a node having four fields as shown in the following:

```
Struct snode{
        Int row,col,val;
       Struct snode *next;
       };
```

row	col	value	link

So a sparse matrix can be represented using a list of such nodes, one per non–zero element of the matrix. For example, consider the sparse matrix shown in Figure 20.11.

0	2	0	0	2	0
0	0	0	1	0	5
0	0	4	0	0	0
0	0	0	0	0	0
0	0	0	0	0	0

FIGURE 20.11 A sparse matrix.

This matrix can be represented using the linked list shown in Figure 20.12.

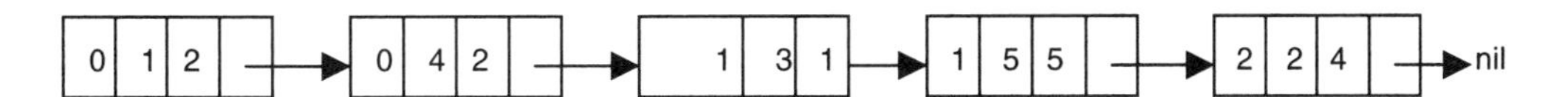

FIGURE 20.12 Linked list representation of sparse matrix of Figure 20.11.

Program

Here is a program for the addition of two sparse matrices:

```
# include <stdio.h>
# include <stdlib.h>
struct  snode
   {
      int row,col,val;
      struct snode *link;
   };

struct snode *insert(struct snode * , int,int,int);
void printlist ( struct  snode *  );
struct snode *sadd(struct snode *, struct snode *);
//struct pnode *sortlist(struct pnode *);

struct snode *insert(struct snode *p , int r,int c,int val)
{
```

```
   struct snode *temp;
    if(p==NULL)
    {
       p=(struct snode *)malloc(sizeof(struct snode));
       if(p==NULL)
       {
    printf("Error\n");
            exit(0);
       }
        p->row = r;
        p->col = c;
        p->val = val;
        p-> link = NULL;
     }
     else
     {
       temp = p;
       while (temp-> link!= NULL)
      temp = temp-> link;
       temp-> link =  (struct snode *)malloc(sizeof(struct snode));
       if(temp -> link == NULL)
       {
    printf("Error\n");
          exit(0);
       }
     temp = temp-> link;
       temp-> row = r;
       temp->col= c;
       temp->val=val;
       temp-> link = NULL;
      }
      return (p);
}

/* A function to add two sparse matrices */
struct snode *sadd(struct snode *p, struct snode *q)
{
    struct snode *r = NULL;
    int  val;
    while((p!=NULL) && (q != NULL))
            {
```

```
                if(p->row < q->row)
                 {
                      r = insert(r,p->row,p->col,p->val);
                      p = p->link;
                  }
                else
                   if(p->row > q->row)
                   {
                      r = insert(r,q->row,q->col,q->val);
                      q = q->link;
                     }
                  else
    if( p->col < q->col)
                       {
      r = insert(r,p->row,p->col,p->val);
                         p = p->link;
                        }
    else
             if(p->col > q->col)
                {
                  r = insert(r,q->row,q->col,q->val);
                  q = q->link;
                }
              else
              {
                 val = p->val  +  q->val;
                 r = insert( r , p->row,p->col,val);
                 p = p->link;
                 q = q->link;
               }
  }
while(p != NULL)
         {
                r = insert( r , p->row ,p->col,p->val);
                p = p->link;
         }
        while(q!=NULL)
        {
                 r = insert( r , q->row ,q->col,q->val);
                 q = q->link;
        }
return(r);
}
```

```
void printlist ( struct snode *p  )
{
   printf("The resultant sparse matrix is\n");
   while (p!= NULL)
        {
    printf("%d  %d  % d\n",p-> row,p->col,p->val);
              p = p-> link;
        }
}
void main()
{
      int r,n,c,val;
      struct  snode *s1 = NULL ;
      struct  snode *s2=NULL;
      struct snode *result = NULL;
      printf("Enter the number of non-zero terms in the sparse matrix1
\n");
      scanf("%d",&n);
     printf("Enter the terms in the sparse matrix1 in the increasing
order of row indices and for the same row index in the increasing order of
row indices and for the same row index in the increasing order of column
indices \n");
             while ( n-- > 0 )
      {
    printf( "Enter the row number, column number, and value\n");
            scanf("%d %d%d",&r,&c,&val);
            s1 = insert ( s1,r,c,val);
      }
   printf("Enter the number of non-zero terms in the sparse matrix1 \n");
      scanf("%d",&n);
     printf("Enter the terms in the sparse matrix2 in the increasing
order of row indices and for the same row index in the increasing order of
row indices and for the same row index in the increasing order of column
indices \n");
      while ( n-- > 0 )
      {
    printf( "Enter the row number, column number, and value\n");
            scanf("%d %d%d",&r,&c,&val);
            s2 = insert ( s2,r,c,val);
      }
```

```
    result = sadd(s1,s2);
    printf("The result of addition is\n");
    printlist ( result );

}
```

Explanation

1. In order to add two sparse matrices represented using the sorted linked lists as shown in the preceding program, the lists are traversed until the end of one of the lists is reached.
2. In the process of traversal, the row indices stored in the nodes of these lists are compared. If they don't match, a new node is created and inserted into the resultant list by copying the contents of a node with a lower value of row index. The pointer in the list containing a node with a lower value of row index is advanced to make it point to the next node.
3. If the row indices match, column indices for the corresponding row positions are compared. If they don't match, a new node is created and inserted into the resultant list by copying the contents of a node with a lower value of column index. The pointer in the list containing a node with a lower value of column index is advanced to make it point to the next node.
4. If the column indices match, a new node is created and inserted into the resultant list by copying the row and column indices from any of the nodes and the value equal to the sum of the values in the two nodes.
5. After this, the pointers in both the lists are advanced to make them point to the next nodes in the respective lists. This process is repeated in each iteration. After reaching the end of any one of the lists, the iterations come to an end and the remaining nodes in the list whose end has not been reached are copied, as it is in the resultant list.

Example

Consider the following sparse matrices:

0	0	5	0
0	1	0	0
0	0	0	0
0	0	2	0

Sparse matrix a

0	0	0	0
0	3	0	1
0	0	0	0
0	0	5	0

Sparse matrix b

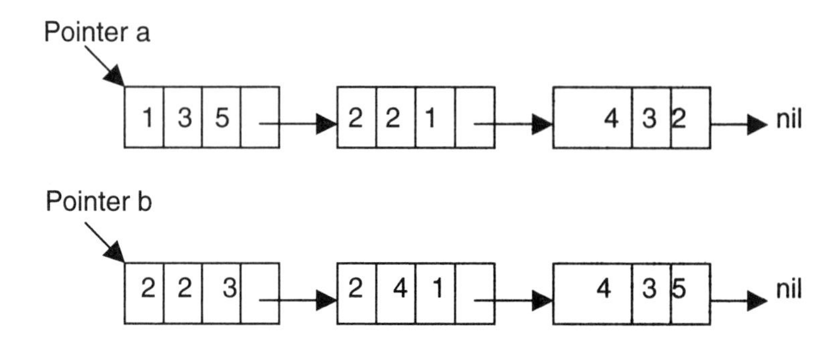

If the procedure sadd is applied to the above linked list representations then we get the resultant list, as shown in Figure 20.13.

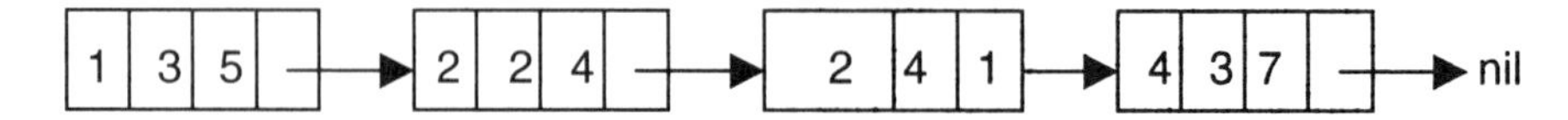

FIGURE 20.13 Result of application of the procedure sadd.

This resultant list represents the matrix shown below:

0	0	5	0
0	4	0	1
0	0	0	0
0	0	7	0

This matrix is an addition of the matrices of a and b, respectively.

Points to Remember

1. If the sparse matrices to be added have n and m non-zero terms, respectively, then the linked list representation of these sparse matrices contains m and n terms, respectively.
2. Since sadd traverses each of these lists sequentially, the maximum number of iterations that sadd will make will not be more than m+n. So the computation time of sadd is $O(m+n)$.

CIRCULAR LINKED LISTS

Introduction

A circular list is a list in which the link field of the last node is made to point to the start/first node of the list, as shown in Figure 20.14.

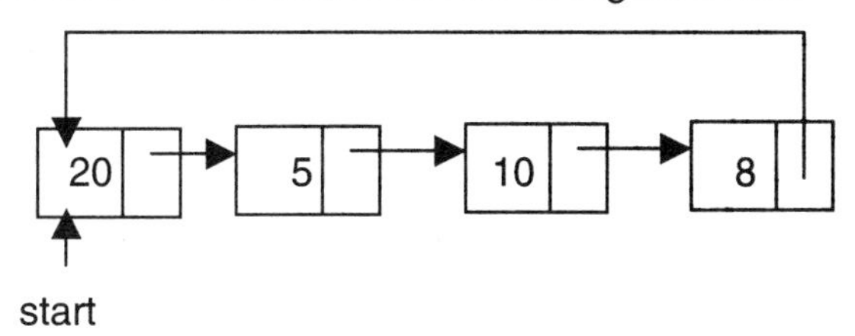

FIGURE 20.14 A circular list.

In the case of circular lists, the empty list also should be circular. So to represent a circular list that is empty, it is required to use a header node or a head-node whose data field contents are irrelevant, as shown in Figure 20.15.

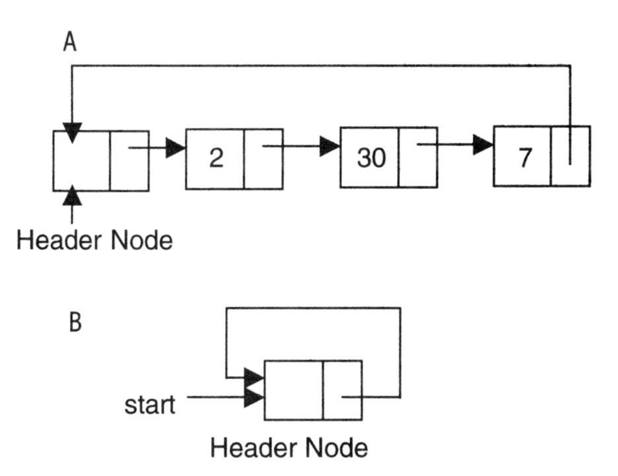

FIGURE 20.15 (A) A circular list with head node, (B) an empty circular list.

Program

Here is a program for building and printing the elements of the circular linked list.

```
# include <stdio.h>
# include <stdlib.h>
struct node
```

```
{
    int data;
    struct node *link;
};
struct node *insert(struct node *p , int n)
{
    struct node *temp;
    /* if the existing list is empty then insert a new node as the
starting node */
    if(p==NULL)
    {
        p=(struct node *)malloc(sizeof(struct node)); /* creates new
node data value passes
    as  parameter */
        if(p==NULL)
        {
                printf("Error\n");
                exit(0);
        }
        p-> data = n;
        p-> link = p; /* makes the pointer pointing to itself because it
is a circular list*/
    }
    else
    {
        temp = p;
  /* traverses the existing list to get the pointer to the last node of
it */
        while (temp-> link != p)
                temp = temp-> link;
        temp-> link =  (struct node *)malloc(sizeof(struct node)); /*
creates new node using
        data value passes as
        parameter  and puts its
        address in the link field
        of last node of the
        existing list*/
        if(temp -> link == NULL)
        {
                printf("Error\n");
          exit(0);
        }
```

```
        temp = temp-> link;
        temp-> data = n;
        temp-> link = p;
       }
       return (p);
  }
  void printlist ( struct node *p  )
  {
      struct node *temp;
      temp = p;
      printf("The data values in the list are\n");
     if(p!= NULL)
     {
         do
        {
                 printf(%d\t",temp->data);
                 temp=temp->link;
         } while (temp!= p)
      }
      else
         printf("The list is empty\n");
   }

  void main()
  {
        int n;
        int x;
        struct node *start = NULL ;
        printf("Enter the nodes to be created \n");
        scanf("%d",&n);
        while ( n– > 0 )
        {
                 printf( "Enter the data values to be placed in a
node\n");
           scanf("%d",&x);
           start = insert ( start , x );
        }
        printf("The created list is\n");
        printlist ( start );
  }
```

Explanation

The program appends a new node to the existing list (that is, it inserts a new node in the existing list at the end), and it makes the link field of the newly inserted node point to the start or first node of the list. This ensures that the link field of the last node always points to the starting node of the list.

SPLITTING A LIST WITH 2*N* NODES INTO TWO SEPARATE AND EQUAL LISTS

Introduction

If the circular linked list has 10 nodes, then the two lists have 5 nodes each. The procedure for splitting a circular list with $2n$ nodes into two equal circular lists is given here:

Program

```
# include <stdio.h>
# include <stdlib.h>
struct node
{
int data;
struct node *link;
};
void split(struct node *p, struct node **q, int n)
   {
      struct node *temp;
     int  i =1;
   temp = p;
          while( i < n)

                        {
             temp  =  temp->link;
             i++;
                        }
   *q = temp->link;
   temp->link  =  p;
          temp = *q;
```

```
            while(temp->link != p)
          temp = temp ->link;
       temp->link = *q;
    }

    struct node *insert(struct node *p , int n)
    {
    struct node *temp;
     /* if the existing list is empty then insert a new node as the
starting node */
     if(p==NULL)
       {
         p=(struct node *)malloc(sizeof(struct node)); /* creates new node
data value passes
    as  parameter */

         if(p==NULL)
         {
       printf("Error\n");
              exit(0);
         }
          p-> data = n;
          p-> link = p; /* makes the pointer point to itself because it is
a circular list*/
       }
       else
       {
         temp = p;
     /* traverses the existing list to get the pointer to the last node of
it */
     while (temp-> link != p)
        temp = temp-> link;
         temp-> link =  (struct node *)malloc(sizeof(struct node)); /*
creates new node using
          data value passes as
            parameter  and puts its
           address in the link field
           of last node of the
           existing list*/
         if(temp -> link == NULL)
         {
       printf("Error\n");
            exit(0);
```

```
      }
      temp = temp-> link;
      temp-> data = n;
      temp-> link = p;
     }
     return (p);
}
void printlist ( struct node *p  )
{
 struct node *temp;
  temp = p;
 printf("The data values in the list are\n");
    if(p!= NULL)
           do
             {
             printf("%d\t",temp->data);
             temp=temp->link;
              } while (temp!= p);

   else
         printf("The list is empty\n");
 }

void main()
{
      int n,num;
      int x;
      struct node *start = NULL ;
      struct node *start1=NULL;
      printf("Enter the value of n \n");
      scanf("%d",&n);
      num = n;
      n*=2;
      /* this will create a circular list with 2n nodes*/
   while ( n-- > 0 )
      {
    printf( "Enter the data values to be placed in a node\n");
            scanf("%d",&x);
            start = insert ( start , x );
      }
      printf("The created list is\n");
      printlist ( start );
```

```
        split(start,&start1,num);
        printf("The first list is:\n");
        printlist(start);
        printf("The second list is:\n");
        printlist(start1);
}
```

Explanation

Consider a circular list containing $2n$ nodes, as shown in Figure 20.16.

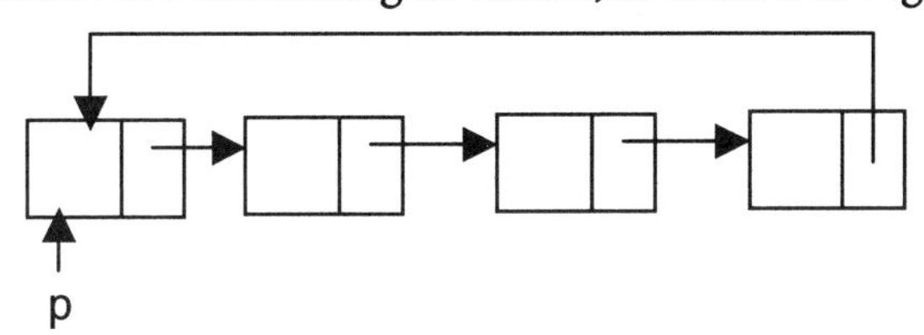

FIGURE 20.16 List containing $2n$ nodes.

To split this list into two equal lists, it is required to traverse the list up to the nth node and store the link of the nth node, which is the address of $(n+1)^{th}$ node in the pointer, say q. After this, make the link field of the nth node point to the first node pointed to by p. Then traverse the list starting from the node pointed to by q up to the end. Then make the link field of the last node point to the node pointed to by q. The result of this is shown in Figure 20.17.

FIGURE 20.17 Splitting of a circular list.

MERGING OF TWO CIRCULAR LISTS

Introduction

You can merge two lists into one list. The following program merges two circular lists.

Program

```
# include <stdio.h>
# include <stdlib.h>
struct node
{
int data;
struct node *link;
};

struct node *insert(struct node *p , int n)
{
struct node *temp;
 /* if the existing list is empty then insert a new node as the
starting node */
  if(p==NULL)
    {
       p=(struct node *)malloc(sizeof(struct node)); /* creates new node
data value passes
   as  parameter */

       if(p==NULL)
       {
     printf("Error\n");
            exit(0);
       }
        p-> data = n;
        p-> link = p; /* makes the pointer pointing to itself because it
is a circular list*/
     }
     else
     {
       temp = p;
  /* traverses the existing list to get the pointer to the last node of
it */
  while (temp-> link != p)
      temp = temp-> link;
       temp-> link =  (struct node *)malloc(sizeof(struct node)); /*
creates new node using
        data value passes as
          parameter  and puts its
         address in the link field
         of last node of the
```

```
        existing list*/
      if(temp -> link == NULL)
      {
    printf("Error\n");
         exit(0);
      }
      temp = temp-> link;
      temp-> data = n;
      temp-> link = p;
     }
     return (p);
}
void printlist ( struct node *p  )
{
 struct node *temp;
  temp = p;
 printf("The data values in the list are\n");
    if(p!= NULL)
    {
       do
             {
             printf("%d\t",temp->data);
             temp=temp->link;
             } while (temp!= p);
      }
   else
         printf("The list is empty\n");
 }
struct node *merge(struct node *p, struct node *q)
     {
       struct node *temp=NULL;
       struct node *r=NULL;
       r = p;
       temp = p;
       while(temp->link != p)
            temp = temp->link;
       temp->link = q;
       temp = q;
       while( temp->link != q)
             temp = temp->link;
       temp->link = r;
      return(r);
```

```
}

void main()
{
      int n;
      int x;
      struct node *start1=NULL ;
      struct node *start2=NULL;
      struct node *start3=NULL;

     /* this will create the first circular list nodes*/
      printf("Enter the number of nodes in the first list \n");
      scanf("%d",&n);
      while ( n-- > 0 )
      {
    printf( "Enter the data value to be placed in a node\n");
           scanf("%d",&x);
           start1 = insert ( start1 , x );
      }
      printf("The first list is\n");
      printlist ( start1 );

      /* this will create the second circular list nodes*/
      printf("Enter the number of nodes in the second list \n");
      scanf("%d",&n);
      while ( n-- > 0 )
      {
    printf( "Enter the data value to be placed in a node\n");
           scanf("%d",&x);
           start2 = insert ( start2, x );
      }
      printf("The second list is:\n");
      printlist ( start2 );

      start3 = merge(start1,start2);
      printf("The resultant list is:\n");
      printlist(start3);
}
```

Explanation

In order to merge or concatenate the two non-empty circular lists pointed to by p and q, it is required to make the start of the resultant list p. Then the list pointed to by p is required to be traversed until its end, and the link field of the last node must become the pointer q. After that, the list pointed to by q is required to be traversed until its end, and the link field of the last node is required to be made p.

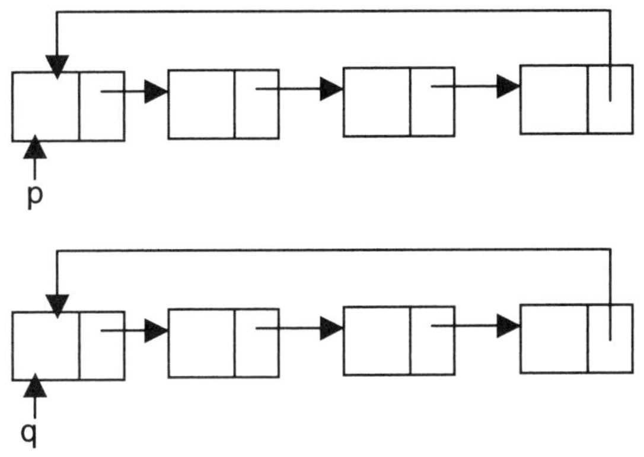

The given lists

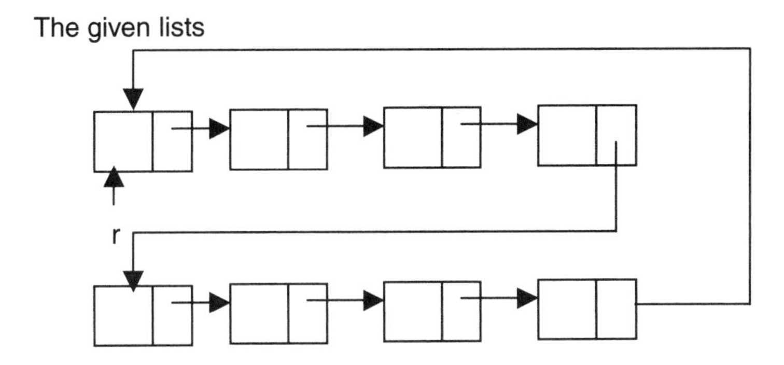

The resultant list

REVERSING THE DIRECTION OF LINKS IN A SINGLY LINKED CIRCULAR LIST

Introduction

You can reverse the direction of links in the circular list. If you do so, each link should be reversed.

Program

```
# include <stdio.h>
# include <stdlib.h>
struct node
{
int data;
struct node *link;
};

/* A function to reverse a singly linked circular list */
struct node *reverselist(struct node *p)
  {
    struct node *temp;
    struct node *prev = NULL;
    struct node *curr;
    if(p != NULL)
    {
    curr = p;
    temp = curr;
    while(curr->link != p)
    {
       curr = curr->link;
               temp ->link = prev;
               prev = temp;
               temp = curr;
            }
    temp ->link = prev;
    p->link = temp;
    p= temp;
    }
    return(p);
}

struct node *insert(struct node *p , int n)
{
struct node *temp;
 /* if the existing list is empty then insert a new node as the
starting node */
 if(p==NULL)
   {
      p=(struct node *)malloc(sizeof(struct node)); /* creates new node
```

```
data value passes
   as  parameter */

         if(p==NULL)
         {
      printf("Error\n");
             exit(0);
         }
          p-> data = n;
          p-> link = p; /* makes the pointer point to itself because it is
a circular list*/
        }
        else
        {
          temp = p;
    /* traverses the existing list to get the pointer to the last node of
it */
     while (temp-> link != p)
        temp = temp-> link;
         temp-> link =  (struct node *)malloc(sizeof(struct node)); /*
creates new node using
          data value passes as
            parameter  and puts its
           address in the link field
           of last node of the
           existing list*/
         if(temp -> link == NULL)
         {
      printf("Error\n");
            exit(0);
         }
         temp = temp-> link;
         temp-> data = n;
         temp-> link = p;
        }
        return (p);
   }
   void printlist ( struct node *p  )
   {
    struct node *temp;
     temp = p;
    printf("The data values in the list are\n");
```

```
    if(p!= NULL)
    {
       do
             {
             printf("%d\t",temp->data);
             temp=temp->link;
             } while (temp!= p);
      }
   else
         printf("The list is empty\n");
 }
void main()
{
      int n;
      int x;
      struct node *start = NULL ;
      struct node *start1=NULL;
     /* this will create at circular list */
      printf("Enter the number of nodes in the list \n");
      scanf("%d",&n);
      while ( n-- > 0 )
      {
   printf( "Enter the data value to be placed in a node\n");
           scanf("%d",&x);
           start = insert ( start , x );
      }
      printf("The list is\n");
      printlist ( start );
      start1 = reverselist(start);
      printf("The reversed list is:\n");
      printlist(start1);
}
```

Explanation

To reverse the links of a singly linked circular list, the list is required to be traversed from the start node until a node is encountered whose link points to the start node (that is, the last node in the list). For this, it is required to maintain the pointers to the current node and the previous node. An additional temporary pointer pointing to the current node is also required to be maintained. Initially, the current, temporary, and previous pointers are set as follows:

1. Set the current as well as the temporary pointer to the start pointer.
2. Set the previous pointer to NULL.
3. The pointers are manipulated in each iteration as follows:
   (i) Advance the current pointer to make it point to the next node.
   (ii) Set the link field of the node pointed to by the temporary pointer to the value of the previous pointer.
   (iii) Make the previous pointer point to the node pointed to by the temporary pointer.
   (iv) Make the temporary pointer point to the node pointed to by the current pointer.
   (v) When the last node is encountered, its link field is made to point to the previous node. After that, the link field of the node pointed to by the start pointer (first node) is made to point to this last node. And the start pointer is made to point to this last node. These manipulations are shown in the following diagrams.

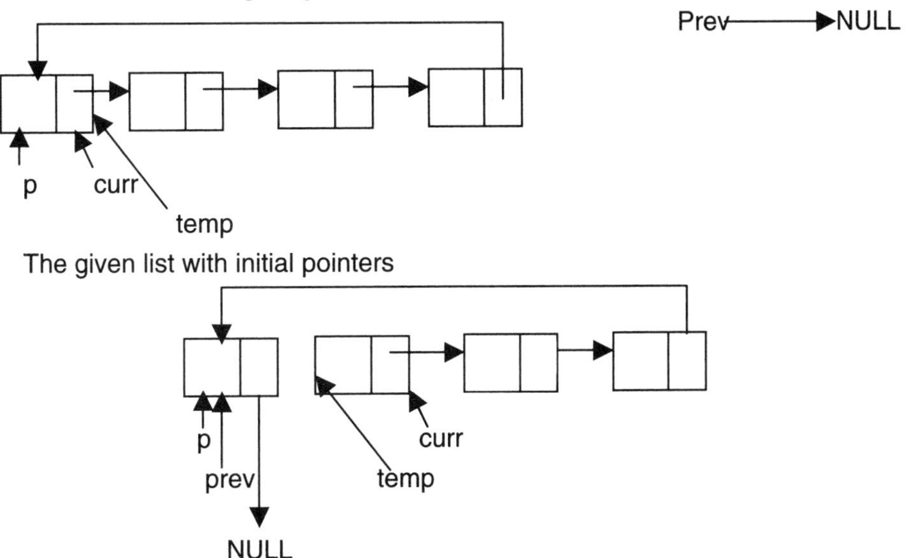

The given list with initial pointers

After the first iteration

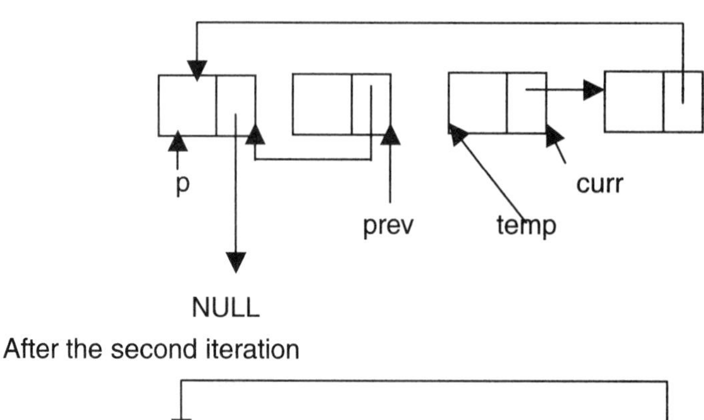

After the second iteration

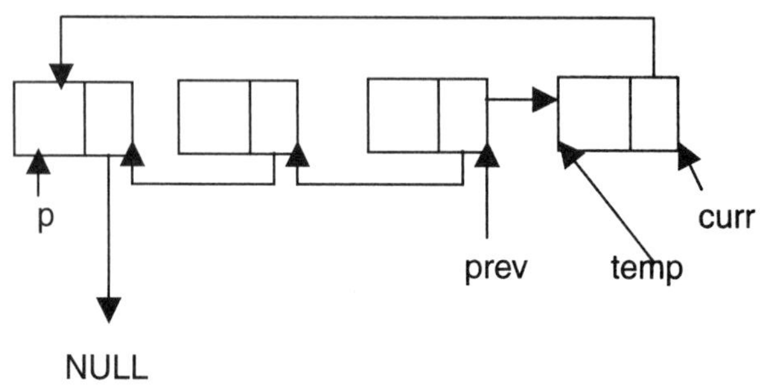

After the third iteration

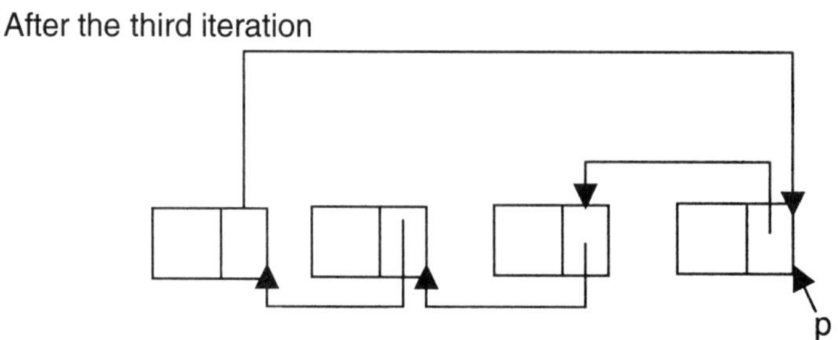

DOUBLY LINKED LISTS

Introduction

The following are problems with singly linked lists:

1. A singly linked list allows traversal of the list in only one direction.
2. Deleting a node from a list requires keeping track of the previous node, that is, the node whose link points to the node to be deleted.

3. If the link in any node gets corrupted, the remaining nodes of the list become unusable.

These problems of singly linked lists can be overcome by adding one more link to each node, which points to the previous node. When such a link is added to every node of a list, the corresponding linked list is called a doubly linked list. Therefore, a doubly linked list is a linked list in which every node contains two links, called left link and right link, respectively. The left link of the node points to the previous node, whereas the right points to the next node. Like a singly linked list, a doubly linked list can also be a chain or it may be circular with or without a header node. If it is a chain, the left link of the first node and the right link of the last node will be NULL, as shown in Figure 20.18.

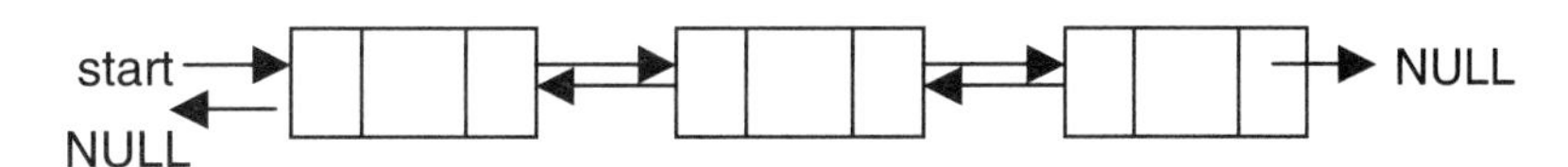

FIGURE 20.18 A doubly linked list maintained as chain.

If it is a circular list without a header node, the right link of the last node points to the first node. The left link of the first node points to the last node, as shown in Figure 20.19.

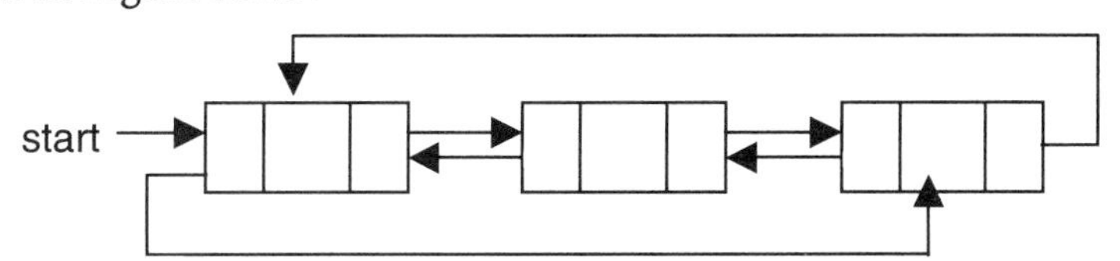

FIGURE 20.19 A doubly linked list maintained as a circular list.

If it is a circular list with a header node, the left link of the first node and the right link of the last node point to the header node. The right link of the header node points to the first node and the left link of the header node points to the last node of the list, as shown in Figure 20.20.

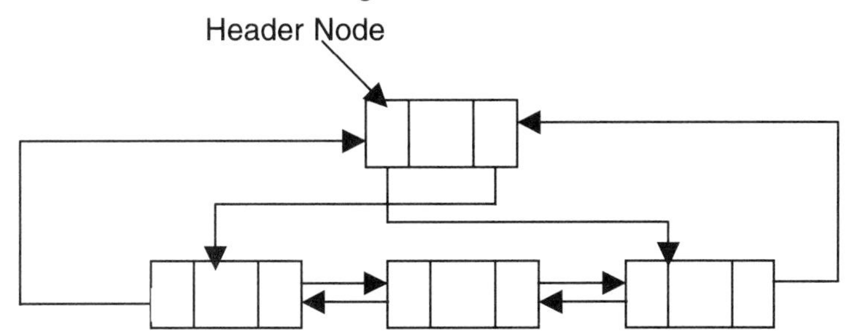

FIGURE 20.20 A doubly linked list maintained as a circular list with a header node.

Therefore, the following representation is required to be used for the nodes of a doubly linked list.

```
struct dnode
  {
    int data;
    struct dnode *left,*right;
  };
```

Program

A program for building and printing the elements of a doubly linked list follows:

```
# include <stdio.h>
# include <stdlib.h>
struct dnode
{
int data;
struct dnode *left, *right;
};
struct dnode *insert(struct dnode *p , struct dnode **q, int n)
{
struct dnode *temp;
 /* if the existing list is empty then insert a new node as the
starting node */
 if(p==NULL)
   {
      p=(struct dnode *)malloc(sizeof(struct dnode)); /* creates new
node data value
      passed as  parameter */

      if(p==NULL)
      {
    printf("Error\n");
          exit(0);
      }
      p->data = n;
      p-> left = p->right =NULL;
      *q =p;
    }
    else
    {
```

```
          temp =  (struct dnode *)malloc(sizeof(struct dnode)); /* creates
new node using
          data value passed as
            parameter  and puts its
           address in the temp
     */
         if(temp == NULL)
         {
       printf("Error\n");
            exit(0);
         }
          temp->data = n;
          temp->left = (*q);
          temp->right = NULL;
          (*q)->right = temp;
         (*q) = temp;
     }
        return (p);
    }
    void printfor( struct dnode *p  )
    {
       printf("The data values in the list in the forward order are:\n");
       while (p!= NULL)
          {
             printf("%d\t",p-> data);
             p = p->right;
          }
    }

    void printrev( struct dnode *p  )
    {
       printf("The data values in the list in the reverse order are:\n");
       while (p!= NULL)
          {
             printf("%d\t",p->data);
             p = p->left;
          }
    }
    void main()
    {
```

```
        int n;
        int x;
        struct dnode *start = NULL ;
        struct dnode *end = NULL;
        printf("Enter the nodes to be created \n");
        scanf("%d",&n);
        while ( n-- > 0 )
        {
      printf( "Enter the data values to be placed in a node\n");
           scanf("%d",&x);
           start = insert ( start , &end,x );
        }
        printf("The created list is\n");
        printfor ( start );
        printrev(end);
}
```

Explanation

1. This program uses a strategy of inserting a node in an existing list to create it. For this, an insert function is used. The insert function takes a pointer to an existing list as the first parameter.
2. The pointer to the last node of a list is the second parameter. A data value with which the new node is to be created is the third parameter. This creates a new node using the data value, appends it to the end of the list, and returns a pointer to the first node of the list. Initially, the list is empty, so the pointer to the start node is NULL. When insert is called the first time, the new node created by the insert becomes the start node.
3. Subsequently, insert creates a new node that stores the pointer to the created node in a temporary pointer. Then the left link of the node pointed to by the temporary pointer becomes the last node of the existing list, and the right link points to NULL. After that, it updates the value of the end pointer to make it point to this newly appended node.
4. The main function reads the value of the number of nodes in the list, and calls insert that many times by going in a while loop, in order to get a doubly linked list with the specified number of nodes created.

INSERTION OF A NODE IN A DOUBLY LINKED LIST

Introduction

The following program inserts the data in a doubly linked list.

Program

```
# include <stdio.h>
# include <stdlib.h>
struct dnode
{
int data;
struct node *left, *right;
};
struct dnode *insert(struct dnode *p , struct dnode **q, int n)
{
struct dnode *temp;
 /* if the existing list is empty then insert a new node as the
starting node */
  if(p==NULL)
    {
       p=(struct dnode *)malloc(sizeof(struct dnode)); /* creates new
node data value
      passed as  parameter */

       if(p==NULL)
       {
     printf("Error\n");
            exit(0);
       }
        p-> data = n;
        p-> left = p->right =NULL;
        *q =p
     }
     else
     {
        temp =  (struct dnode *)malloc(sizeof(struct dnode)); /* creates
new node using
        data value passed as
```

```
         parameter  and puts its
        address in the temp
 */
      if(temp == NULL)
      {
    printf("Error\n");
         exit(0);
      }
       temp-> data = n;
       temp->left = (*q);
       temp->right = NULL;
       (*q) = temp;
 }
     return (p);
}
void printfor( struct dnode *p  )
{
   printf("The data values in the list in the forward order are:\n");
   while (p!= NULL)
      {
         printf("%d\t",p-> data);
         p = p-> right;
      }
}
/* A function to count the number of nodes in a doubly linked list */
int  nodecount (struct dnode *p )
{
   int count=0;
    while (p != NULL)
    {
        count ++;
        p = p->right;
    }
      return(count);
}


   /* a function which inserts a newly created node after the specified
node in a doubly
          linked list */
```

```
struct node * newinsert ( struct dnode *p, int node_no, int value )
      {
      struct dnode *temp , * temp1;
      int i;
      if ( node_no <= 0 || node_no > nodecount (p))
      {
      printf("Error! the specified node does not exist\n");
      exit(0);
      }
      if ( node_no == 0)
      {
      temp = ( struct dnode * )malloc ( sizeof ( struct dnode ));
      if ( temp == NULL )
      {
      printf( " Cannot allocate \n");
      exit (0);
      }
      temp -> data = value;
      temp -> right = p;
      temp->left = NULL
      p = temp ;
      }
      else
      {
      temp = p ;
      i = 1;
      while ( i < node_no )
      {
      i = i+1;
      temp = temp-> right ;
      }
      temp1 = ( struct dnode * )malloc ( sizeof(struct dnode));
      if ( temp == NULL )
      {
      printf("Cannot allocate \n");
      exit(0);
      }
      temp1 -> data = value ;
      temp1 -> right = temp -> right;
      temp1 -> left = temp;
      temp1->right->left = temp1;
      temp1->left->right = temp1
```

```
        }
        return (p);
}
void main()
{
        int n;
        int x;
        struct dnode *start = NULL ;
        struct dnode *end = NULL;
        printf("Enter the nodes to be created \n");
        scanf("%d",&n);
        while ( n-- > 0 )
        {
     printf( "Enter the data values to be placed in a node\n");
           scanf("%d",&x);
           start = insert ( start , &end,x );
        }
        printf("The created list is\n");
        printfor ( start );
        printf("enter the node number after which the new node is to be
inserted\n");
        scanf("%d",&n);
        printf("enter the data value to be placed in the new node\n");
        scanf("%d",&x);
        start=newinsert(start,n,x);
        printfor(start);
}
```

Explanation

1. To insert a new node in a doubly linked chain, it is required to obtain a pointer to the node in the existing list after which a new node is to be inserted.
2. To obtain this pointer, the node number after which the new node is to be inserted is given as input. The nodes are assumed to be numbered as 1,2,3,..., etc., starting from the first node.
3. The list is then traversed starting from the start node to obtain the pointer to the specified node. Let this pointer be x. A new node is then created with the required data value, and the right link of this node is made to point to

the node to the right of the node pointed to by x. And the left link of the newly created node is made to point to the node pointed to by x. The left link of the node which was to the right of the node pointed to by x is made to point to the newly created node. The right link of the node pointed to by x is made to point to the newly created node.

DELETING A NODE FROM A DOUBLY LINKED LIST

Introduction

The following program deletes a specific node from the linked list.

Program

```
# include <stdio.h>
# include <stdlib.h>
struct dnode
{
int data;
struct dnode *left, *right;
};

struct dnode *insert(struct dnode *p , struct dnode **q, int n)
{
struct dnode *temp;
 /* if the existing list is empty then insert a new node as the
starting node */
 if(p==NULL)
   {
      p=(struct dnode *)malloc(sizeof(struct dnode)); /* creates new
node data value
   passed as  parameter */

      if(p==NULL)
      {
    printf("Error\n");
           exit(0);
      }
```

```
            p-> data = n;
            p-> left = p->right =NULL;
            *q =p;
         }
         else
         {
            temp =  (struct dnode *)malloc(sizeof(struct dnode)); /* creates
new node using
            data value passed as
              parameter  and puts its
             address in the temp
      */
           if(temp == NULL)
           {
        printf("Error\n");
              exit(0);
           }
            temp-> data = n;
            temp->left = (*q);
            temp->right = NULL;
            (*q)->right = temp;
           (*q) = temp;
      }
          return (p);
    }
    void printfor( struct dnode *p  )
    {
       printf("The data values in the list in the forward order are:\n");
       while (p!= NULL)
          {
             printf("%d\t",p-> data);
             p = p-> right;
          }
    }
    /* A function to count the number of nodes in a doubly linked list */
    int  nodecount (struct dnode *p )
    {
       int count=0;
        while (p != NULL)
        {
```

```
           count ++;
           p = p->right;
        }
          return(count);
    }


    /* a function which inserts a newly created node after the specified
node in a doubly
           linked list */
    struct dnode * delete( struct dnode *p, int node_no, int *val)
          {
          struct dnode *temp ,*prev=NULL;
          int i;
          if ( node_no <= 0 || node_no > nodecount (p))
          {
          printf("Error! the specified node does not exist\n");
          exit(0);
          }
          if ( node_no == 0)
          {
           temp = p;
           p = temp->right;
           p->left = NULL;
           *val = temp->data;
           return(p);
          }
           else
          {
          temp = p ;
           i = 1;
          while ( i < node_no )
          {
          i = i+1;
           prev = temp;
          temp = temp-> right ;
          }
           prev->right = temp->right;
           if(temp->right != NULL)
             temp->right->left = prev;
           *val = temp->data;
           free(temp);
```

```
}
        return (p);
}

void main()
{
        int n;
        int x;
        struct dnode *start = NULL ;
        struct dnode *end = NULL;
        printf("Enter the nodes to be created \n");
        scanf("%d",&n);
        while ( n-- > 0 )
        {
     printf( "Enter the data values to be placed in a node\n");
            scanf("%d",&x);
            start = insert ( start , &end,x );
        }
        printf("The created list is\n");
        printfor ( start );
        printf("enter the number of the node which is to be deleted\n");
        scanf("%d",&n);
       start=delete(start,n,&x);
       printf("The data value of the node deleted from list is :
%d\n",x);
       printf("The list after deletion of the specified node is :\n");
        printfor(start);
}
```

Explanation

1. To delete a node from a doubly linked chain, it is required to obtain a pointer to the node in the existing list that appears to the left of the node which is to be deleted.
2. To obtain this pointer, the node number which is to be deleted is given as input. The nodes are assumed to be numbered 1,2,3,..., etc., starting from the first node.
3. The list is then traversed starting from the start node to obtain the pointer to the specified node. Let this pointer be x. A pointer to the node to the right of the node x is also obtained. Let this be pointer y (this is a pointer to the

node to be deleted). The right link of the node pointed to by x is the node pointing to the node to which the right link of the node pointed to by y points. The left link of the node to the right of the node pointed to by y is made to point to x. The node pointed to by y is then freed.

APPLICATION OF DOUBLY LINKED LISTS TO MEMORY MANAGEMENT

Introduction

A doubly linked list is used to maintain both the list of allocated blocks and the list of free blocks by the memory manager of the operating system. To keep track of the allocated and free portions of memory, the memory manager is required to maintain a linked list of allocated and free segments. Each node of this list contains a starting address, size, and status of the segment. This list is kept sorted by the starting address field to facilitate the updating, because when a process terminates, the memory segment allocated to it becomes free, and so if any of the segments are freed, then they can be merged with the adjacent segment, if the adjacent segment is already free. This requires traversal of the list both ways to find out whether any of the adjacent segments are free. So this list is required to be maintained as a doubly linked list. For example, at a particular point in time, the list may be as shown in Figure 20.21.

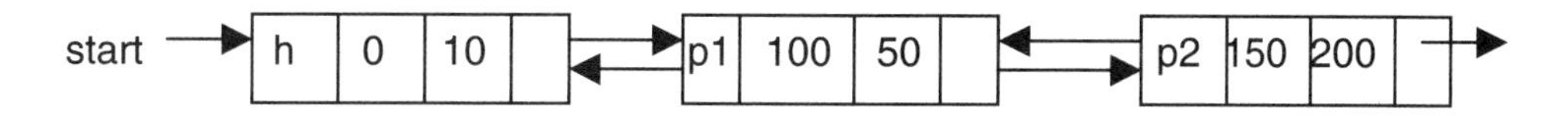

FIGURE 20.21 Before termination of process p1.

If the process p1 terminates, it is required to be modified as shown in Figure 20.22.

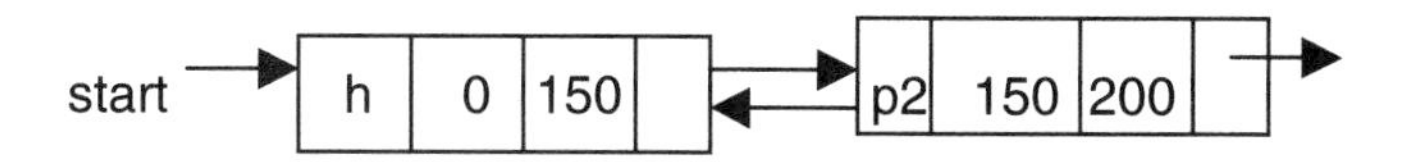

FIGURE 20.22 After termination of process p1.

General Comments on Linked Lists

1. A linked list is a dynamic data structure that can grow and shrink based on need.
2. The elements are not necessarily at a fixed distance apart.
3. In a linked list, the elements are placed in non-contiguous blocks of memory, and each block is linked to its previous block.
4. To link the next element to the previous element, the address of the next element is stored in the previous element itself.
5. Insertion or deletion at any arbitrary position in the linked list can be done easily, since it requires adjustment of only a few pointers.
6. Linked lists can be used for manipulation of symbolic polynomials.
7. A linked list is suitable for representation of sparse matrices.
8. A circular list is a list in which the link field of the last node is made to point to the start/first node of the list
9. A doubly linked list (DLL) is a linked list in which every node contains two links, called the left link and right link, respectively.
10. The left link of the node in a DLL is made to point to the previous node, whereas the right link is made to point to the next node.
11. A DLL can be traversed in both directions.
12. Having two pointers in a DLL provides safety, because even if one of the pointers get corrupted, the node still remains linked.
13. Deleting a particular node from a list, therefore, does not require keeping track of the previous node in a DLL.

Exercises

1. Write a C program to delete a node with the minimum value from a singly linked list.
2. Write a C program that will remove a specified node from a given doubly linked list and insert it at the end of the list.
3. Write a C program to transform a circular list into a chain.
4. Write a C program to merge two given lists A and B to form C in the following manner:

The first element of C is the first element of A and the second element of C is the first element of B. The second elements of A and B become the third and fourth elements of C, and so on. If either A or B gets exhausted, the remaining elements of the other are to be copied to C.

5. Write a C program to delete all occurrences of x from a given singly linked list.

21 Trees

THE CONCEPT OF TREES

Introduction

Trees are used to impose a hierarchical structure on a collection of data items. For example, we need to impose a hierarchical structure on a collection of data items while preparing organizational charts and geneologies to represent the syntactic structure of a source program in compilers. So the study of trees as one of the data structures is important.

Definition of a Tree

A tree is a set of one or more nodes T such that:

(i) there is a specially designated node called a root

(ii) The remaining nodes are partitioned into *n disjointed* set of nodes T_1, T_2,...,Tn, each of which is a tree.

A tree strucutre is shown in Figure 21.1.

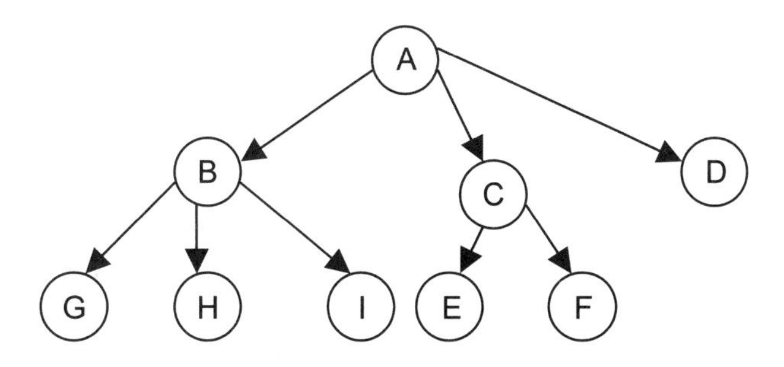

FIGURE 21.1 A tree structure.

This is a tree because it is a set of nodes {A,B,C,D,E,F,G,H,I}, with node A as a root node and the remaining nodes partitioned into three disjointed sets {B,G,H,I}, { C,E,F} and {D}, respectively. Each of these sets is a tree because each satisfies the aforementioned definition properly.

Shown in Figure 21.2 is a structure that is not a tree.

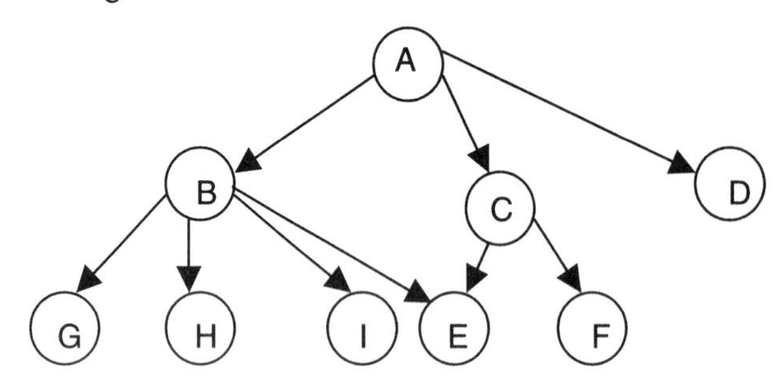

FIGURE 21.2 A non-tree structure.

Even though this is a set of nodes {A,B,C,D,E,F,G,H,I}, with node A as a root node, this is not a tree because the fact that node E is shared makes it impossible to partition nodes B through I into disjointed sets.

Degree of a Node of a Tree

The *degree of a node of a tree* is the number of subtrees having this node as a root. In other words, the degree is the number of descendants of a node. If the degree is zero, it is called a terminal or leaf node of a tree.

Degree of a Tree

The *degree of a tree* is defined as the maximum of degree of the nodes of the tree, that is, degree of tree = max (degree(node i) for I = 1 to n)

Level of a Node

We define the level of the node by taking the level of the root node as 1, and incrementing it by 1 as we move from the root towards the subtrees. So the level of all the descendants of the root nodes will be 2. The level of their descendants will be 3, and so on. We then define the depth of the tree to be the maximum value of the level of the node of the tree.

BINARY TREE AND ITS REPRESENTATION

Introduction

A *binary tree* is a special case of tree as defined in the preceding section, in which no node of a tree can have a degree of more than 2. Therefore, a binary tree is a set of zero or more nodes T such that:

(i) there is a specially designated node called the root of the tree

(ii) the remaining nodes are partitioned into two disjointed sets, T_1 and T_2, each of which is a binary tree. T_1 is called the left subtree and T_2 is called right subtree, or vice-versa.

A binary tree is shown in Figure 21.3.

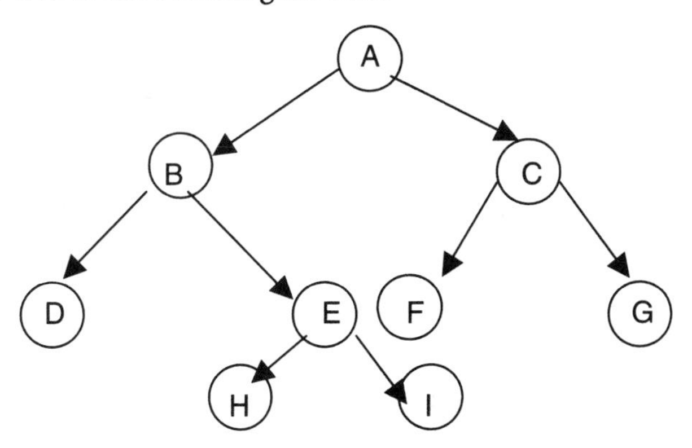

FIGURE 21.3 Binary tree structure.

So, for a binary tree we find that:

(i) The maximum number of nodes at level i will be 2^{i-1}

(ii) If k is the depth of the tree then the maximum number of nodes that the tree can have is

$2^k - 1 = 2^{k-1} + 2^{k-2} + ... + 2^0$

Also, there are skewed binary trees, such as the one shown in Figure 21.4.

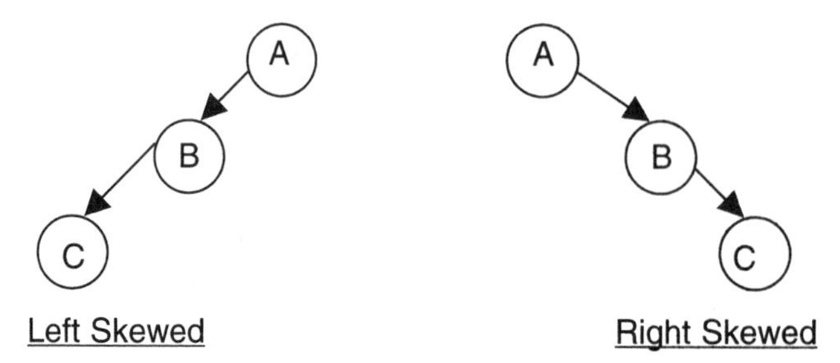

FIGURE 21.4 Skewed trees.

A full binary tree is a binary of depth k having $2^k - 1$ nodes. If it has $< 2^k - 1$, it is not a full binary tree. For example, for k = 3, the number of nodes = $2^k - 1 = 2^3 - 1 = 8 - 1 = 7$. A full binary tree with depth k = 3 is shown in Figure 21.5.

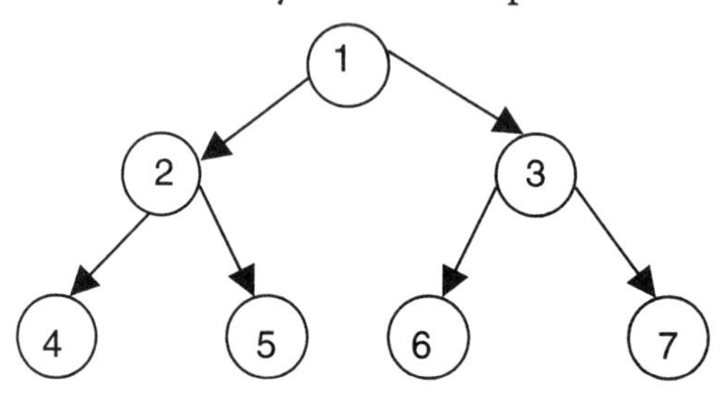

FIGURE 21.5 A full binary tree.

We use numbers from 1 to $2^k - 1$ as labels of the nodes of the tree.

If a binary tree is full, then we can number its nodes sequentially from 1 to 2^{k-1}, starting from the root node, and at every level numbering the nodes from left to right.

A complete binary tree of depth k is a tree with n nodes in which these n nodes can be numbered sequentially from 1 to n, as if it would have been the first n nodes in a full binary tree of depth k.

A complete binary tree with depth $k = 3$ is shown in Figure 21.6.

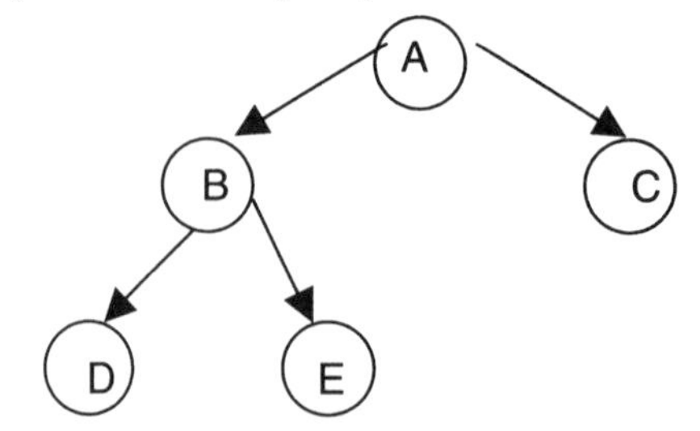

FIGURE 21.6 A complete binary tree.

Representation of a Binary Tree

If a binary tree is a *complete binary tree*, it can be represented using an array capable of holding n elements where n is the number of nodes in a complete binary tree. If the tree is an array of n elements, we can store the data values of the ith node of a complete binary tree with n nodes at an index i in an array tree. That means we can map node i to the ith index in the array, and the parent of node i will get mapped at an index i/2, whereas the left child of node i gets mapped at an index 2i and the right child gets mapped at an index 2i + 1. For example, a complete binary tree with depth $k = 3$, having the number of nodes $n = 5$, can be represented using an array of 5 as shown in Figure 21.7.

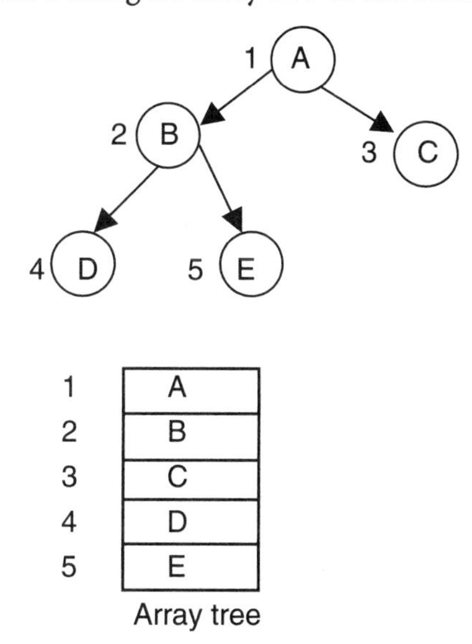

FIGURE 21.7 An array representation of a complete binary tree having 5 nodes and depth 3.

Shown in Figure 21.8 is another example of an array representation of a complete binary tree with depth $k = 3$, with the number of nodes $n = 4$.

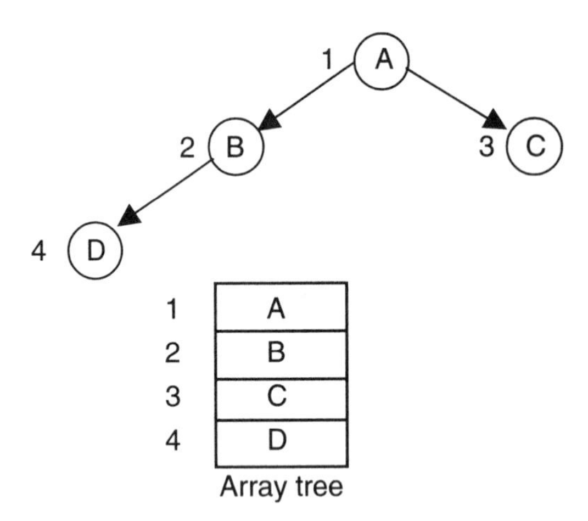

FIGURE 21.8 An array representation of a complete binary tree with 4 nodes and depth 3.

In general, any binary tree can be represented using an array. We see that an array representation of a complete binary tree does not lead to the waste of any storage. But if you want to represent a binary tree that is not a complete binary tree using an array representation, then it leads to the waste of storage as shown in Figure 21.9.

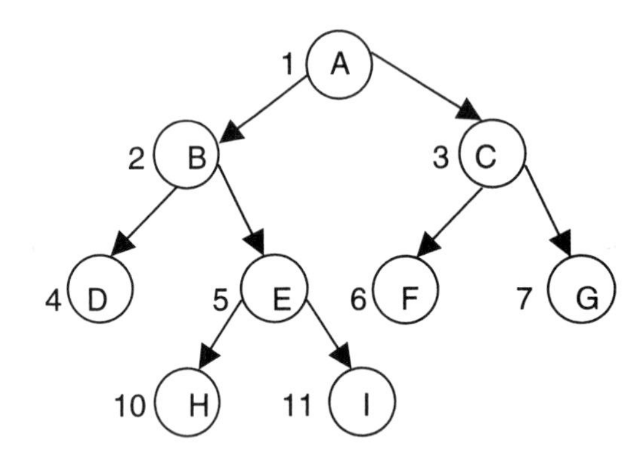

1	A
2	B
3	C
4	D
5	E
6	F
7	G
8	
9	
10	H
11	I

Array tree

FIGURE 21.9 An array representation of a binary tree.

An array representation of a binary tree is not suitable for frequent insertions and deletions, even though no storage is wasted if the binary tree is a complete binary tree. It makes insertion and deletion in a tree costly. Therefore, instead of using an array representation, we can use a linked representation, in which every node is represented as a structure with three fields: one for holding data, one for linking it with the left subtree, and the third for linking it with right subtree as shown here:

leftchild	data	rightchild

We can create such a structure using the following C declaration:

```
struct tnode
{
   int data
   struct tnode *lchild,*rchild;
};
```

A tree representation that uses this node structure is shown in Figure 21.10.

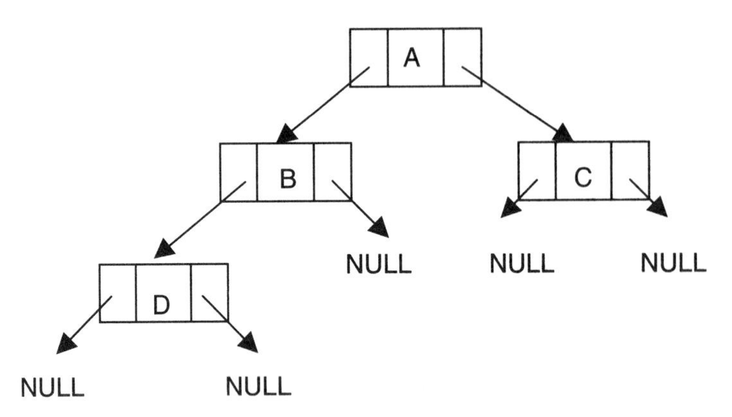

FIGURE 21.10 Linked representation of a binary tree.

BINARY TREE TRAVERSAL

Introduction

Order of Traversal of Binary Tree

The following are the possible orders in which a binary tree can be traversed:

LDR
LRD
DLR
RDL
RLD
DRL

where L stands for traversing the left subtree, R stands for traversing the right subtree, and D stands for processing the data of the node. Therefore, the order LDR is the order of traversal in which we start with the root node, visit the left subtree, process the data of the root node, and then visit the right subtree. Since the left and right subtrees are also the binary trees, the same procedure is used recursively while visiting the left and right subtrees.

The order LDR is called as inorder; the order LRD is called as postorder; and the order DLR is called as preorder. The remaining three orders are not used. If the processing that we do with the data in the node of tree during the traversal is simply printing the data value, then the output generated for a tree is given in Figure 21.11, using inorder, preorder and postorder as shown.

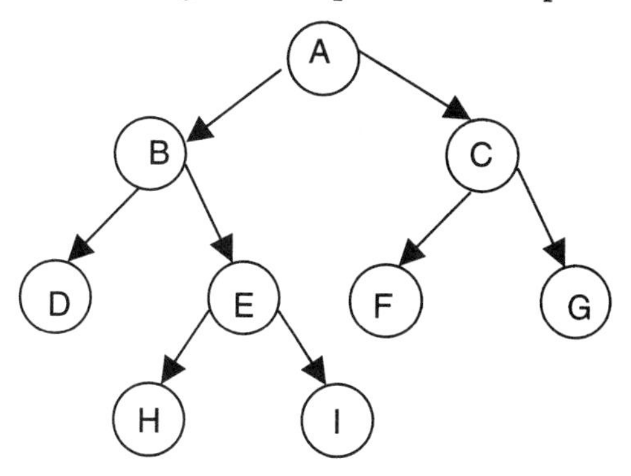

Inorder : DBHEIAFCG

Preorder : ABDEHICFG

Postorder : DHIEBFGCA

FIGURE 21.11 A binary tree along with its inorder, preorder and postorder.

If an expression is represented as a binary tree, the inorder traversal of the tree gives us an infix expression, whereas the postorder traversal gives us a postfix expression as shown in Figure 21.12.

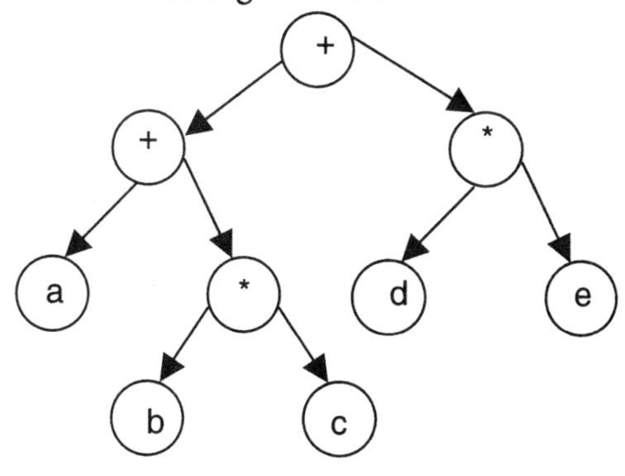

Inorder : a + b * c + d * e

postorder : abc*+de*+

FIGURE 21.12 A binary tree of an expression along with its inorder and postorder.

Given an order of traversal of a tree, it is possible to construct a tree; for example, consider the folowing order:

Inorder = DBEAC

We can construct the binary trees shown in Figure 21.13 by using this order of traversal:

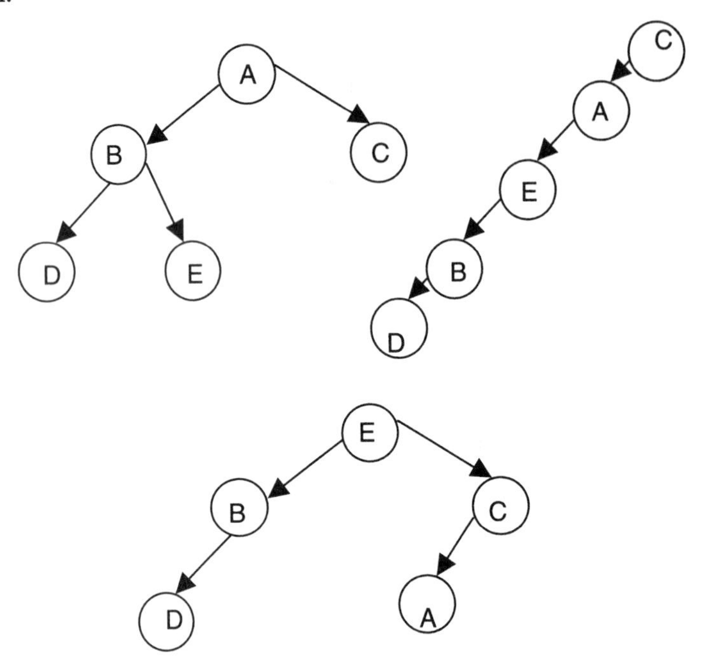

FIGURE 21.13 Binary trees constructed using the given inorder.

Therefore, we conclude that given only one order of traversal of a tree, it is possible to construct a number of binary trees; a unique binary tree cannot be constructed with only one order of traversal. For construction of a unique binary tree, we require two orders, in which one has to be inorder; the other can be preorder or postorder. For example, consider the following orders:

Inorder = DBEAC

Postorder = DEBCA

We can construct the unique binary tree shown in Figure 21.14 by using these orders of traversal:

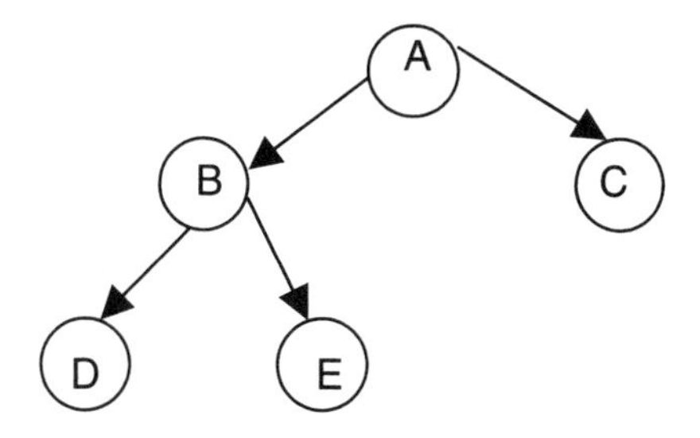

FIGURE 21.14 A unique binary tree constructed using its inorder and postorder.

BINARY SEARCH TREE

Introduction

A *binary search tree* is a binary tree that may be empty, and every node must contain an identifier. An identifier of any node in the left subtree is less than the identifier of the root. An identifier of any node in the right subtree is greater than the identifier of the root. Both the left subtree and right subtree are binary search trees.

A binary search tree is shown in Figure 21.15.

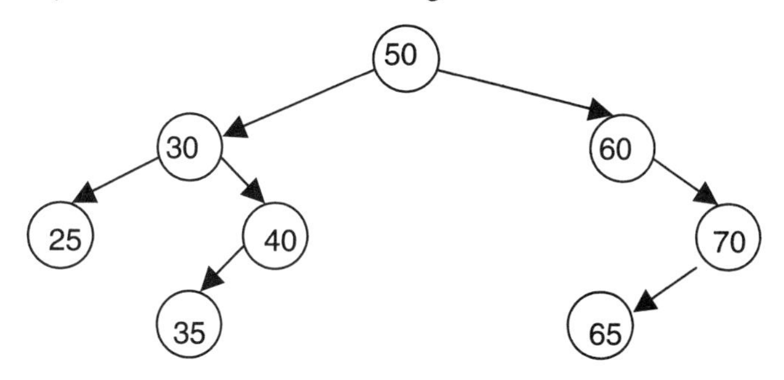

FIGURE 21.15 The binary search tree.

The binary search tree is basically a binary tree, and therefore it can be traversed in inorder, preorder, and postorder. If we traverse a binary search tree in inorder and print the identifiers contained in the nodes of the tree, we get a sorted list of identifiers in ascending order.

A binary search tree is an important search structure. For example, consider the problem of searching a list. If a list is ordered, searching becomes faster if we use a contiguous list and perform a binary search. But if we need to make changes in the list, such as inserting new entries and deleting old entries, using a contiguous list would be much slower, because insertion and deletion in a contiguous list requires moving many of the entries every time. So we may think of using a linked list because it permits insertions and deletions to be carried out by adjusting only a few pointers. But in an n-linked list, there is no way to move through the list other than one node at a time, permitting only sequential access. Binary trees provide an excellent solution to this problem. By making the entries of an ordered list into the nodes of a binary search tree, we find that we can search for a key in $O(n \log n)$ steps.

Program: Creating a Binary Search Tree

We assume that every node of a binary search tree is capable of holding an integer data item and that the links can be made to point to the root of the left subtree and the right subtree, respectively. Therefore, the structure of the node can be defined using the following declaration:

```
struct tnode
   {
        int data;
        struct tnode *lchild,*rchild;
   };
```

A complete C program to create a binary search tree follows:

```
#include <stdio.h>
#include <stdlib.h>
struct tnode
{
  int data;
  struct tnode *lchild, *rchild;
};

struct tnode *insert(struct tnode *p,int val)
{
   struct tnode *temp1,*temp2;
   if(p == NULL)
   {
      p = (struct tnode *) malloc(sizeof(struct tnode)); /* insert the
new node as root node*/
```

```
          if(p == NULL)
            {
         printf("Cannot allocate\n");
         exit(0);
            }
          p->data = val;
          p->lchild=p->rchild=NULL;
     }
    else
    {
     temp1 = p;
    /* traverse the tree to get a pointer to that node whose child will be
the newly created node*/
    while(temp1 != NULL)
    {
      temp2 = temp1;
      if( temp1 ->data > val)
           temp1 = temp1->lchild;
      else
           temp1 = temp1->rchild;
    }
    if( temp2->data > val)
    {
    temp2->lchild = (struct tnode*)malloc(sizeof(struct tnode));/*inserts
the newly created node as left child*/
       temp2 = temp2->lchild;
       if(temp2 == NULL)
            {
        printf("Cannot allocate\n");
        exit(0);
            }
       temp2->data = val;
       temp2->lchild=temp2->rchild = NULL;
    }
    else
    {
       temp2->rchild = (struct tnode*)malloc(sizeof(struct tnode));/
*inserts the newly created node
    as left child*/
       temp2 = temp2->rchild;
       if(temp2 == NULL)
            {
```

```
    printf("Cannot allocate\n");
    exit(0);
        }
   temp2->data = val;
   temp2->lchild=temp2->rchild = NULL;
}
}
return(p);
}
/* a function to binary tree in inorder */
void inorder(struct tnode *p)
    {
       if(p != NULL)
         {
           inorder(p->lchild);
           printf("%d\t",p->data);
          inorder(p->rchild);
          }
    }
void main()
{
   struct tnode *root = NULL;
   int n,x;
   printf("Enter the number of nodes\n");
   scanf("%d",&n);
   while( n - > 0)
     {
        printf("Enter the data value\n");
        scanf("%d",&x);
        root = insert(root,x);
    }
     inorder(root);
  }
```

Explanation

1. To create a binary search tree, we use a function called insert, which creates a new node with the data value supplied as a parameter to it, and inserts it into an already existing tree whose root pointer is also passed as a parameter.
2. The function accomplishes this by checking whether the tree whose root pointer is passed as a parameter is empty. If it is empty, then the newly created node is inserted as a root node. If it is not empty, then it copies the

root pointer into a variable temp1. It then stores the value of temp1 in another variable, temp2, and compares the data value of the node pointed to by temp1 with the data value supplied as a parameter. If the data value supplied as a parameter is smaller than the data value of the node pointed to by temp1, it copies the left link of the node pointed to by temp1 into temp1 (goes to the left); otherwise it copies the right link of the node pointed to by temp1 into temp1 (goes to the right).

3. It repeats this process until temp1 reaches 0. When temp1 becomes 0, the new node is inserted as a left child of the node pointed to by temp2, if the data value of the node pointed to by temp2 is greater than the data value supplied as a parameter. Otherwise, the new node is inserted as a right child of the node pointed to by temp2. Therefore the insert procedure is:

   Input: 1. The number of nodes that the tree to be created should have

   2. The data values of each node in the tree to be created

   Output: The data value of the nodes of the tree in inorder

Example

Input: 1. The number of nodes that the created tree should have = 5

2. The data values of the nodes in the tree to be created are: 10, 20, 5, 9, 8

Output : 5 8 9 10 20

Program

A function for inorder traversal of a binary tree:

```
void inorder(struct tnode *p)
    {
        if(p != NULL)
         {
 inorder(p->lchild);
 printf("%d\t",p->data);
     inorder(p->rchild);
}
```

A non-recursive/iterative function for traversing a binary tree in inorder is given here for the purpose of doing the analysis.

```
void inorder(struct tnode *p)
{
 struct tnode *stack[100];
```

```
 int top;
 top = -1;
if(p != NULL)
 {
    top++;
    stack[top] = p;
    p = p->lchild;
    while(top >= 0)
       {
          while ( p!= NULL)/* push the left child onto stack*/
           {
       top++;
                 stack[top] =p;
        p = p->lchild;
    }
           p = stack[top];
    top—;
printf("%d\t",p->data);
    p = p->rchild;
    if ( p != NULL) /* push right child*/
       {
          top++;
                          stack[top] = p;
           p = p->lchild;
       }
    }
  }
}
```

A function for preorder traversal of a binary tree:

```
void preorder(struct tnode *p)
    {
       if(p != NULL)
        {
printf("%d\t",p->data);
            preorder(p->lchild);
preorder(p->rchild);
          }
```

A function for postorder traversal of a binary tree:

```
void postorder(struct node *p)
    {
       if(p != NULL)
        {
```

```
 postorder(p->lchild);
     postorder(p->rchild);
printf("%d\t",p->data);
          }
```

Explanation

Consider the iterative version of the inorder just given. If the binary tree to be traversed has n nodes, the number of NULL links are $n+1$. Since every node is placed on the stack once, the statements `stack[top]:=p` and `p:=stack[top]` are executed n times. The test for NULL links will be done exactly $n+1$ times. So every step will be executed no more than some small constant times n. So the order of the algorithm is $O(n)$. A similar analysis can be done to obtain the estimate of the computation time for preorder and postorder.

Constructing a Binary Tree Using the Preorder and Inorder Traversals

To obtain the binary tree, we reverse the preorder traversal and take the first node that is a root node. We then search for this node in the inorder traversal. In the inorder traversal, all the nodes to the left of this node will be the part of the left subtree, and all the nodes to the right of this node will be the part of the right subtree. We then consider the next node in the reversed preorder. If it is a part of the left subtree, then we make it the left child of the root; if it is part of the right subtree, we make it part of right subtree. This procedure is repeated recursively to get the tree as shown in Figure 21.16.

For example, for the preorder and inorder traversals of a binary tree, the binary tree and its postorder traversal are as follows:

Z,A,Q,P,Y,X,C,B = Preorder

Q,A,Z,Y,P,C,X,B = Inorder

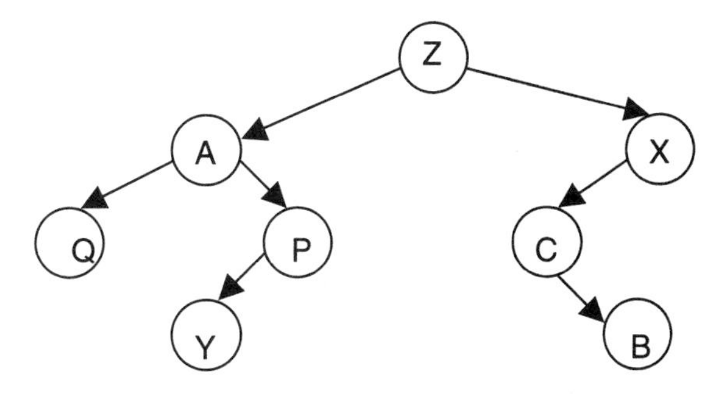

FIGURE 21.16 A unique binary tree constructed using the inorder and postorder.

The postorder for this tree is:

Z,A,P,X,B,C,Y,Q

COUNTING THE NUMBER OF NODES IN A BINARY SEARCH TREE

Introduction

To count the number of nodes in a given binary tree, the tree is required to be traversed recursively until a leaf node is encountered. When a leaf node is encountered, a count of 1 is returned to its previous activation (which is an activation for its parent), which takes the count returned from both the children's activation, adds 1 to it, and returns this value to the activation of its parent. This way, when the activation for the root of the tree returns, it returns the count of the total number of the nodes in the tree.

Program

A complete C program to count the number of nodes is as follows:

```
#include <stdio.h>
#include <stdlib.h>
struct tnode
{
  int data;
  struct tnode *lchild, *rchild;
};
int count(struct tnode *p)
   {
              if( p == NULL)
      return(0);
              else
      if( p->lchild == NULL && p->rchild == NULL)
            return(1);
      else
            return(1 + (count(p->lchild) + count(p->rchild)));
    }

struct tnode *insert(struct tnode *p,int val)
```

```
{
   struct tnode *temp1,*temp2;
   if(p == NULL)
   {
      p = (struct tnode *) malloc(sizeof(struct tnode)); /* insert the
new node as root node*/
      if(p == NULL)
        {
     printf("Cannot allocate\n");
     exit(0);
        }
    p->data = val;
    p->lchild=p->rchild=NULL;
 }
 else
 {
  temp1 = p;
 /* traverse the tree to get a pointer to that node whose child will be
the newly created node*/
 while(temp1 != NULL)
 {
   temp2 = temp1;
   if( temp1 ->data > val)
        temp1 = temp1->lchild;
   else
        temp1 = temp1->rchild;
 }
 if( temp2->data > val)
 {
    temp2->lchild = (struct tnode*)malloc(sizeof(struct tnode)); /
*inserts the newly created node
 as left child*/
    temp2 = temp2->lchild;
    if(temp2 == NULL)
        {
     printf("Cannot allocate\n");
     exit(0);
        }
    temp2->data = val;
    temp2->lchild=temp2->rchild = NULL;
 }
 else
```

```
{
   temp2->rchild = (struct tnode*)malloc(sizeof(struct tnode));/
*inserts the newly created node
as left child*/
   temp2 = temp2->rchild;
   if(temp2 == NULL)
       {
    printf("Cannot allocate\n");
    exit(0);
       }
   temp2->data = val;
   temp2->lchild=temp2->rchild = NULL;
}
}
return(p);
}
/* a function to binary tree in inorder */
void inorder(struct tnode *p)
   {
      if(p != NULL)
       {
         inorder(p->lchild);
         printf("%d\t",p->data);
        inorder(p->rchild);
        }
   }
void main()
{
  struct tnode *root = NULL;
  int n,x;
  printf("Enter the number of nodes\n");
  scanf("%d",&n);
  while( n - > 0)
    {
      printf("Enter the data value\n");
      scanf("%d",&x);
      root = insert(root,x);
   }
    inorder(root);
   printf("\nThe number of nodes in tree are :%d\n",count(root));
}
```

Explanation

Input: 1. The number of nodes that the tree to be created should have
2. The data values of each node in the tree to be created

Output: 1. The data value of the nodes of the tree in inorder
2. The count of number of node in a tree.

Example

Input: 1. The number of nodes the created tree should have = 5
2. The data values of the nodes in the tree to be created are: 10, 20, 5, 9, 8

Output: 1. 5 8 9 10 20
2. The number of nodes in the tree is 5

SWAPPING OF LEFT AND RIGHT SUBTREES OF A GIVEN BINARY TREE

Introduction

An elegant method of swapping the left and right subtrees of a given binary tree makes use of a recursive algorithm, which recursively swaps the left and right subtrees, starting from the root.

Program

```
#include <stdio.h>
#include <stdlib.h>
struct tnode
{
  int data;
  struct tnode *lchild, *rchild;
};

struct tnode *insert(struct tnode *p,int val)
{
   struct tnode *temp1,*temp2;
   if(p == NULL)
   {
```

```
        p = (struct tnode *) malloc(sizeof(struct tnode)); /* insert the
new node as root node*/
        if(p == NULL)
          {
      printf("Cannot allocate\n");
      exit(0);
          }
       p->data = val;
       p->lchild=p->rchild=NULL;
   }
  else
  {
   temp1 = p;
  /* traverse the tree to get a pointer to that node whose child will be
the newly created node*/
  while(temp1 != NULL)
  {
    temp2 = temp1;
    if( temp1 ->data > val)
         temp1 = temp1->lchild;
    else
         temp1 = temp1->rchild;
  }
  if( temp2->data > val)
  {
     temp2->lchild = (struct tnode*)malloc(sizeof(struct tnode));/
*inserts the newly created node
  as left child*/
     temp2 = temp2->lchild;
     if(temp2 == NULL)
         {
      printf("Cannot allocate\n");
      exit(0);
         }
     temp2->data = val;
     temp2->lchild=temp2->rchild = NULL;
  }
  else
  {
     temp2->rchild = (struct tnode*)malloc(sizeof(struct tnode));/
*inserts the newly created node
  as left child*/
```

```
   temp2 = temp2->rchild;
   if(temp2 == NULL)
       {
    printf("Cannot allocate\n");
    exit(0);
       }
   temp2->data = val;
   temp2->lchild=temp2->rchild = NULL;
}
}
return(p);
}
/* a function to binary tree in inorder */
void inorder(struct tnode *p)
    {
       if(p != NULL)
         {
            inorder(p->lchild);
            printf("%d\t",p->data);
         inorder(p->rchild);
         }
    }

struct tnode *swaptree(struct tnode *p)
{
   struct tnode *temp1=NULL, *temp2=NULL;
    if( p != NULL)
    {  temp1= swaptree(p->lchild);
                    temp2 = swaptree(p->rchild);
                  p->rchild = temp1;
                   p->lchild = temp2;

       }
        return(p);
 }

void main()
{
  struct tnode *root = NULL;
  int n,x;
  printf("Enter the number of nodes\n");
  scanf("%d",&n);
  while( n - > 0)
```

```
    {
       printf("Enter the data value\n");
       scanf("%d",&x);
       root = insert(root,x);
    }
      printf("The created tree is :\n");
      inorder(root);
      printf("The tree after swapping is :\n");
      root = swaptree(root);
      inorder(root);
      printf("\nThe original tree is \n");
      root = swaptree(root);
      inorder(root);

}
```

Explanation

Input: 1. The number of nodes that the tree to be created should have

2. The data values of each node in the tree to be created

Output: 1. The data value of the nodes of the tree in inorder before interchanging the left and right subtrees

2. The data value of the nodes of the tree in inorder after interchanging the left and right subtrees

Example

Input: 1. The number of nodes that the created tree should have = 5

2. The data values of the nodes in the tree to be created are: 10, 20, 5, 9, 8

Output: 1. 5 8 9 10 20

2. 20 10 9 8 5

SEARCHING FOR A TARGET KEY IN A BINARY SEARCH TREE

Introduction

Data values are given which we call a key and a binary search tree. To search for the key in the given binary search tree, start with the root node and compare the key with the data value of the root node. If they match, return the root pointer. If the key is less than the data value of the root node, repeat the process by using the left subtree. Otherwise, repeat the same process with the right subtree until either a match is found or the subtree under consideration becomes an empty tree.

Program

A complete C program for this search is as follows:

```
#include <stdio.h>
#include <stdlib.h>
struct tnode
{
  int data;
  struct tnode *lchild, *rchild;
};
/* A function to serch for a given data value in a binary search tree*/
struct tnode *search( struct tnode *p,int key)
  {
    struct tnode *temp;
     temp = p;
     while( temp != NULL)
      {
        if(temp->data == key)
            return(temp);
        else
    if(temp->data > key)
       temp = temp->lchild;
    else
       temp = temp->rchild;
         }
return(NULL);
}

/*an iterative function to print the binary tree in inorder*/
void inorder1(struct tnode *p)
{
 struct tnode *stack[100];
 int top;
 top = -1;
if(p != NULL)
```

```
{
    top++;
    stack[top] = p;
    p = p->lchild;
    while(top >= 0)
       {
          while ( p!= NULL)/* push the left child onto stack*/
           {
                     top++;
               stack[top] =p;
               p = p->lchild;
    }
    p = stack[top];
    top--;
    printf("%d\t",p->data);
    p = p->rchild;
    if ( p != NULL) /* push right child*/
      {
         top++;
                      stack[top] = p;
          p = p->lchild;
            }
       }
  }
}

/* A function to insert a new node in binary search tree to
get a tree created*/
struct tnode *insert(struct tnode *p,int val)
{
   struct tnode *temp1,*temp2;
   if(p == NULL)
   {
      p = (struct tnode *) malloc(sizeof(struct tnode)); /* insert the
new node as root node*/
      if(p == NULL)
        {
    printf("Cannot allocate\n");
    exit(0);
        }
     p->data = val;
     p->lchild=p->rchild=NULL;
```

```
     }
    else
    {
     temp1 = p;
    /* traverse the tree to get a pointer to that node whose child will be
the newly created node*/
    while(temp1 != NULL)
    {
      temp2 = temp1;
      if( temp1 ->data > val)
           temp1 = temp1->lchild;
      else
           temp1 = temp1->rchild;
    }
    if( temp2->data > val)
    {
       temp2->lchild = (struct tnode*)malloc(sizeof(struct tnode));/
*inserts the newly created node
    as left child*/
       temp2 = temp2->lchild;
       if(temp2 == NULL)
            {
        printf("Cannot allocate\n");
        exit(0);
            }
       temp2->data = val;
       temp2->lchild=temp2->rchild = NULL;
    }
    else
    {
       temp2->rchild = (struct tnode*)malloc(sizeof(struct tnode));/
*inserts the newly created node
    as left child*/
       temp2 = temp2->rchild;
       if(temp2 == NULL)
            {
        printf("Cannot allocate\n");
        exit(0);
            }
       temp2->data = val;
       temp2->lchild=temp2->rchild = NULL;
    }
```

```
}
return(p);
}
void main()
{
  struct tnode *root = NULL, *temp = NULL;
  int n,x;
  printf("Enter the number of nodes in the tree\n");
  scanf("%d",&n);
  while( n - > 0)
      {
       printf("Enter the data value\n");
       scanf("%d",&x);
       root = insert(root,x);
      }
     printf("The created tree is :\n");
     inorder1(root);
     printf("\n Enter the value of the node to be searched\n");
     scanf("%d",&n);
     temp=search(root,n);
     if(temp != NULL)
          printf("The data value is present in the tree \n");
     else
          printf("The data value is not present in the tree \n");

}
```

Explanation

Input:
1. The number of nodes that the tree to be created should have
2. The data values of each node in the tree to be created
3. The key value

Output: If the key is present and appears in the created tree, then a message "The data value is present in the tree" appears. Otherwise the message "The data value is not present in the tree" appears.

Example

Input:
1. The number of nodes that the created tree should have = 5
2. The data values of the nodes in the tree to be created are: 10, 20, 5, 9, 8
3. The key value = 9

Output: The data is present in the tree

DELETION OF A NODE FROM BINARY SEARCH TREE

Introduction

To delete a node from a binary search tree, the method to be used depends on whether a node to be deleted has one child, two children, or no children.

Deletion of a node with two children

Consider the binary search tree shown in Figure 21.17.

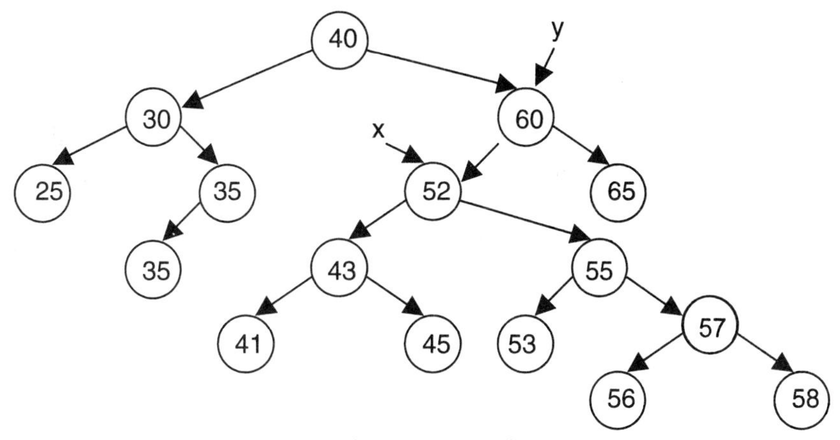

FIGURE 21.17 A binary tree before deletion of a node pointed to by x.

To delete a node printed to by *x*, we start by letting *y* be a pointer to the node that is the root of the node pointed to by *x*. We store the pointer to the left child of the node pointed to by *x* in a temporary pointer temp. We then make the left child of the node pointed to by *y* the left child of the node pointed to by *x*. We then traverse the tree with the root as the node pointed to by temp to get its right leaf, and make the right child of this right leaf the right child of the node pointed to by *x*, as shown in Figure 21.18.

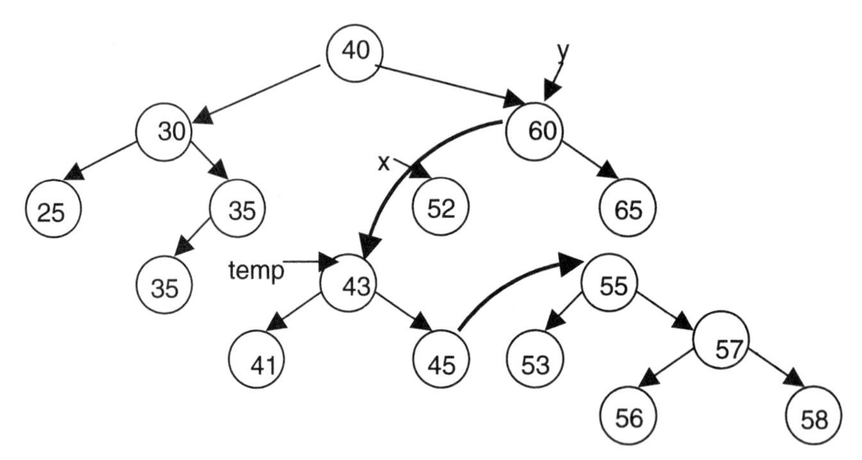

FIGURE 21.18 A binary tree after deletion of a node pointed to by x.

Another method is to store the pointer to the right child of the node pointed to by *x* in a temporary pointer temp. We then make the left child of the node pointed by *y* to be the right child of the node pointed to by *x*. We then traverse the tree with the root as the node pointed to by temp to get its left leaf, and make the left child of this left leaf the left child of the node pointed to by *x*, as shown in Figure 21.19.

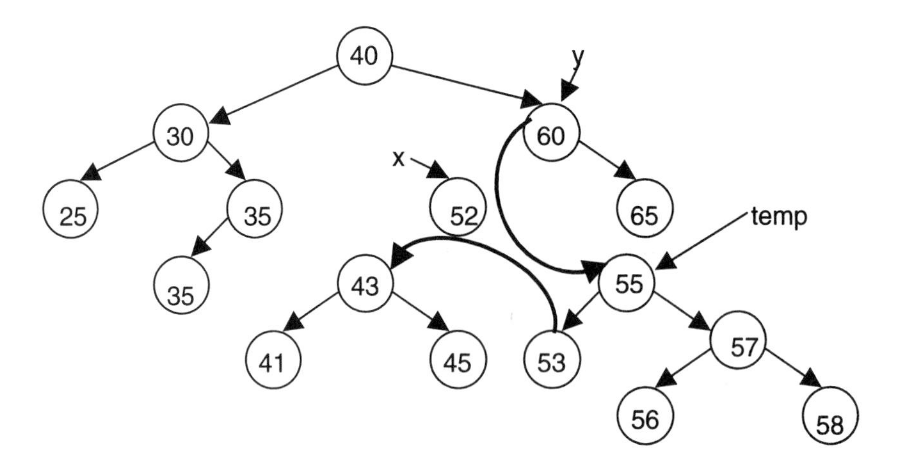

FIGURE 21.19 A binary tree after deletion of a node pointed to by x.

Deletion of a Node with One Child

Consider the binary search tree shown in Figure 21.20.

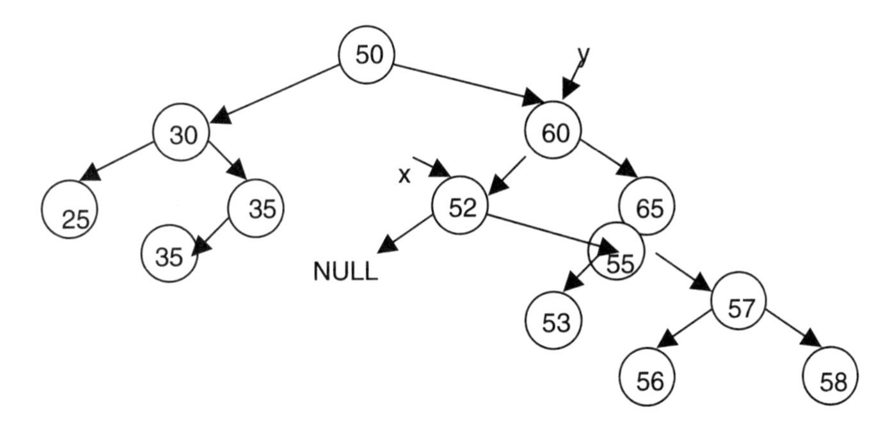

FIGURE 21.20 A binary tree before deletion of a node pointed to by x.

If we want to delete a node pointed to by *x*, we can do that by letting *y* be a pointer to the node that is the root of the node pointed to by *x*. Make the left child of the node pointed to by *y* the right child of the node pointed to by *x*, and dispose of the node pointed to by *x*, as shown in Figure 21.21.

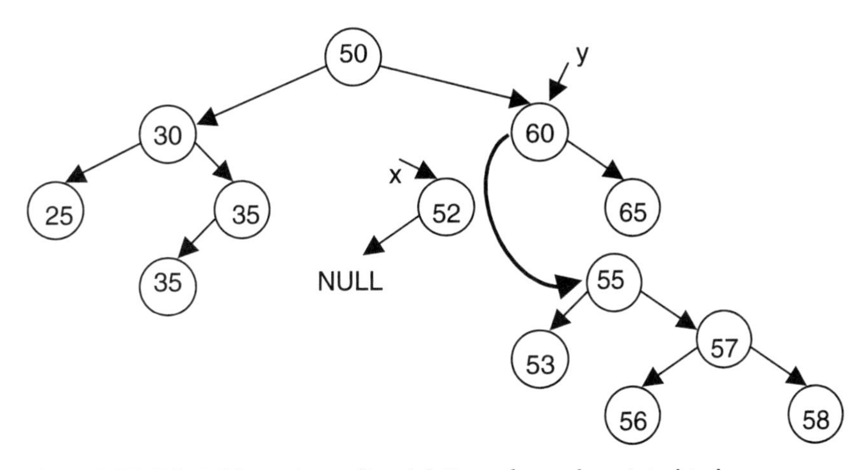

FIGURE 21.21 A binary tree after deletion of a node pointed to by x.

Deletion of a Node with No Child

Consider the binary search tree shown in Figure 21.22.

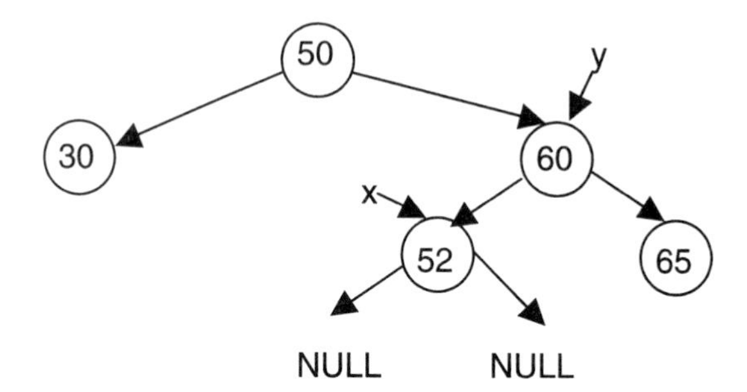

FIGURE 21.22 A binary tree before deletion of a node pointed to by x.

Set the left child of the node pointed to by *y* to NULL, and dispose of the node pointed to by *x*, as shown in Figure 21.23.

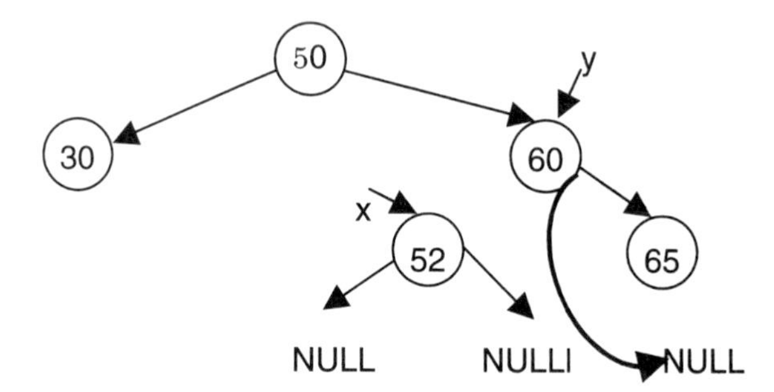

FIGURE 21.23 A binary tree after deletion of a node pointed to by x.

Program

A complete C program to delete a node, where the data value of the node to be deleted is known, is as follows:

```
#include <stdio.h>
#include <stdlib.h>
struct tnode
{
  int data;
  struct tnode *lchild, *rchild;
};
/* A function to get a pointer to the node whose data value is given
    as well as the pointer to its root */
struct tnode *getptr(struct tnode *p, int key, struct tnode **y)
```

```
{
 struct tnode *temp;
    if( p == NULL)
      return(NULL);
    temp = p;
    *y = NULL;
     while( temp != NULL)
      {
        if(temp->data == key)
            return(temp);
        else
    {
            *y = temp; /*store this pointer as root */
    if(temp->data > key)
       temp = temp->lchild;
    else
       temp = temp->rchild;
       }
      }
     return(NULL);
 }

/* A function to delete the node whose data value is given */
struct tnode *delete(struct tnode *p,int val)
  {
   struct tnode *x , *y, *temp;
   x = getptr(p,val,&y);
   if( x == NULL)
   {
     printf("The node does not exists\n");
     return(p);
   }
   else
   {
   /* this code is for deleting root node*/
   if( x == p)
     {
      temp = x->lchild;
      y = x->rchild;
      p = temp;
      while(temp->rchild != NULL)
    temp = temp->rchild;
```

```
        temp->rchild=y;
        free(x);
        return(p);
       }
   /* this code is for deleting node having both children */
     if( x->lchild != NULL && x->rchild != NULL)
        {

         if(y->lchild == x)
         {
    temp = x->lchild;
    y->lchild = x->lchild;
    while(temp->rchild != NULL)
        temp = temp->rchild;
    temp->rchild=x->rchild;
    x->lchild=NULL;
    x->rchild=NULL;
         }
         else
         {
           temp = x->rchild;
           y->rchild = x->rchild;
            while(temp->lchild != NULL)
        temp = temp->lchild;
     temp->lchild=x->lchild;
    x->lchild=NULL;
     x->rchild=NULL;
         }

    free(x);
      return(p);
     }
      /* this code is for deleting a node with only one child*/
      if( x->lchild == NULL && x->rchild != NULL)
        {
          if(y->lchild == x)
     y->lchild = x->rchild ;
          else
            y->rchild = x->rchild;
          x->rchild = NULL;
          free(x);
          return(p);
        }
```

```
        if( x->lchild != NULL && x->rchild == NULL)
          {
            if(y->lchild == x)
               y->lchild = x->lchild ;
            else
               y->rchild = x->lchild;
            x->lchild = NULL;
            free(x);
            return(p);
          }
        /* this code is for deleting a node with no child*/
        if(x->lchild == NULL && x->rchild == NULL)
         {
            if(y->lchild == x)
               y->lchild = NULL ;
            else
               y->rchild = NULL;
            free(x);
            return(p);
          }
      }

}
/*an iterative function to print the binary tree in inorder*/
void inorder1(struct tnode *p)
{
 struct tnode *stack[100];
 int top;
 top = -1;
if(p != NULL)
 {
    top++;
    stack[top] = p;
    p = p->lchild;
    while(top >= 0)
       {
          while ( p!= NULL)/* push the left child onto stack*/
           {
                      top++;
                stack[top] =p;
                p = p->lchild;
    }
```

```
        p = stack[top];
        top--;
        printf("%d\t",p->data);
        p = p->rchild;
        if ( p != NULL) /* push right child*/
           {
              top++;
           stack[top] = p;
              p = p->lchild;
        }
    }
      }
    }

    /* A function to insert a new node in binary search tree to get a tree
created*/
    struct tnode *insert(struct tnode *p,int val)
    {
       struct tnode *temp1,*temp2;
       if(p == NULL)
       {
          p = (struct tnode *) malloc(sizeof(struct tnode)); /* insert the
new node as root node*/
          if(p == NULL)
            {
        printf("Cannot allocate\n");
        exit(0);
            }
         p->data = val;
         p->lchild=p->rchild=NULL;
     }
    else
    {
     temp1 = p;
    /* traverse the tree to get a pointer to that node whose child will be
the newly created node*/
    while(temp1 != NULL)
    {
      temp2 = temp1;
      if( temp1 ->data > val)
           temp1 = temp1->lchild;
      else
```

```
        temp1 = temp1->rchild;
    }
    if( temp2->data > val)
    {
       temp2->lchild = (struct tnode*)malloc(sizeof(struct tnode));/
*inserts the newly created node
    as left child*/
       temp2 = temp2->lchild;
       if(temp2 == NULL)
           {
       printf("Cannot allocate\n");
       exit(0);
           }
       temp2->data = val;
       temp2->lchild=temp2->rchild = NULL;
    }
    else
    {
       temp2->rchild = (struct tnode*)malloc(sizeof(struct tnode));/
*inserts the newly created node
    as left child*/
       temp2 = temp2->rchild;
       if(temp2 == NULL)
           {
       printf("Cannot allocate\n");
       exit(0);
           }
       temp2->data = val;
       temp2->lchild=temp2->rchild = NULL;
    }
    }
    return(p);
    }

    void main()
    {
      struct tnode *root = NULL;
      int n,x;
      printf("Enter the number of nodes in the tree\n");
      scanf("%d",&n);
      while( n - > 0)
          {
```

```
        printf("Enter the data value\n");
        scanf("%d",&x);
        root = insert(root,x);
       }
      printf("The created tree is :\n");
      inorder1(root);
      printf("\n Enter the value of the node to be deleted\n");
      scanf("%d",&n);
      root=delete(root,n);
      printf("The tree after deletion is \n");
      inorder1(root);
}
```

Explanation

This program first creates a binary tree with a specified number of nodes with their respective data values. It then takes the data value of the node to be deleted, obtains a pointer to the node containing that data value, and obtains another pointer to the root of the node to be deleted. Depending on whether the node to be deleted is a root node, a node with two children a node with only one child, or a node with no children, it carries out the manipulations as discussed in the section on deleting a node. After deleting the specified node, it returns the pointer to the root of the tree.

Input: 1. The number of nodes that the tree to be created should have

2. The data values of each node in the tree to be created

3. The data value in the node to be deleted

Output: 1. The data values of the nodes in the tree in inorder before deletion

2. The data values of the nodes in the tree in inorder after deletion

Example

Input: 1. The number of nodes taht the created tree should have = 5

2. The data values of the nodes in the tree to be created are: 10, 20, 5, 9, 8

3. The data value in the node to be deleted = 9

Output: 1. 5 8 9 10 20

2 5 8 10 20

Applications of Binary Search Trees

One of the applications of a binary search tree is the implementation of a dynamic dictionary. This application is appropriate because a dictionary is an ordered list that is required to be searched frequently, and is also required to be updated (insertion and deletion mode) frequently. So it can be implemented by making the entries in a dictionary into the nodes of a binary search tree. A more efficient implementation of a dynamic dictionary involves considering a key to be a sequence of characters, and instead of searching by comparison of entire keys, we use these characters to determine a multi-way branch at each step. This will allow us to make a 26-way branch according to the first letter, followed by another branch according to the second letter and so on.

General Comments on Binary Trees

1. Trees are used to organize a collection of data items into a hierarchical structure.
2. A tree is a collection of elements called nodes, one of which is distinguished as the root, along with a relation that places a hierarchical structure on the node.
3. The degree of a node of a tree is the number of descendants that node has.
4. A leaf node of a tree is a node with a degree equal to 0.
5. The degree of a tree is the maximum of the degree of the nodes of the tree.
6. The level of the root node is 1, and as we descend the tree, we increment the level of each node by 1.
7. Depth of a tree is the maximum value of the level for the nodes in the tree.
8. A binary tree is a special case of tree, in which no node can have degree greater than 2.
9. The maximum number of nodes at level i in a binary tree is 2^{i-1}.
10. The maximum number of nodes in a binary tree of depth k is 2^{k-1}.
11. A complete binary tree of depth k is a tree with n nodes in which these n nodes can be numbered sequentially from 1 to n.
12. If a binary tree is a complete binary tree, it can be represented by an array capable of holding n elements where n is the number of nodes in a complete binary tree.

13. Inorder, preorder, and postorder are the three commonly used traversals that are used to traverse a binary tree.
14. In inorder traversal, we start with the root node, visit the left subtree first, then process the data of the root node, followed by that of the right subtree.
15. In preorder traversal, we start with the root node. First we process the data of the root node, then visit the left subtree, then the right subtree.
16. In postorder traversal, we start with the root node, visit the left subtree first, then visit the right subtree, and then process the data of the root node.
17. To construct a unique binary tree, we require two orders of traversal, in which one has to be inorder; the other could be preorder or postorder.
18. A binary search tree is an important search structure that is dynamic and allows a search by using $O(\log_2^n)$ steps.

Exercises

1. Write a C program to count the number of non-leaf nodes of a binary tree.
2. Write a C program to delete all the leaf nodes of a binary tree.
3. How many binary trees are possible with three nodes?
4. Write a C program to construct a binary tree with inorder and preorder traversals. Test it for the following inorder and preorder traversals:

   Inorder: 5, 1, 3, 11, 6, 8, 4, 2, 7

   Preorder: 6, 1, 5, 11, 3, 4, 8, 7, 2

22 Graphs

GRAPHS

Introduction

Graphs are natural models that are used to represent arbitrary relationships among data objects. We often need to represent such arbitrary relationships among the data objects while dealing with problems in computer science, engineering, and many other disciplines. Therefore, the study of graphs as one of the basic data structures is important.

Basic Definitions and Terminology

A graph is a structure made of two components: a set of vertices V, and a set of edges E. Therefore, a graph is G = (V,E), where G is a graph. The graph may be directed or undirected. In a *directed graph*, every edge of the graph is an ordered pair of vertices connected by the edge, whereas in an *undirected graph*, every edge is an unordered pair of vertices connected by the edge. Figure 22.1 shows an undirected and a directed graph.

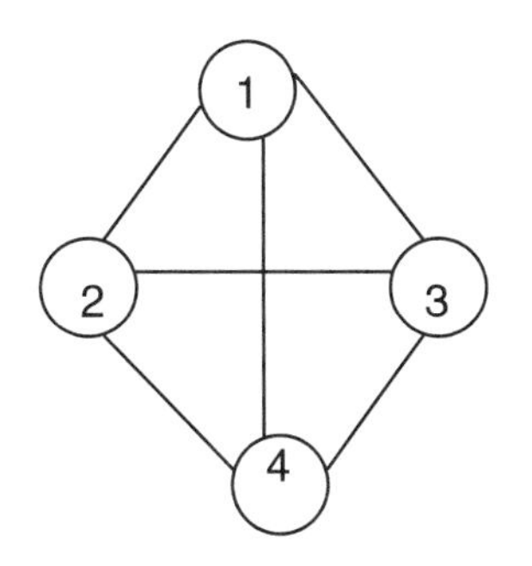

Undirected Graph G_1

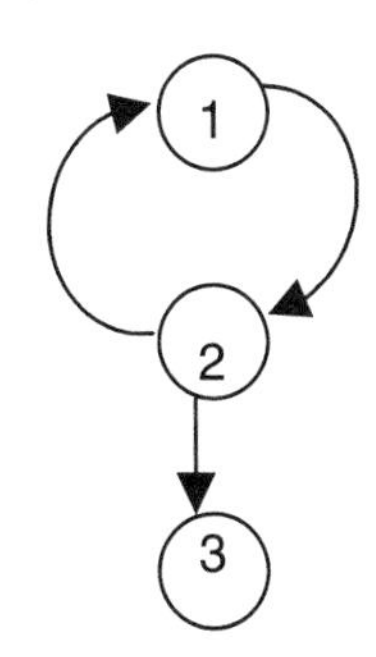

Directed Graph G_2

FIGURE 22.1 Graphs.

Incident edge: If (v_i,v_j) is an edge, then edge(v_i,v_j) is said to be incident to vertices v_i and v_j. For example, in graph G_1 shown in Figure 22.1, the edges incident on vertex 1 are (1,2), (1,4), and (1,3), whereas in G_2, the edges incident on vertex 1 are (1,2).

Degree of vertex: The number of edges incident onto the vertex. For example, in graph G_1, the degree of vertex 1 is 3, because 3 edges are incident onto it. For a directed graph, we need to define indegree and outdegree. *Indegree* of a vertex vi is the number of edges incident onto vi, with vi as the head. *Outdegree* of vertex vi is the number of edges incident onto vi, with vi as the tail. For graph G_2, the indegree of vertex 2 is 1, whereas the outdegree of vertex 2 is 2.

Directed edge: A directed edge between the vertices vi and vj is an ordered pair. It is denoted by <vi,vj>.

Undirected edge: An undirected edge between the vertices v_i and v_j is an unordered pair. It is denoted by (v_i,v_j).

Path: A path between vertices v_p and v_q is a sequence of vertices $v_p, v_{i1}, v_{i2},\ldots, v_{in},v_q$ such that there exists a sequence of edges $(v_p, v_{i1}), (v_{i1}, v_{i2}), \ldots, (v_{in}, v_q)$. In case of a directed graph, a path between the vertices v_p and v_q is a sequence of vertices $v_p, v_{i1}, v_{i2},\ldots, v_{in}, v_q$ such that there exists a sequence of edges $<v_p, v_{i1}>$, $<v_{i1}, v_{i2}>, \ldots, <v_{in}, v_q>$. If there exists a path from vertex v_p to v_q in an undirected graph, then there always exists a path from v_q to v_p also. But, in the case of a directed graph, if there exists a path from vertex v_p to v_q, then it does not necessarily imply that there exists a path from v_q to v_p also.

Simple path: A simple path is a path given by a sequence of vertices in which all vertices are distinct except the first and the last vertices. If the first and the last vertices are same, the path will be a cycle.

Maximum number of edges: The maximum number of edges in an undirected graph with n vertices is $n(n-1)/2$. In a directed graph, it is $n(n-1)$.

Subgraph: A *subgraph* of a graph G = (V,E) is a graph G where V(G) is a subset of V(G). E(G) consists of edges (v1,v2) in E(G), such that both v1 and v2 are in V(G). [Note: If G = (V,E) is a graph, then V(G) is a set of vertices of G and E(G) is a set of edges of G.]

If E(G) consists of all edges (v1,v2) in E(G), such that both v1 and v2 are in V(G), then G is called an induced subgraph of G. For example, the graph shown in Figure 22.2 is a subgraph of the graph G2.

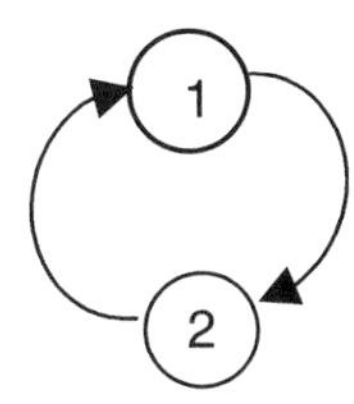

FIGURE 22.2 The subgraph of graph G2.

For the graph shown in Figure 22.3, one of the induced subgraphs is shown in Figure 22.4.

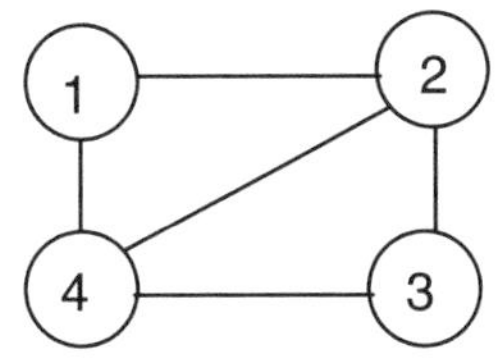

FIGURE 22.3 Graph G.

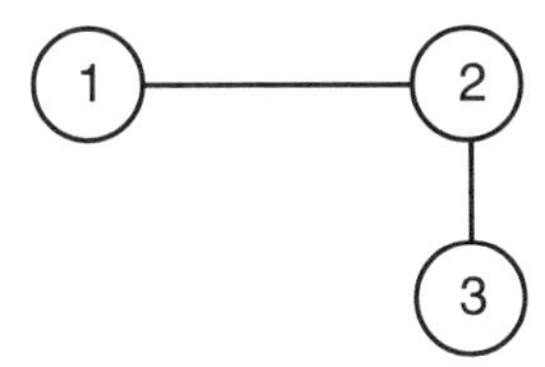

FIGURE 22.4 Induced subgraph of Graph G of Figure 22.3.

In the undirected graph G, the two vertices v_1 and v_2 are said to be connected if there exists a path in G from v_1 to v_2 (being an undirected graph, there exists a path from v_2 to v_1 also).

Connected graph: A graph G is said to be connected if for every pair of distinct vertices (v_i,v_j), there is a path from v_i to v_j. A connected graph is shown in Figure 22.5.

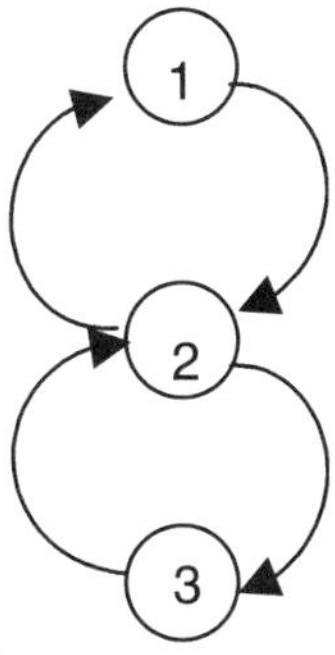

FIGURE 22.5 A connected graph.

Completely connected graph: A graph G is completely connected if, for every pair of distinct vertices (v_i, v_j), there exists an edge. A completely connected graph is shown in Figure 22.6.

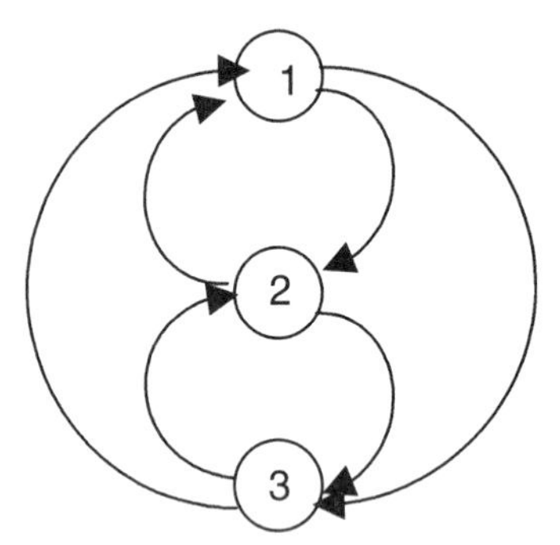

FIGURE 22.6 A completely connected graph.

REPRESENTATIONS OF A GRAPH

Array Representation

One way of representing a graph with n vertices is to use an n^2 matrix (that is, a matrix with *n* rows and *n* columns—that means there is a row as well as a column corresponding to every vertex of the graph). If there is an edge from v_i to v_j then the entry in the matrix with row index as v_i and column index as v_j is set to 1 ($adj[v_i, v_j] = 1$, if (v_i, v_j) is an edge of graph G). If e is the total number of edges in the graph, then there will 2e entries which will be set to 1, as long as G is an undirected graph. Whereas if G were a directed graph, only e entries would have been set to 1 in the adjacency matrix. The adjacency matrix representation of an undirected as well as a directed graph is show in Figure 22.7.

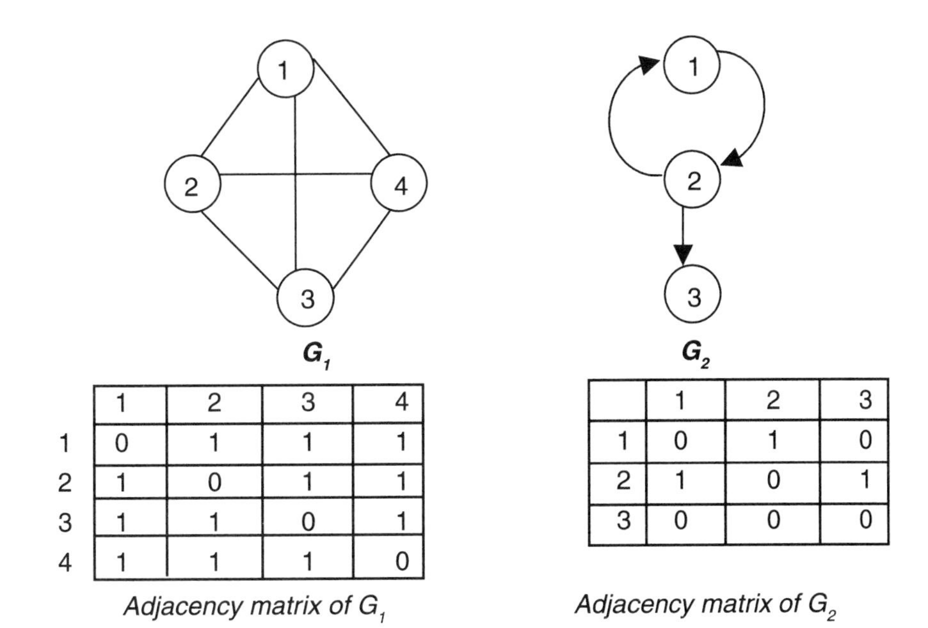

	1	2	3	4
1	0	1	1	1
2	1	0	1	1
3	1	1	0	1
4	1	1	1	0

	1	2	3
1	0	1	0
2	1	0	1
3	0	0	0

FIGURE 22.7 Adjacency matrices.

Example

The adjacency matrix representation of the following diagraph(directed graph), along with the indegree and outdegree of each node is shown here:

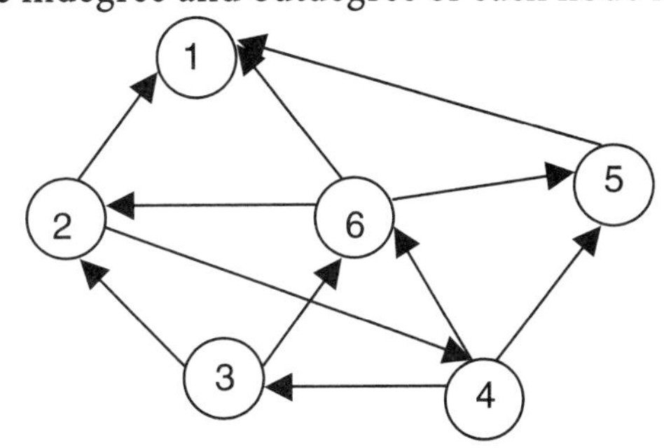

The adjacency matrix representation of the above diagraph is shown here:

	1	2	3	4	5	6
1	0	0	0	0	0	0
2	1	0	0	1	0	0
3	0	1	0	0	0	1
4	0	0	1	0	1	1
5	1	0	0	0	0	0
6	1	1	0	0	1	0

The indegree and outdegree of each node is shown here:

	Indegree	Outdegree
1	3	0
2	2	2
3	1	2
4	1	3
5	2	1
6	2	3

Linked List Representation

Another way of representing a graph G is to maintain a list for every vertex containing all vertices adjacent to that vertex, as shown in Figure 22.8.

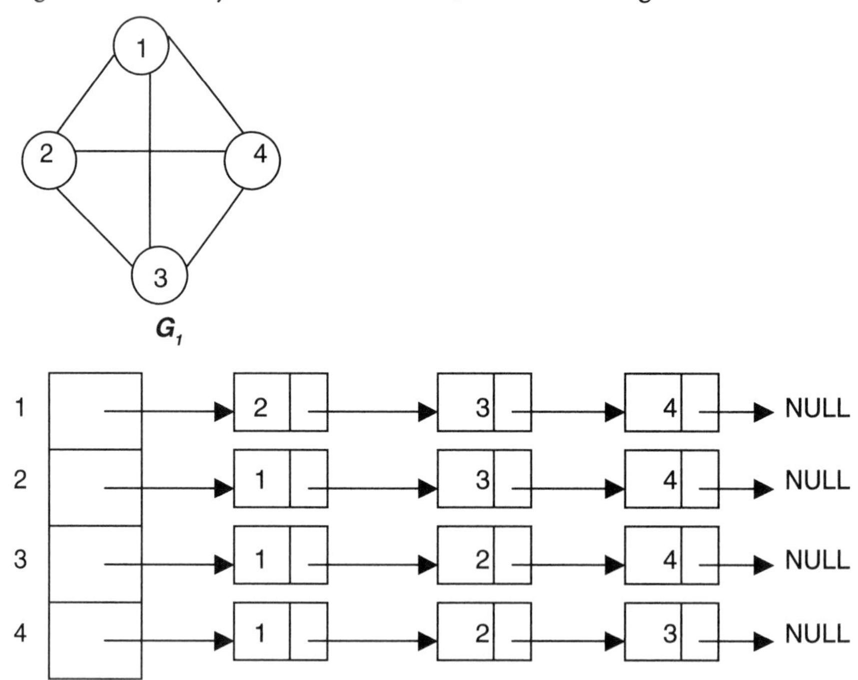

FIGURE 22.8 Adjacency list of G_1.

COMPUTING INDEGREE AND OUTDEGREE OF A NODE OF A GRAPH USING ADJACENCY MATRIX REPRESENTATION

Introduction

To compute the indegree of a node n by using the adjacency matrix representation of a graph, use the node number *n* as a column index in the adjacency matrix and count the number of 1's in that column of the adjacency matrix. This count is the indegree of node *n*. Similarly, to compute the outdegree of a node *n* of a graph, use the node number *n* as the row index in the adjacency matrix and count the number of 1's in that row of the adjacency matrix. This is the outdegree of the node *n*. A complete C program to compute the indegree and outdegree of each node of a graph using the adjacency matrix representation of a graph follows.

Program: Computing the indegree and outdegree

```
#include <stdio.h>
#define MAX 10
/* a function to build an adjacency matrix of the graph*/
void buildadjm(int adj[][MAX], int n)
   {
     int i,j;
     for(i=0;i<n;i++)
         for(j=0;j<n;j++)
         {
         printf("Enter 1 if there is an edge from %d to %d, otherwise
enter 0 \n",
   i,j);
         scanf("%d",&adj[i][j]);
         }
   }

/* a function to compute outdegree of a node*/
int outdegree(int adj[][MAX],int x,int n)
    {
       int i, count =0;
       for(i=0;i<n;i++)
           if( adj[x][i] ==1) count++;
      return(count);
    }
```

```
/* a function to compute indegree of a node*/
int indegree(int adj[][MAX],int x,int n)
    {
       int i, count =0;
       for(i=0;i<n;i++)
           if( adj[i][x] ==1) count++;
      return(count);
    }
void main()
   {
     int adj[MAX][MAX],node,n,i;
     printf("Enter the number of nodes in graph maximum = %d\n",MAX);
      scanf("%d",&n);
     buildadjm(adj,n);
     for(i=0;i<n;i++)
     {
        printf("The indegree of the node %d is
%d\n",i,indegree(adj,i,n));
      printf("The outdegree of the node %d is %d\n",
i,outdegree(adj,i,n));
   }
}
```

Explanation

1. This program uses the adjacency matrix representation of a directed graph to compute the indegree and outdegree of each node of the graph.
2. It first builds an adjacency matrix of the graph by calling a `buildadjm` function, then goes in a loop to compute the indegree and outdegree of each node by calling the `indegree` and `outdegree` functions, respectively.
3. The indegree function counts the number of 1's in a column of an adjacency matrix using the node number whose indegree is to be computed as a column index.
4. The `outdegree` function counts the number of 1's in a row of an adjacency matrix by using the node number whose outdegree is to be computed as a row index.

Input: 1. The number of nodes in a graph

2. Information about edges, in the form of values, to be stored in adjacency matrix 1, if there is an edge from node i to node j; 0 otherwise.

Output: The indegree and outdegree of each node.

Example

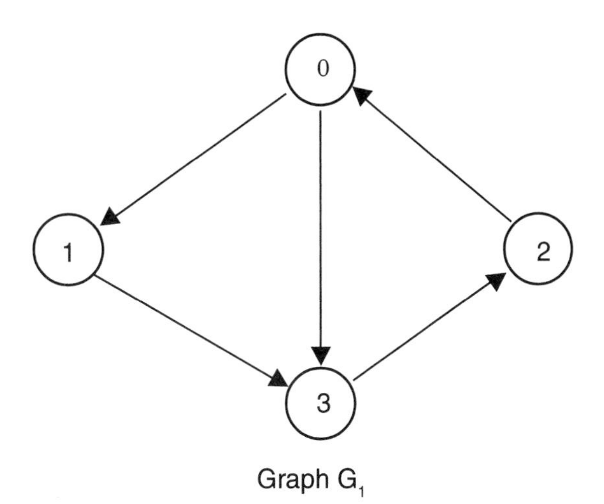

Graph G_1

The adjacency matrix for graph G_1 is:

	0	1	2	3
0	0	1	0	1
1	0	0	0	1
2	1	0	0	0
3	0	0	1	0

For this graph as the input, the output is:

The indegree of node 0 is 1

The outdgree of node 0 is 2

The indegree of node 1 is 1

The outdgree of node 1 is 1

The indegree of node 2 is 1

The outdgree of node 2 is 1

The indegree of node 3 is 2

The outdgree of node 3 is 1

DEPTH-FIRST TRAVERSAL

Introduction

A graph can be traversed either by using the *depth-first traversal* or *breadth-first traversal*. When a graph is traversed by visiting the nodes in the forward (deeper) direction as long as possible, the traversal is called depth-first traversal. For example, for the graph shown in Figure 22.9, the depth-first traversal starting at the vertex 0 visits the node in the orders:

(i) 0 1 2 6 7 8 5 3 4

(ii) 0 4 3 5 8 6 7 2 1

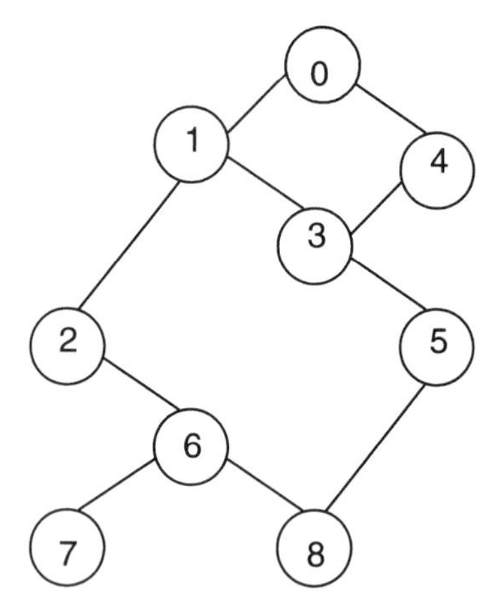

FIGURE 22.9 Graph G and its depth first traversals starting at vertex 0.

A complete C program for depth-first traversal of a graph follows. It makes use of an array visited of n elements where n is the number of vertices of the graph, and the elements are Boolean. If visited[i] = 1 then it means that the ith vertex is visited. Initially we set visited[i] = 0.

Program

```
#include <stdio.h>
#define max 10

/* a function to build adjacency matrix of a graph */
void buildadjm(int adj[][max], int n)
   {
     int i,j;
```

```
        for(i=0;i<n;i++)
            for(j=0;j<n;j++)
             {
        printf("enter 1 if there is an edge from %d to %d, otherwise enter
0 \n",
   i,j);
              scanf("%d",&adj[i][j]);
              }
  }

/* a function to visit the nodes in a depth-first order */
void dfs(int x,int visited[],int adj[][max],int n)
{
    int j;
    visited[x] = 1;
    printf("The node visited id %d\n",x);
    for(j=0;j<n;j++)
       if(adj[x][j] ==1 && visited[j] ==0)
            dfs(j,visited,adj,n);
}
void main()
   {
     int adj[max][max],node,n;
     int i, visited[max];
     printf("enter the number of nodes in graph maximum = %d\n",max);
     scanf("%d",&n);
     buildadjm(adj,n);
     for(i=0; i<n; i++)
       visited[i] =0;
     for(i=0; i<n; i++)
      if(visited[i] ==0)
          dfs(i,visited,adj,n);
     }
```

Explanation

1. Initially, all the elements of an array named `visited` are set to 0 to indicate that all the vertices are unvisited.
2. The traversal starts with the first vertex (that is, vertex 0), and marks it visited by setting `visited[0]` to 1. It then considers one of the unvisited vertices adjacent to it and marks it `visited`, then repeats the process by considering one of its unvisited adjacent vertices.

3. Therefore, if the following adjacency matrix that represents the graph of Figure 22.9 is given as input, the order in which the nodes are visited is given here:

   Input: 1. The number of nodes in a graph

   2. Information about edges, in the form of values to be stored in adjacency matrix 1 if there is an edge from node i to node j , 0 otherwise

   Output: Depth-first ordering of the nodes of the graph starting from the initial vertex, which is vertex 0, in our case.

Example

Input

	0	1	2	3	4	5	6	7	8
0	0	1	0	0	1	0	0	0	0
1	1	0	1	1	0	0	0	0	0
2	0	1	0	0	0	0	1	0	0
3	0	1	0	0	1	1	0	0	0
4	1	0	0	1	0	0	0	0	0
5	0	0	0	1	0	0	0	0	1
6	0	0	1	0	0	0	0	0	1
7	0	0	0	0	0	0	1	0	1
8	0	0	0	0	0	1	1	0	0

Output

0, 1, 2, 6, 8, 5, 3, 4, 7

Analysis

1. If the graph G to which the depth-first search (dfs) is applied is represented using adjacency lists, then the vertices y adjacent to x can be determined by following the list of adjacent vertices for each vertex.
2. Therefore, the for loop searching for adjacent vertices has the total cost of $d_1 + d_2 + \ldots + d_n$, where d_i is the degree of vertex v_i, because the number of nodes in the adjacency list of vertex v_i are d_i.

3. If the graph G has n vertices and e edges, then the sum of the degree of each vertex ($d_1 + d_2 + ... + d_n$) is 2e. Therefore, there are total of 2e list nodes in the adjacency lists of G. If G is a directed graph, then there are a total of e list nodes only.
4. The algorithm examines each node in the adjacency lists once, at most. So the time required to complete the search is O(e), provided $n <= e$. Instead of using adjacency lists, if an adjacency matrix is used to represent a graph G, then the time required to determine all adjacent vertices of a vertex is $O(n)$, and since at most n vertices are visited, the total time required is $O(n^2)$.

BREADTH-FIRST TRAVERSAL

Introduction

When a graph is traversed by visiting all the adjacent nodes/vertices of a node/vertex first, the traversal is called breadth-first traversal. For example, for a graph in which the breadth-first traversal starts at vertex v_1, visits to the nodes take place in the order shown in Figure 22.10.

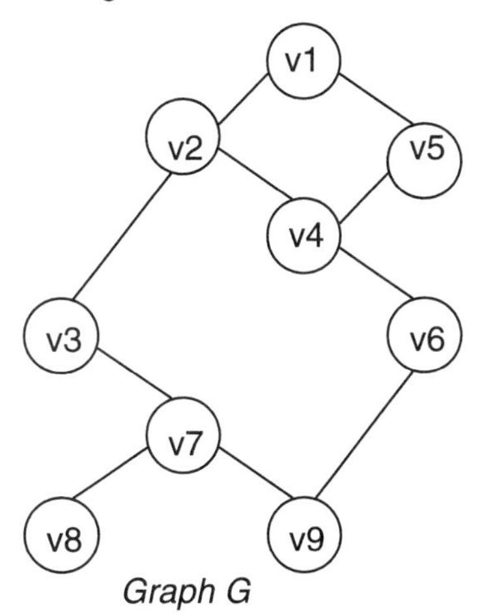

Graph G

breadth-first traversal order = v1 v2 v5 v3 v4 v7 v6 v8 v9

FIGURE 22.10 Breadth-first traversal of graph G starting at vertex v1.

Program

A complete C program for breadth-first traversal of a graph appears next. The program makes use of an array of n visited elements where n is the number of vertices of the graph. If `visited[i] = 1`, it means that the ith vertex is visited. The program also makes use of a queue and the procedures `addqueue` and `deletequeue` for adding a vertex to the queue and for deleting the vertex from the queue, respectively. Initially, we set `visited[i] = 0`.

```
#include <stdio.h>
#include <stdlib.h>
#define MAX 10
struct node
   {
     int data;
     struct node *link;
   };
void buildadjm(int adj[][MAX], int n)
   {
     int i,j;
     printf("enter adjacency matrix  \n",i,j);
     for(i=0;i<n;i++)
         for(j=0;j<n;j++)
                scanf("%d",&adj[i][j]);
    }

/* A function to insert a new node in queue*/
struct node *addqueue(struct node *p,int val)
{
   struct node *temp;
   if(p == NULL)
   {
      p = (struct node *) malloc(sizeof(struct node)); /* insert the
new node first node*/
        if(p == NULL)
          {
               printf("Cannot allocate\n");
               exit(0);
          }
        p->data = val;
        p->link=NULL;
  }
  else
  {
```

```
 temp= p;
while(temp->link != NULL)
{
  temp = temp->link;
}
   temp->link = (struct node*)malloc(sizeof(struct node));
   temp = temp->link;
   if(temp == NULL)
        {
              printf("Cannot allocate\n");
              exit(0);
        }
   temp->data = val;
   temp->link = NULL;
}
return(p);
}
struct node *deleteq(struct node *p,int *val)
{  struct node *temp;
   if(p == NULL)
   {
        printf("queue is empty\n");
              return(NULL);
   }
    *val =  p->data;
    temp = p;
    p = p->link;
    free(temp);
    return(p);
 }


void bfs(int adj[][MAX], int x,int visited[], int n, struct node **p)
{
    int y,j,k;
    *p = addqueue(*p,x);
     do{

          *p = deleteq(*p,&y);
           if(visited[y] == 0)
                {
                      printf("\nnode visited = %d\t",y);
             visited[y] = 1;
```

```
                    for(j=0;j<n;j++)
              if((adj[y][j] ==1) && (visited[j] == 0))
                      *p = addqueue(*p,j);
              }

        }while((*p) != NULL);
}
void main()
   {
     int adj[MAX][MAX];

     int n;
     struct node *start=NULL;
     int i, visited[MAX];
     printf("enter the number of nodes in graph maximum = %d\n",MAX);
     scanf("%d",&n);
     buildadjm(adj,n);
     for(i=0; i<n; i++)
       visited[i] =0;
     for(i=0; i<n; i++)
      if(visited[i] ==0)
           bfs(adj,i,visited,n,&start);

     }
```

Example

Input and Output

Enter the number of nodes in graph maximum = 10

9

Enter adjacency matrix

0 1 0 0 1 0 0 0 0
1 0 1 1 0 0 0 0 0
0 1 0 0 0 0 1 0 0
0 1 0 0 1 1 0 0 0
1 0 0 1 0 0 0 0 0
0 0 0 1 0 0 0 0 1
0 0 1 0 0 0 0 1 1
0 0 0 0 0 0 1 0 0
0 0 0 0 0 1 1 0 0

node visited = 0
node visited = 1
node visited = 4
node visited = 2
node visited = 3
node visited = 6
node visited = 5
node visited = 7
node visited = 8

CONNECTED COMPONENT OF A GRAPH

Introduction

The *connected component* of a graph is a maximal subgraph of a given graph, which is connected. For example, consider the graph that follows.

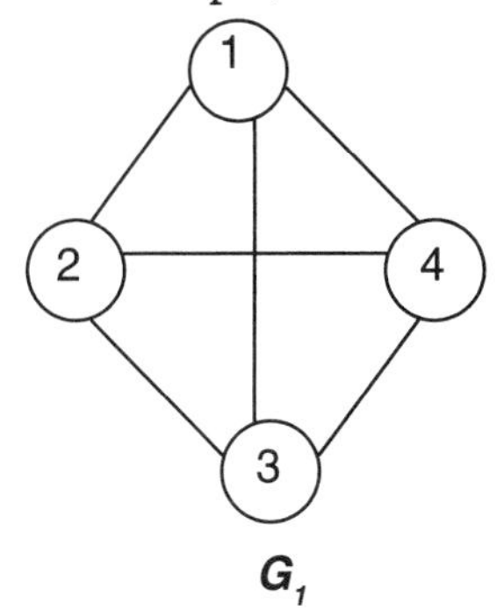

The connected component of the graph G1 is shown in Figure 22.11.

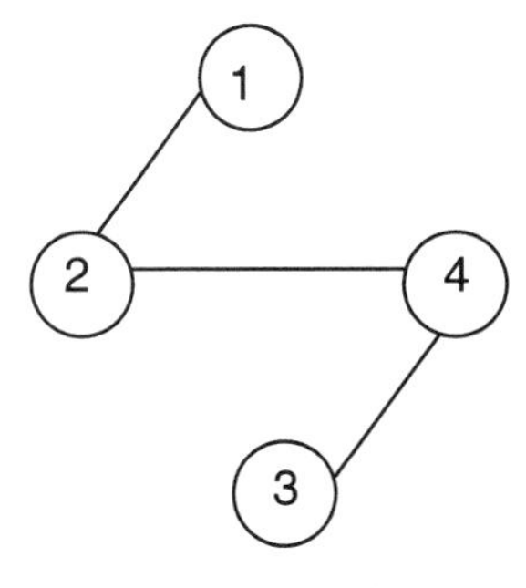

FIGURE 22.11 Connected component of G_1.

Strongly Connected Component

For a diagraph (directed graph), a *strongly connected component* is that component of the graph in which, for every pair of distinct vertices vi and vj, there is a directed path from vi to vj, and also a directed path from vj to vi. For example, consider the diagraph shown in Figure 22.12.

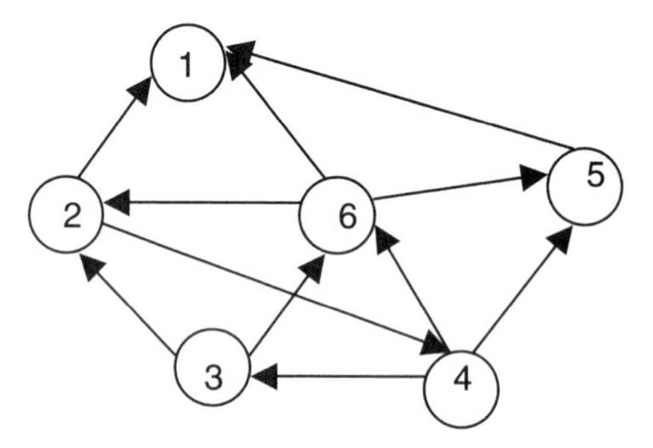

FIGURE 22.12 A diagraph.

The strongly connected components of the graph are shown in Figure 22.13.

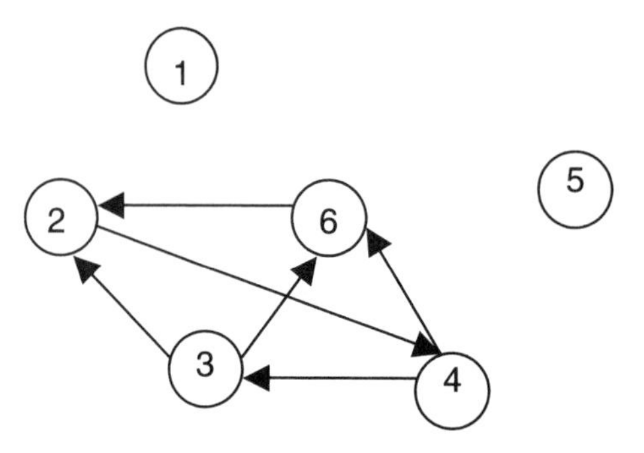

FIGURE 22.13 Strongly connected components of the graph shown in Figure 22.12.

Example

Is the following diagraph strongly connected?

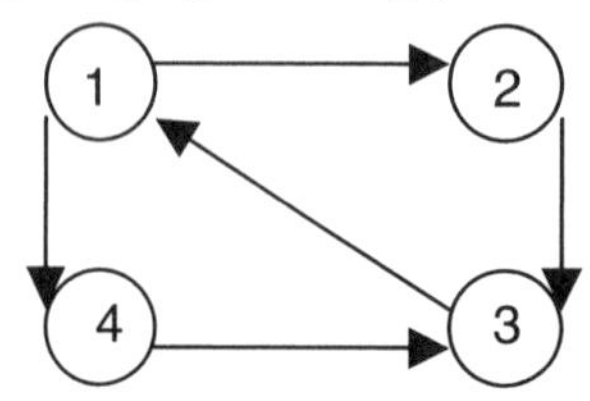

This table shows the possible pairs of vertices, and the forward and backward paths between them, for the previous graph:

PAIR OF VERTICES	FORWARD PATH	BACKWARD PATH
<1, 2>	1-2	2-3-4
<1,3>	1-2-3	3-1
<1,4>	1-4	4-3-1
<2,3>	2-3	3-1-2
<2,4>	2-3-1-4	4-3-1-2
<3,4>	3-1-4	4-3

Therefore, we see that between every pair of distinct vertices of the given graph there exists a forward as well as a backward path, so it is strongly connected.

Program

Write a function dfs(v) to traverse a graph in a depth-first manner starting from vertex v. Use this function to find connected components in the graph. Modify dfs() to produce a list of newly visited vertices. The graph is represented as adjacency lists.

```
#include <stdio.h>

#define MAXVERTICES 20
#define MAXEDGES    20

typedef enum {FALSE, TRUE, TRISTATE} bool;
typedef struct node node;

struct node {
    int dst;
    node *next;
};

void printGraph( node *graph[], int nvert ) {
    /*
     * prints the graph.
     */
    int i, j;
```

```
        for( i=0; i<nvert; ++i ) {
            node *ptr;
            for( ptr=graph[i]; ptr; ptr=ptr->next )
                printf( "[%d] ", ptr->dst );
            printf( "\n" );
        }
    }

    void insertEdge( node **ptr, int dst ) {
        /*
         * insert a new node at the start.
         */
        node *newnode = (node *)malloc( sizeof(node) );
        newnode->dst = dst;
        newnode->next = *ptr;
        *ptr = newnode;
    }

    void buildGraph( node *graph[], int edges[2][MAXEDGES], int nedges ) {
        /*
         * fills graph as adjacency list from array edges.
         */
        int i;
        for( i=0; i<nedges; ++i ) {
            insertEdge( graph+edges[0][i], edges[1][i] );
            insertEdge( graph+edges[1][i], edges[0][i] );   // undirected
graph.
        }
    }

    void dfs( int v, int *visited, node *graph[] ) {
        /*
         * recursively traverse graph from v using visited.
         * and mark all the vertices that come in dfs path to TRISTATE.
         */
        node *ptr;

        visited[v] = TRISTATE;
        //printf( "%d \n", v );
        for( ptr=graph[v]; ptr; ptr=ptr->next )
            if( visited[ ptr->dst ] == FALSE )
                dfs( ptr->dst, visited, graph );
    }
```

```
void printSetTristate( int *visited, int nvert ) {
    /*
     * prints all vertices of visited which are TRISTATE.
     * and set them to TRUE.
     */
    int i;

    for( i=0; i<nvert; ++i )
        if( visited[i] == TRISTATE ) {
            printf( "%d ", i );
            visited[i] = TRUE;
        }
    printf( "\n\n" );
}

void compINC(node *graph[], int nvert) {
    /*
     * prints all connected components of graph represented using INC
lists.
     */
    int *visited;
    int i;

    visited = (int *)malloc( nvert*sizeof(int) );
    for( i=0; i<nvert; ++i )
        visited[i] = FALSE;

    for( i=0; i<nvert; ++i )
        if( visited[i] == FALSE ) {
            dfs( i, visited, graph );
            // print all vertices which are TRISTATE.
            // and mark them to TRUE.
            printSetTristate( visited, nvert );
        }
    free( visited );
}

int main() {
    int edges[][MAXEDGES] = { {0,2,4,5,5,4},
                              {1,1,3,4,6,6}
                            };
    int nvert = 7;      // no of vertices.
    int nedges = 6; // no of edges in the graph.
```

```
    node **graph = (node **)calloc( nvert, sizeof(node *) );

    buildGraph( graph, edges, nedges );
    printGraph( graph, nvert );
    compINC( graph, nvert );

    return 0;
}
```

Explanation

1. The graph is represented as adjacency lists. The graph contains an array of n pointers where n is the number of vertices in the graph. Each entry i in the array contains a list of vertices to which i is connected. For example, if the graph is as shown in the first diagram, the adjacency lists for the graph are as shown in the subsequent diagram.

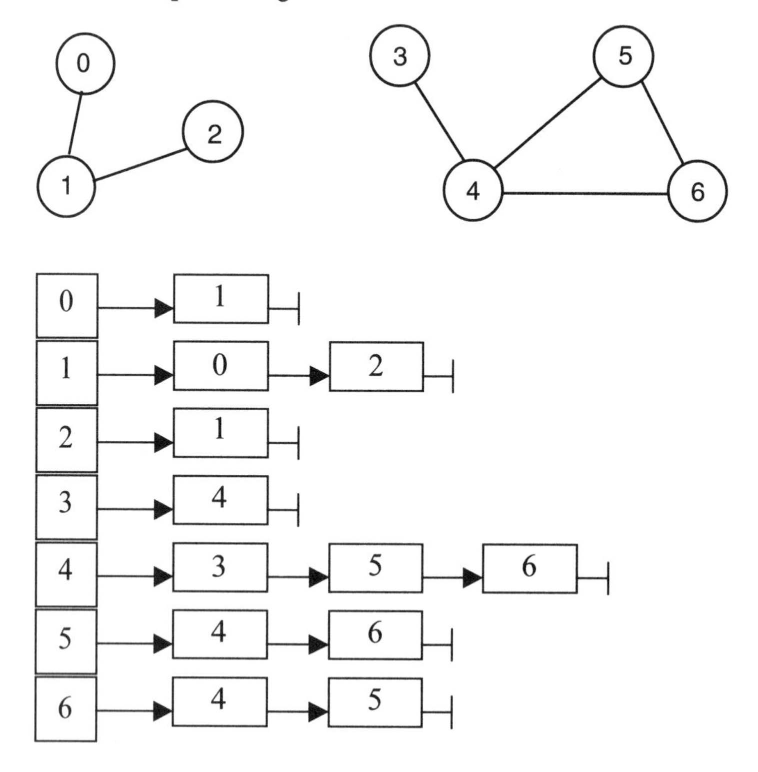

Each node in list i contains a vertex to which i is connected.

2. dfs(v) is implemented recursively. A Boolean vector visited[] is maintained whose entries are initially all FALSE. dfs(v) marks v as visited by making visited[v] = TRUE. It then finds all the adjacent nodes of v and starts dfs() from those nodes that have not yet been visited. For example, if dfs(v) is called with v = 0, it marks 0 and then it traverses the adjacency list graph[0] and calls dfs(1). This marks 1 and traverses the adjacency list graph[1]. But since 0 is already marked, dfs(2) is called. It marks 2 and starts traversal of graph[2]. But since 1 is marked, it returns. All the previous invocations return as there are no nodes being considered in the lists. Thus, the marked vertices are {0, 1, 2}.
3. compINC() is a function that finds all the connected components of a graph. It maintains a local copy of the vector visited[] and passes it as a parameter to dfs(v). compINC() passes that vertex as a parameter to dfs( )that has not yet been visited. Thus each invocation of dfs() finds one connected component of the graph.
4. In order to modify dfs() to produce a list of newly visited vertices, we tag the vertices visited using dfs() as TRISTATE. In compINC(), all these TRISTATE vertices will form one connected component. This status is then converted to TRUE. The next invocation to dfs() returns another set of vertices tagged as TRISTATE, which forms another connected component and so on.

   For example, in the previous program, first all vertices are tagged as FALSE. After the invocation of dfs(0), the vertices tagged as TRISTATE are {0, 1, 2}. These are output and their tags are changed from TRISTATE to TRUE. The next invocation of dfs(3) tags vertices {3, 4, 5, 6} as TRISTATE. These are then output and their tags are changed from TRISTATE to TRUE. Since there is no vertex remaining whose tag is FALSE, the algorithm stops.

Points to Remember

1. All the reachable vertices can be traversed from a source vertex by using depth-first search.
2. The data representation (a graph in this case) should be such that it should make algorithms operate on the data efficiently. Being represented as adjacency lists, we could easily traverse the list to get the vertices adjacent to a particular vertex.
3. Note how a simple recursive procedure solves the problem of finding all the reachable vertices from a vertex.

4. Note the use of descriptive words such as FALSE, TRUE and TRISTATE, rather than integers 0, 1 and 2. It makes the program easily understandable.

DEPTH-FIRST SPANNING TREE AND BREADTH-FIRST SPANNING TREE

Introduction

If graph G is connected, the edges of G can be partitioned into two disjointed sets. One is a set of tree edges, which we denote by set T, and the other is a set of back edges, which we denote by B. The tree edges are precisely those edges that are followed during the depth-first traversal or during the breadth-first traversal of graph G. If we consider only the tree edges, we get a subgraph of G containing all the vertices of G, and this subgraph is a tree called *spanning tree* of the graph G. For example, consider the graph shown in Figure 22.14.

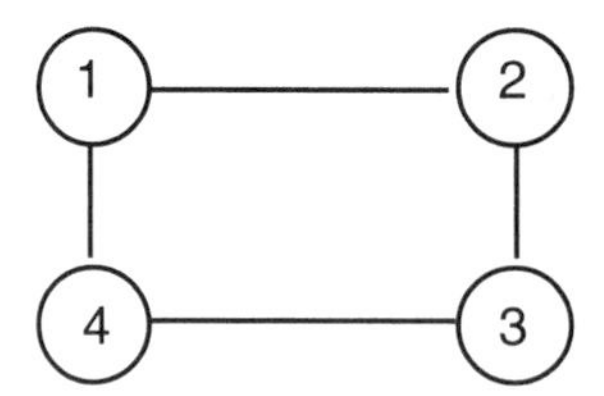

FIGURE 22.14 Graph G.

One of the depth-first traversal orders for this tree is 1-2-3-4, so the tree edges are (1,2), (2,3) and (3,4). Therefore, one of the spanning trees obtained by using depth-first traversal of the graph of Figure 22.14 is shown in Figure 22.15.

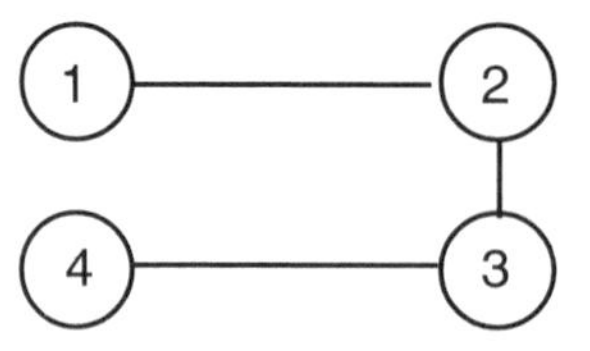

FIGURE 22.15 Depth first spanning tree of the graph of Figure 22.14.

Similarly, one of the breadth-first traversal orders for this tree is 1-2-4-3, so the tree edges are (1,2), (1,4) and (4,3). Therefore, one of the spanning trees obtained using breadth-first traversal of the graph of Figure 22.14 is shown in Figure 22.16.

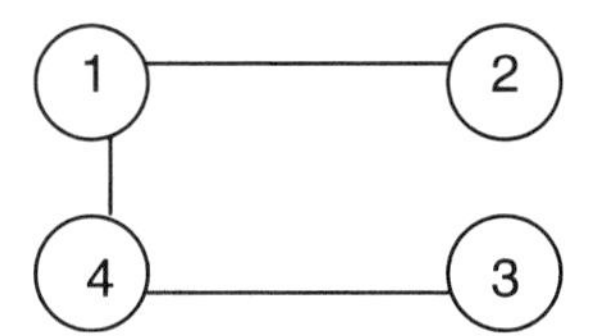

FIGURE 22.16 Breadth-first spanning tree of the graph of Figure 22.14.

The algorithm for obtaining the depth-first spanning tree (dfst) appears next.

```
T = f; {initially set of tree nodes is empty}
dfst( v : node);
{
    if (visited[v] = false)
    {
        visited[v] = true;
        for every adjacent i of v do
        {
                T =  T  È {(v,i)}
                dfst(i);
        }
    }
}
```

If a graph G is not connected, the tree edges, which are precisely those edges followed during the depth-first, traversal of the graph G, constitute the depth-first *spanning forest*. The depth-first spanning forest will be made of trees, each of which is one of the connected components of graph G.

When a graph G is directed, the tree edges, which are precisely those edges followed during the depth-first traversal of the graph G, form a depth-first spanning forest for G. In addition to this, there are three other types of edges. These are called *back edges*, *forward edges*, and *cross edges*. An edge A →B is called a back edge, if B is an ancestor of A in the spanning forest. A non-tree edge that goes from a vertex to a proper descendant is called a forward edge. An edge which goes from a vertex to another vertex that is neither an ancestor nor a descendant is called a cross edge. An edge from a vertex to itself is a back edge. For example, consider the directed graph G shown in Figure 22.17.

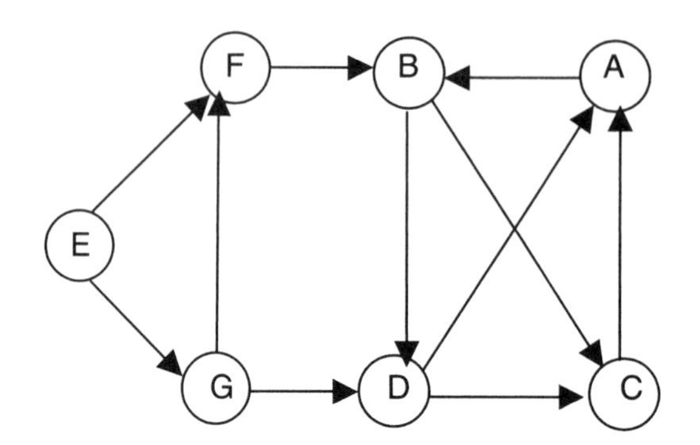

FIGURE 22.17 A directed graph G.

The depth-first spanning forest for graph G of Figure 22.17 is shown in Figure 22.18.

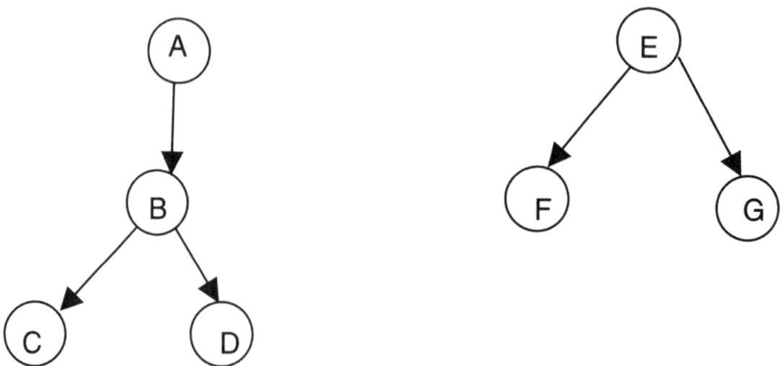

FIGURE 22.18 Depth-first spanning forest for the graph G of Figure 22.17.

In graph G of Figure 22.17, the edges such as C → A and D → A are the back edges, the edges such as D → C and G → D are cross edges.

Example

Consider the graph shown in Figure 22.19.

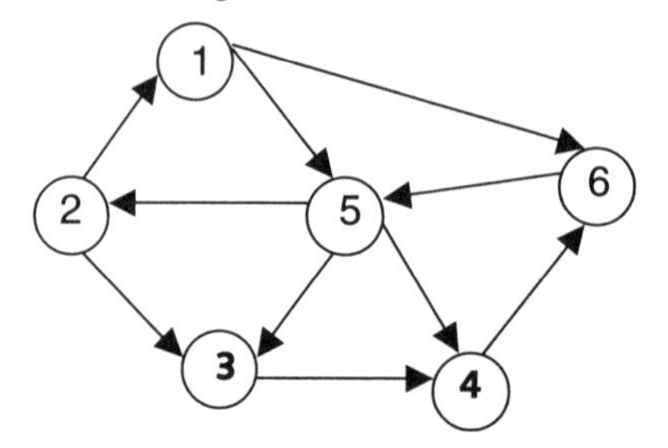

FIGURE 22.19 A graph G.

If we apply the procedure dfst to this graph, one of the depth-first spanning trees that we get by starting with vertex 1 is shown in Figure 22.20.

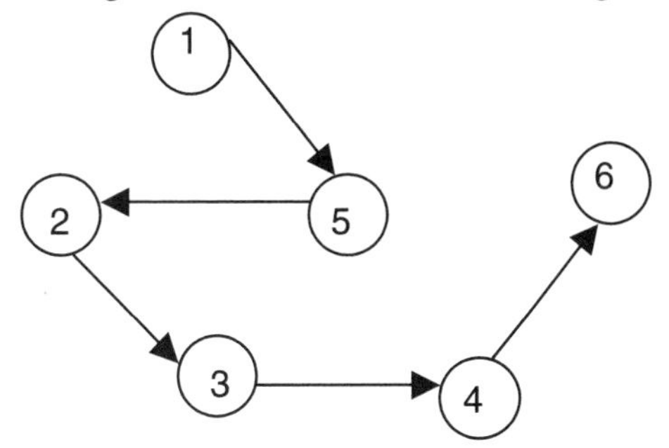

FIGURE 22.20 Depth-first spanning tree of the graph G of Figure 22.19.

MINIMUM-COST SPANNING TREE

Introduction

When the edges of the graph have weights representing the cost in some suitable terms, we can obtain that spanning tree of a graph whose cost is minimum in terms of the weights of the edges. For this, we start with the edge with the minimum-cost/weight, add it to set T, and mark it as visited. We next consider the edge with minimum-cost that is not yet visited, add it to T, and mark it as visited. While adding an edge to the set T, we first check whether both the vertices of the edge are visited; if they are, we do not add to the set T, because it will form a cycle. For example, consider the graph shown in Figure 22.21.

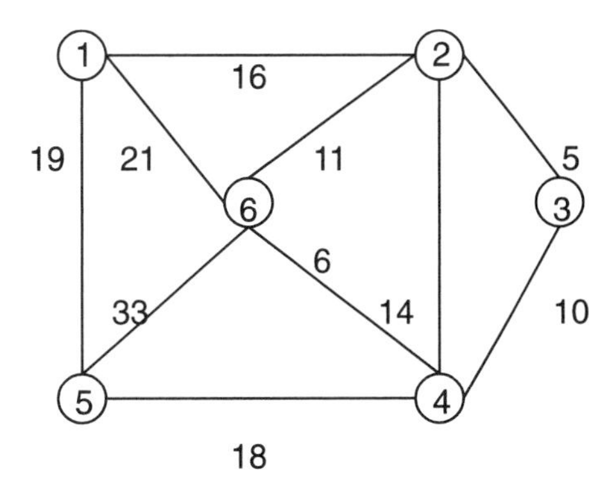

FIGURE 22.21 A graph G.

The *minimum-cost spanning tree* of the graph of Figure 22.21 is shown in Figure 22.22.

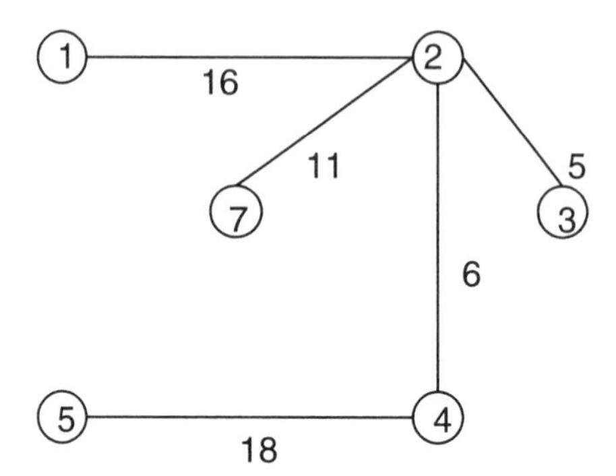

FIGURE 22.22 The minimum-cost spanning tree of graph G of Figure 22.21.

MST Property

Let G = (V,E) be a connected graph with a cost function defined on the edges. Let U be some proper subset of the set of vertices V. If (u,v) is an edge of lowest cost such that u is in U, and v is in V-U, there is a minimum-cost spanning tree that includes edge (u,v). Many of the methods of constructing a minimum-cost spanning tree use the following properties.

Prim's Algorithm

Let G = (V,E) be a weighted graph, and suppose V = {1,2,...,n}. The Prim's algorithm begins with a set U initialized to {1}, and at each stage finds the shortest edge (u,v) that connects u in U and v in V-U, and then adds v to U. It repeats this step until U = V.

```
mcost(G is a  graph; T is a  set of edges)
   U is a  set of vertices
   u,v are vertices;
{
   T= ô
   U = {1}
   while U  '" V
   {
```

```
    Find the lowest-cost edge (u,v) such that u is in U and v is in V-U
        add (u,v) to T
        add v to U
    }
}
```

Program

The following program can be used to find the minimum spanning tree of a graph.

```
#include <stdio.h>

#define MAXVERTICES 10
#define MAXEDGES 20
typedef enum {FALSE, TRUE} bool;

int getNVert(int edges[][3], int nedges) {
    /*
     * returns no of vertices = maxvertex + 1;
     */
    int nvert = -1;
    int j;

    for( j=0; j<nedges; ++j ) {
        if( edges[j][0] > nvert )
                nvert = edges[j][0];

        if( edges[j][1] > nvert )
                nvert = edges[j][1];
    }
    return ++nvert;          // no of vertices = maxvertex + 1;
}

bool isPresent(int edges[][3], int nedges, int v) {
    /*
     * checks whether v has been included in the spanning tree.
     * thus we see whether there is an edge incident on v which has
     * a negative cost. negative cost signifies that the edge has been
     * included in the spanning tree.
     */

    int j;
```

```
        for(j=0; j<nedges; ++j)
            if(edges[j][2] < 0 && (edges[j][0] == v || edges[j][1] == v))
                    return TRUE;

        return FALSE;
    }

    void spanning(int edges[][3], int nedges) {
        /*
         * finds a spanning tree of the graph having edges.
         * uses kruskal's method.
         * assumes all costs to be positive.
         */
        int i, j;
        int tv1, tv2, tcost;
        int nspanedges = 0;
        int nvert = getNVert(edges, nedges);

        // sort edges on cost.
        for(i=0; i<nedges-1; ++i)
            for(j=i; j<nedges; ++j)
                    if(edges[i][2] > edges[j][2]) {
                            tv1 = edges[i][0]; tv2 = edges[i][1]; tcost =
edges[i][2];
                            edges[i][0] = edges[j][0]; edges[i][1] =
edges[j][1]; edges[i][2] = edges[j][2];
                            edges[j][0] = tv1; edges[j][1] = tv2; edges[j][2]
= tcost;
                    }

        for(j=0; j<nedges-1; ++j) {
            // consider edge j connecting vertices v1 and v2.
            int v1 = edges[j][0];
            int v2 = edges[j][1];

            // check whether it forms a cycle in the up until now formed
spanning tree.
            // checking can be done easily by checking whether both v1 and
v2 are in
            // the current spanning tree!
            if(isPresent(edges, nedges, v1) && isPresent(edges, nedges, v2))
// cycle.
```

```
                printf("rejecting: %d %d %d...\n", edges[j][0],
edges[j][1], edges[j][2]);
            else {
                edges[j][2] = -edges[j][2];
                printf("%d %d %d.\n", edges[j][0], edges[j][1], -
edges[j][2]);
                if(++nspanedges == nvert-1)
                    return;
            }
        }

        printf("No spanning tree exists for the graph.\n");
    }

    main() {
        int edges[][3] = {
                                        {0,1,16},
                                        {0,4,19},
                                        {0,5,21},
                                        {1,2,5},
                                        {1,3,6},
                                        {1,5,11},
                                        {2,3,10},
                                        {3,4,18},
                                        {3,5,14},
                                        {4,5,33}
                                    };
        int nedges = sizeof(edges)/3/sizeof(int);
        spanning(edges, nedges);

        return 0;
    }
```

Explanation

1. A tree consisting solely of edges in a graph G, and including all vertices in G, is called a spanning tree. A minimum spanning tree of a weighted graph is the spanning tree with minimum total cost of its edges.

Example:

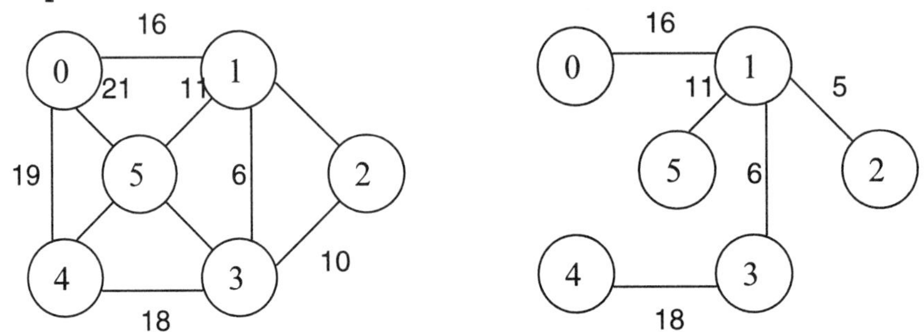

An example graph and its minimum spanning tree.

2. The graph is represented as an array of edges. Each entry in the array is a triplet representing an edge consisting of source vertex, destination vertex, and the cost associated with the edge. The method used in finding a minimum spanning tree is that given by Kruskal. In this approach, a minimum spanning tree T is built edge by edge. Edges are considered for inclusion in T in non-decreasing order of their costs. An edge is included if it does not form a cycle with the edges that are already in T. Since graph G is connected and has $n > 0$ vertices, exactly $n - 1$ edges will be selected for inclusion in T.
3. Kruskal's algorithm is as follows:

```
T={};        // empty set.
while T contains less than n-1 edges and E not empty do
   choose an edge (v, w) from E of lowest cost.
   delete (v, w) from E.
   if (v, w) does NOT create a cycle in T
         add (v, w) to T.
   else
         discard (v, w).
endwhile.
if T contains less than n-1 edges
   print("no spanning tree exists for this graph.");
```

4. In order for the choice of the lowest-cost edge from E to become efficient, we sort the edge array over the cost of edge. To check whether an edge (v, w) forms a cycle, we simply need to check whether both v and w appear in any of the previously added edges in T. We assume that all the costs are positive and we make them negative to signify that the edge has been included in T.

5. **Example:**

   For the example graph in item 1, the run of the algorithm goes as follows:

STEP	EDGE	COST	ACTION	SPANNING-TREE
0	—	—	—	{}.
1	(1, 2)	5	accept	{(1, 2)}.
2	(1, 3)	6	accept	{(1, 2), (1, 3)}.
3	(2, 3)	10	reject	{(1, 2), (1, 3)}.
4	(1, 5)	11	accept	{(1, 2), (1, 3), (1, 5)}.
5	(3, 5)	14	reject	{(1, 2), (1, 3), (1, 5)}.
6	(0, 1)	16	accept	{(1, 2), (1, 3), (1, 5), (0, 1)}.
7	(3, 4)	18	accept	{(1, 2), (1, 3), (1, 5), (0, 1), (3, 4)}.

Points to Remember

1. A minimum spanning tree of a weighted graph G is a tree that consists of edges solely from the edges of G, which covers all the vertices in G, and which has the minimum combined cost of its edges.
2. The complexity of Kruskal's method used for finding the minimum spanning tree of a graph G is O(eloge) where e is the number of edges in G.
3. Note that the union and find algorithms for set representation can be used for checking for cycle and inclusion of an edge in a set.
4. There can be multiple minimum spanning trees in a graph.

Application of Minimum-Cost Spanning Tree

A property of a spanning tree of a graph G is that a spanning tree is a minimal connected subgraph of G (by minimal, we mean the one with the fewest number of edges). Therefore, if nodes of G represent cities and the edges represent possible communication links connecting two cities, then the spanning trees of graph G represent all feasible choices of the communication network. If each edge has weight representing cost measured in some suitable terms (such as cost of construction or distance etc.), then the minimum-cost spanning tree of G is the selection of the required communication network.

DIRECTED ACYCLIC GRAPH (DAG)

Concept

A *directed acyclic graph* (DAG) is a directed graph with no cycles. A DAG represents more general relationships than trees but less general than arbitrary directed graphs. An example of a DAG is given in Figure 22.23.

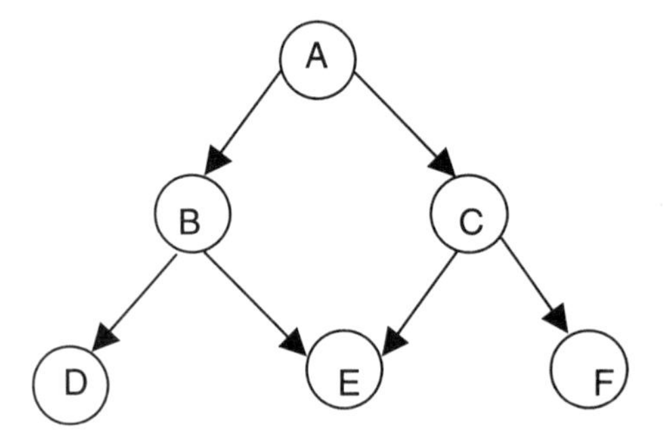

FIGURE 22.23 Directed acyclic graph.

DAGs are useful in representing the syntactic structure of arithmetic expressions with common sub-expressions. For example, consider the following expression:

$(a+b)*c + ((a+b) + e)$

In this expression, the term $(a + b)$ is a common sub-expression, and therefore represented in the DAG by the vertices with more than one incoming edge, as shown in Figure 22.24.

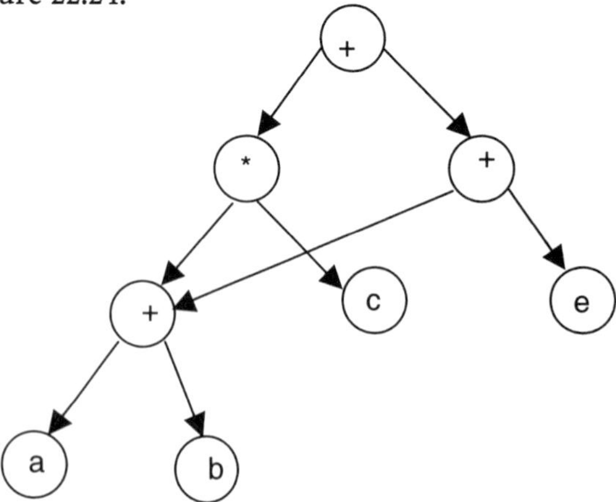

FIGURE 22.24 DAG representation of expression $(a+b)*c + ((a+b) + e)$.

Topological Sort of Directed Graph

A *topological sort* is a method for ordering the nodes of the directed graph G in which the nodes represent tasks or activities, and the edges represent precedence relations between nodes, that is, when the edges of the graph represent dependency among the node/vertices of graph G. It lists the vertices/nodes in such an order that a vertex *vi* gets listed only after all the vertices on which *vi* depends have been listed. For a topological sort to be feasible, it is required that the directed graph G not have any directed cycles. In other words, the graph G should be a DAG. This also means that the precedence relation defined by the edges of G must be irreflexive. The precedence relation defined by the edges of G is certainly transitive and so is a partial order. It starts with a vertex that does not have any predecessor and lists it. After that it logically deletes it from the graph and once again searches for that vertex that does not have any predecessor, and repeats the procedure. It does not give a unique order. For example, consider the directed graph shown in Figure 22.25.

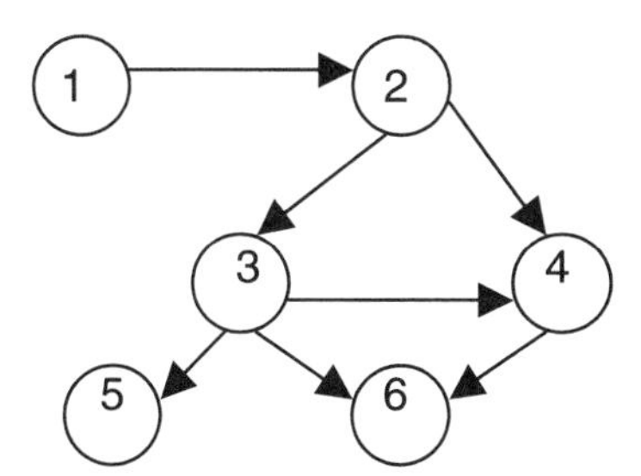

FIGURE 22.25 A graph G.

The topological sort of the graph in figure 22.25 gives the following orders:

1-2-3-4-5-6

1-2-3-5-4-6

1-2-3-4-6-5

The algorithm for a topological sort is presented here:

```
while there exist a node
   {
      select a node which does not have any predecessor
      list the selected node
      delete it from the graph
   }
```

The procedure for a depth first search, with a print statement added to print the nodes, can also be used, as shown in the following program, to perform a topological sort. This algorithm prints the vertices accessible from x in reverse topological sort. This algorithm prints the vertices accessible from x in reverse topological order.

Program

Write a program to find the topological order of a diagraph G represented as adjacency lists.

```
#include <stdio.h>

#define N 11          // no of total vertices in the graph.

typedef enum {FALSE, TRUE} bool;
typedef struct node node;

struct node {
    int count;        // for arraynodes : in-degree.
                      // for listnodes  : vertex no this vertex is
connected to.
                      // if this node is out of graph : -1.
                      // if this has 0 indegree then it occurs in
zerolist.
    node *next;
};

node graph[N];
node *zerolist;

void addToZerolist( int v ) {
    /*
     * adds v to zerolist as v has 0 predecessors.
     */
    node *ptr = (node *)malloc( sizeof(node) );
    ptr->count = v;
    ptr->next  = zerolist;
    zerolist   = ptr;
}

void buildGraph( int a[][2], int edges ) {
    /*
```

```
     * fills global graph with input given in a.
     * a[i][0] is src vertex and a[i][1] is dst vertex.
     */
    int i;

    // init graph.
    for( i=0; i<N; ++i ) {
        graph[i].count = 0;
        graph[i].next  = NULL;
    }

    // now add the list entries.
    for( i=0; i<edges; ++i ) {
        // add new node to src list.
        node *ptr = (node *)malloc( sizeof(node) );
        ptr->count = a[i][1];
        ptr->next = graph[ a[i][0] ].next;
        graph[ a[i][0] ].next = ptr;
        // increase indegree of dst.
        graph[ a[i][1] ].count++;
    }

    // now create list of zero predecessors.
    zerolist = NULL;  // list of vertices having 0 predecessors.
    for( i=0; i<N; ++i )
        if( graph[i].count == 0 ) {
                addToZerolist(i);
        }
}

void printGraph() {
    int i;
    node *ptr;

    for( i=0; i<N; ++i ) {
        node *ptr;
        printf( "%d: pred=%d: ", i, graph[i].count );
        for( ptr=graph[i].next; ptr; ptr=ptr->next )
                printf( "%d ", ptr->count );
        printf( "\n" );
    }
    printf( "zerolist: " );
```

```
        for( ptr=zerolist; ptr; ptr=ptr->next )
            printf( "%d ", ptr->count );
        printf( "\n" );
    }

    int getZeroVertex() {
        /*
         * returns the vertex with zero predecessors.
         * if no such vertex then returns -1.
         */
        int v;
        node *ptr;

        if( zerolist == NULL )
            return -1;
        ptr = zerolist;
        v = ptr->count;
        zerolist = zerolist->next;
        free(ptr);

        return v;
    }

    void removeVertex( int v ) {
        /*
         * deletes vertex v and its outgoing edges from global graph.
         */
        node *ptr;
        graph[v].count = -1;
        // free the list graph[v].next.
        for( ptr=graph[v].next; ptr; ptr=ptr->next ) {
            if( graph[ ptr->count ].count > 0 ) // normal nodes.
                    graph[ ptr->count ].count-;
            if( graph[ ptr->count ].count == 0 )       // this is NOT else
of above if.
                    addToZerolist( ptr->count );
        }
    }

    void topsort( int nvert ) {
        /*
         * finds recursively topological order of global graph.
```

```
     * nvert vertices of graph are needed to be ordered.
     */
    int v;

    if( nvert > 0 ) {
        v = getZeroVertex();
        if( v == -1 ) {       // no such vertex.
                fprintf( stderr, "graph contains a cycle.\n" );
                return;
        }
        printf( "%d.\n", v );
        removeVertex(v);
        topsort( nvert-1 );
    }
}

int main() {
    int a[][2] = {
                          {0,1},
                          {0,3},
                          {0,2},
                          {1,4},
                          {2,4},
                          {2,5},
                          {3,4},
                          {3,5}
                    };
    buildGraph( a, 8 );
    printGraph();
    topsort(N);
}
```

Explanation

1. A linear ordering of vertices of a diagraph G, with the property that if i is a predecessor of j then i precedes j in linear order, is called a topological order of G.
2. The diagraph G is maintained as adjacency lists. In this representation, G is an array `graph[0..n-1]` where each element `graph[i]` is a linked list of vertices. Vertex i is connected to, and n is the number of, vertices in G.

3. We also maintain a zerolist, which is a list of vertices that have zero predecessors. The necessity of this list will be clear in the algorithm shown in number four of this explanation.
4. The algorithm for topological sort is:

```
    topsort(n) {
        if( n > 0 ) {
    if every vertex has a predecessor then
        error( "graph contains a cycle." ).
    pick a vertex v that has no predecessors.                    //
getZeroVertex()
    output v.
    delete v and all the edges leading out of v in the graph.    //
removeVertex()
    topsort(n-1).
        }
    }
```

5. The algorithm topsort() is tail-recursive. From zerolist, it removes a vertex v containing zero predecessors and outputs it. This vertex v has no predecessors in G or all its predecessors have already been output. Thus all the vertices in zerolist are the candidates for the next output. After v is output, all the vertices to which v points may become the candidates for the next output. Thus we remove all the edges starting from v and rerun topsort() over the remaining vertices.
6. See the following example of a diagraph.

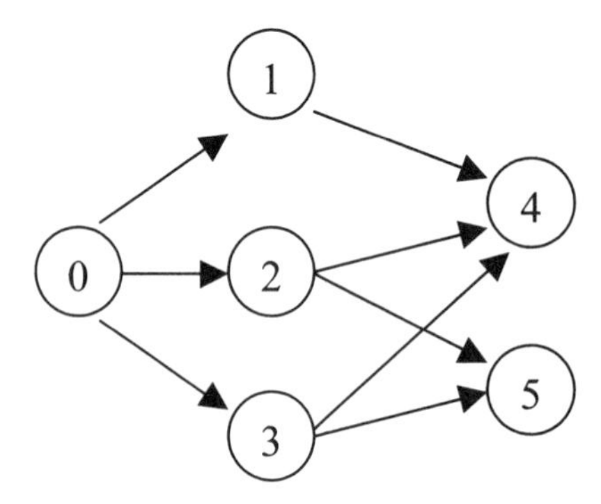

STEP	ZEROLIST	OUTPUT
0	{0}	nil
1	{1, 2, 3}	0
2	{2, 3}	1
3	{3}	2
4	{4,5}	3
5	{5}	4
6	{}	5

Points to Remember

1. A linear ordering of vertices of a diagraph G, with the property that if i is a predecessor of j then i precedes j in the linear ordering, is called a topological order of G.
2. The complexity of topological order is $O(n+e)$ where n is the number of vertices and e is the number of edges in the diagraph.
3. Removal of an edge results in a decrease in the predecessor count of the destination vertex. If this count reaches 0, the vertex should be inserted in the `zerolist`.
4. By maintaining a list of vertices with zero predecessors, the computing time of the algorithm decreases.
5. The algorithm is therefore a total time of $O(e)$, if graph G is represented by an adjacency list, and $O(n^2)$ if graph G is represented by an adjacency matrix.

General Comments on Graphs

1. A graph is a structure that is often used to model the arbitrary relationships among the data objects while solving many problems.
2. A graph is a structure made of two components: a set of vertices V, and the set of edges E. Therefore, a graph is G = (V,E), where G is a graph.
3. The graph may be directed or undirected. When the graph is directed, every edge of the graph is an ordered pair of vertices connected by the edge.
4. When the graph is undirected, every edge of the graph is an unordered pair of vertices connected by the edge.

5. The maximum number of edges in an undirected graph with n vertices is $n(n - 1)/2$, whereas in the case of a directed graph, it is $n(n - 1)$.
6. Adjacency matrices and adjacency lists are used to represent graphs.
7. A graph can be traversed by using either the depth first traversal or the breadth first traversal.
8. When a graph is traversed by visiting in the forward (deeper) direction as long as possible, the traversal is called depth first traversal.
9. When a graph is traversed by visiting all the adjacencies of a node/vertex first, the traversal is called breadth first traversal.
10. A connected component of a graph is a maximal subgraph of a given graph that is connected.
11. A DAG is an important data structure used for representing syntactic structure of expressions with common subexpressions.
12. A DAG is also used in representing partial orders.
13. A topological sort lists the vertices in such an order that if a vertex vi is a predecessor of vertex vj then vi precedes vj in the linear ordering.
14. Topological sort is possible only for a DAG.

Exercises

1. Find out the minimum number of edges in a strongly connected diagraph on *n* vertices.
2. Test the program for obtaining the depth first spanning tree for the following graph:

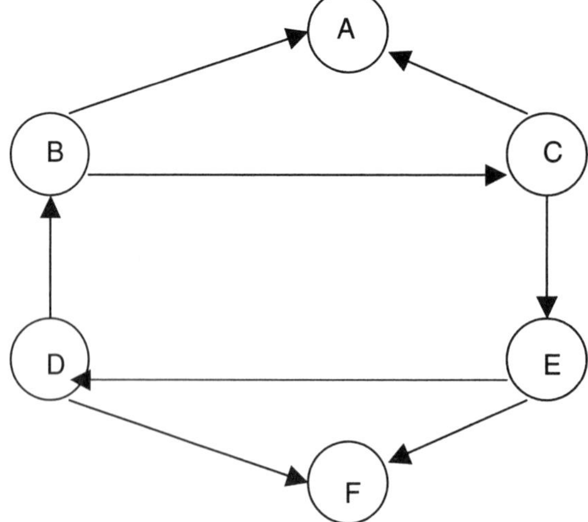

3. Test the program for obtaining the minimum cost spanning tree for the following graph:

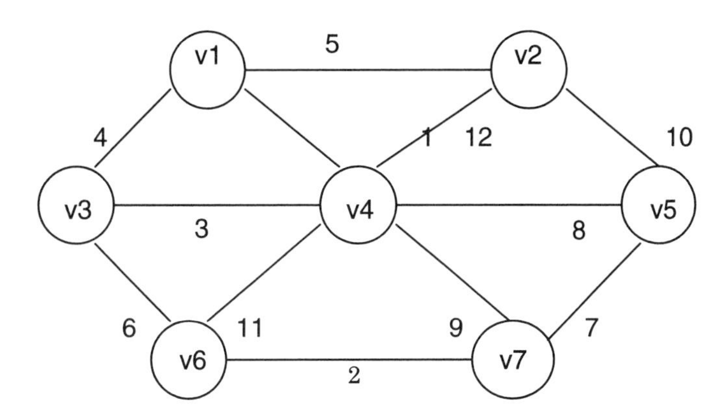

4. Test a program for topological sort for the following DAG:

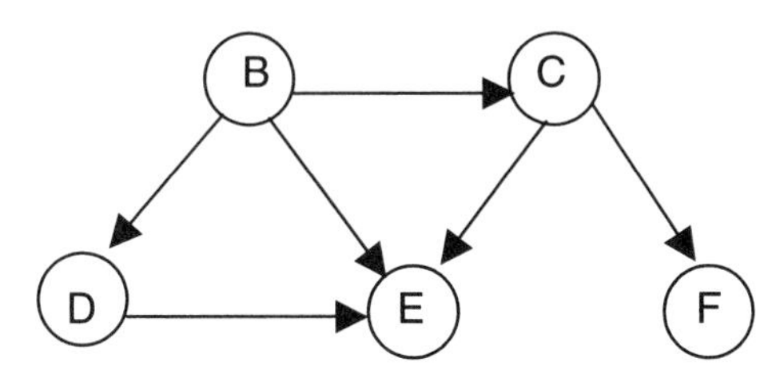

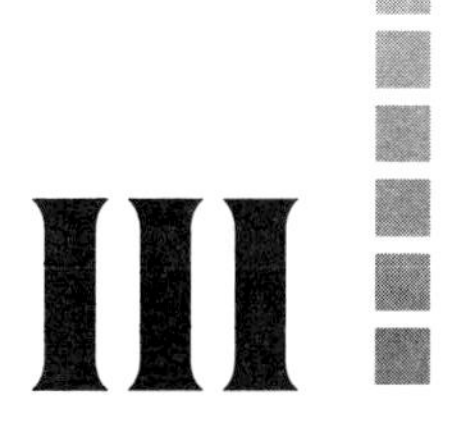

Advanced Problems in Data Structures

23 Problems in Arrays, Searching, Sorting, Hashing

PROBLEM: CALCULATE THE VALUE OF AN $N \times N$ DETERMINANT

Program

```
#include <stdio.h>
#include <math.h>

#define N 3
#define EPSILON 1e-10

typedef enum {FALSE, TRUE} bool;

void print( double a[][N] ) {
    /*
     * print the matrix.
     */
    int i, j;

    for( i=0; i<N; ++i ) {
        for( j=0; j<N; ++j )
                printf( "%8.4g ", a[i][j] );
        printf( "\n" );
    }
    printf( "\n" );
}

void divRow( double a[][N], int row, double divisor ) {
    /*
     * divides row of a by divisor.
     */
    int j;
    for( j=0; j<N; ++j )
```

```
        a[row][j] /= divisor;
}

void subRow( double a[][N], int row1, int row2 ) {
    /*
     * row1 -= row2.
     */
    int j;
    for( j=0; j<N; ++j )
        a[row1][j] -= a[row2][j];
}

bool anyZero( double a[][N] ) {
    /*
     * returns TRUE if any diagonal entry of a is zero (less than
EPSILON).
     */
    int i;
    for( i=0; i<N; ++i )
        if( fabs(a[i][i]) <= EPSILON )
                return TRUE;
    return FALSE;
}

double makeUpper( double a[][N] ) {
    /*
     * makes a an upper-triangular matrix.
     * returns 0 if any of the diagonal entries are 0; 1 otherwise.
     */
    int i, j;
    double factor = 1.0;
    double temp;

    for( i=1; i<N; ++i )     // dont worry about row 0.
        for( j=0; j<i; ++j ) {
                temp = a[i][j];
                if( fabs(temp) > EPSILON ) {
                        printf( "factor=%g dividing row %d by %g...\n",
factor, i, temp );
                        divRow( a, i, temp );
                        print(a);
                        factor *= temp;
```

```
                }
                temp = a[j][j];
                if( fabs(temp) > EPSILON && fabs(temp-1.0) > EPSILON ) {
                      printf( "factor=%g dividing row %d by %g...\n",
factor, j, temp );
                      divRow( a, j, temp );
                      print(a);
                      factor *= temp;
                }
                if( fabs(a[i][j]) > EPSILON ) {
                      printf( "factor=%g row[%d] -= row[%d]...\n",
factor, i, j );
                      subRow( a, i, j );
                      print(a);
                }
                if( anyZero(a) == TRUE )
                      return 0;
          }
      a[N-1][N-1] *= factor;          // all but(?) last element of row
N-1 are zero.

      return 1;
   }

   double multDia( double a[][N] ) {
      /*
       * returns multiplication of diagonal elements.
       */
      int i;
      double factor = 1;

      for( i=0; i<N; ++i )
         factor *= a[i][i];
      return factor;
   }


   int main() {
      double a[N][N] =                        {      {8,0,1},
                                              {2,1,3},
                                              {5,3,9}
                                         };
      double factor;
```

```
    print(a);
    factor = makeUpper(a);
    print(a);
    printf( "determinant = %g.\n", factor*multDia(a) );
}
```

Explanation

1. The usual way of finding the value of an $N \times N$ determinant is to take the first element of the first row and multiply it by the value of the determinant formed by removing that row and that column. This procedure is followed recursively at every step until only a single element remains whose value itself is the value of the 1×1 determinant. The sign of the element being multiplied may change depending on its row and column. In general, if the element has row i and column j, then its sign is determined by the formula (–1)^(i+j).
2. There is another way to solve the problem. We note that if we perform row or column transformations on the determinant, its value does not change. We take advantage of this fact to transform the matrix corresponding to the determinant into an upper or lower triangular matrix and then find its determinant value. It becomes clear that the determinant value of an upper or lower triangular matrix is the product of the diagonal elements. Thus the job of finding a determinant value is basically a matter of making a matrix upper or lower triangular.
3. To make a matrix upper triangular (`makeUpper()`), we perform row transformations on the matrix. Another important observation is that multiplication of a determinant D by a value v is a new determinant in which any of the rows of D are multiplied by v. Multiplication of a row by v means multiplying each element in the row by v. Thus we maintain a variable factor that keeps track of the multiplier as the rows of the determinant are multiplied or divided by various factors in the process of upper-triangulation. We assume all elements of the determinant to be floating-point numbers.
4. In the process of upper-triangulation, if any of the diagonal elements become 0 (`anyZero()`), that means the value of the determinant is 0 (remember that the value of the determinant is the multiplication of diagonal elements (`multDia()`).
5. Since here we assume that the elements are floating-point numbers, the calculations might not be exact. To account for this approximation, we

maintain an error term EPSILON which is set to a sufficiently low value (tending to 0). Any value in the range +EPSILON...0...–EPSILON is considered to be 0.

6. The algorithm for making a matrix upper triangular is as follows:

```
factor = 1.
for i=1 to N-1 do
for j=0 to i-1 do
   divide row i by D[i][j]     // divRow().
   factor *= D[i][j].
   divide row j by D[j][j]     // divRow().
   factor *= D[j][j].
   subtract elements of row j from corresponding elements of row i
// subRow().
   check for any of the diagonal elements of D to be 0.
// anyZero().
    determinant = factor*product of diagonal elements.
```

7. **Example:** Let the determinant be

   | 8 0 1 |
   | 2 1 3 |
   | 5 3 9 |.

   factor = 1.

   Snapshots of the algorithm when run on this determinant are shown here.

I	J	STEP	FACTOR	DETERMINANT
1	0	divide row 1 by 2	2	| 8 0 1 | | 1 0.5 1.5 | | 5 3 9 |
1	0	divide row 0 by 8	16	| 1 0 0.125 | | 1 0.5 1.5 | | 5 3 9 |

I	J	STEP	FACTOR	DETERMINANT
1	0	row 1 -= row 0|	16	| 1 0 0.125 | | 0 0.5 1.375 | | 5 3 9 |
2	0	divide row 2 by 5	80	| 1 0 0.125 | | 0 0.5 1.375 | | 1 0.6 1.8 |
2	0	row 2 -= row 0	80	| 1 0 0.125 | | 0 0.5 1.375 | | 0 0.6 1.675 |
2	1	divide row 2 by 0.6	48	| 1 0 0.125 | | 0 0.5 1.375 | | 0 1 2.792 |
2	1	divide row 1 by 0.5	24	| 1 0 0.125 | | 0 1 2.75 | | 0 1 2.792 |
2	1	row 2 -= row 1	24	| 1 0 0.125 | | 0 1 2.75 | | 0 0 0.0417|

Thus, the determinant = 24*(1*1*0.0417) = 1.

Points to Remember

1. The first way to solve the problem recursively was natural but clumsy, as it requires removal of a row and a column. So, when designing an algorithm, we should try different approaches and then select the most appropriate one.

2. Note how we reduced the problem of finding a determinant value to making a matrix upper triangular. These reductions not only simplify a problem but can also help in reusing the code and analysis.
3. An error term such as EPSILON should be used in floating point computations.

PROBLEM: WRITE A PROGRAM TO FIND THE SADDLE POINT OF A MATRIX, IF IT EXISTS

Program

```
#include <stdio.h>

#define M 3
#define N 3

int findMin( int a[][N], int row ) {
    /*
     * find min value in row of a.
     * return the value.
     */
    int min = a[row][0];
    int j;

    for( j=1; j<N; ++j )
        if( a[row][j] < min )
            min = a[row][j];
    return min;
}

int findMax( int a[][N], int col ) {
    /*
     * find max val in col of a.
     * return the value.
     */
    int max = a[0][col];
    int i;

    for( i=1; i<M; ++i )
       if( a[i][col] > max )
              max = a[i][col];
```

```
    return max;
}

void saddle( int a[][N] ) {
    /*
     * finds ALL saddle points of a if exist.
     */
    int i, j;

    for( i=0; i<M; ++i ) {
        int min = findMin( a, i );

        for( j=0; j<N; ++j ) {
            int max;
            max = findMax( a, j );
            if( min == max )
                printf( "Saddle : (%d,%d).\n", i, j );
        }
    }
}

int main() {
    int a[M][N] = {5,7,3,7,7,9,7,1,2};
    int row, col;

    saddle(a);

    return 0;
}
```

Explanation

1. An M × N matrix is said to have a saddle point if some entry a[i][j] is the smallest in row i and the largest in column j. In the following example matrix, 7 is the saddle point.

```
    |1 2 3|
a = |4 5 6|
    |7 8 9|
```

2. The program is simple and could be completed in O(M × N × M) time by using nested for loops. The algorithm is as follows:

```
for i=0 to M-1 {
```

```
        min = minimum value in row i.             // findMin().
        for j=0 to N-1 {
            max = maximum value in column j.      // findMax().
            if (min == max)
         print "saddle found at row ", i, "and column ", j.
        }
    }
```

3. Finding the minimum in a row is O(N) and finding the maximum in a column is O(M). Thus the complexity of the algorithm becomes O(M*(N+N*M)), or O(M*N*M).

Points to Remember

1. A saddle point is the value which is minimum in the row and maximum in the column.
2. There can be more than one saddle point in a matrix.
3. There may be no saddle point in a matrix.
4. The complexity of the algorithm is $O(M \times M \times N)$, where $M \times N$ is the size of the matrix.

PROBLEM: MULTIPLY TWO SPARSE MATRICES

Write a function `mmult()` to multiply two sparse matrices and store the result in another sparse matrix. Each sparse matrix is represented by an array of triplets. Each triplet consists of a row, a column, and a value.

Program

```
#include <stdio.h>

#define M 10          // the matrices are MxP, PxN and MxN.
#define P 7
#define N 8

void ftrans( int a[][3], int b[][3] ) {
    /*
     * finds fast-transpose of a in b.
```

```
     */
    int t[P+1];
    int i, j, n, terms;
    int temp, temp2;

    n = a[0][1];
    terms = a[0][2];
    b[0][0] = n;
    b[0][1] = a[0][0];
    b[0][2] = terms;

    if( terms <= 0 )
        return;
    for( i=0; i<n; ++i )
        t[i] = 0;
    for( i=1; i<=terms; ++i )
        t[a[i][1]]++;
    temp = t[0];
    t[0] = 1;
    for( i=1; i<n; ++i ) {
        temp2 = t[i], t[i] = t[i-1]+temp, temp = temp2;
    }
    for( i=1; i<=terms; ++i ) {
        j = t[a[i][1]];
        b[j][0] = a[i][1], b[j][1] = a[i][0], b[j][2] = a[i][2];
        t[a[i][1]] = j+1;
    }
}

void printMatrix( int a[][3] ) {
    /*
     * prints the matrix in the form of 3-tuples.
     */
    int i;
    int nterms = a[0][2];

    printf( "rows=%d cols=%d vals=%d.\n", a[0][0], a[0][1], a[0][2] );
    for( i=1; i<=nterms; ++i )
        printf( "a[%d][%d] = %d.\n", a[i][0], a[i][1], a[i][2] );
    putchar( '\n' );
}
```

```
void insert( int c[][3], int row, int col, int val ) {
    /*
     * insert or add the triplet (row,col,val) in c.
     * update c[0][2] if necessary.
     */
    int i, terms = c[0][2];

    for( i=1; i<=terms && c[i][0]<row; ++i )
        ;
    for( ; i<=terms && c[i][1]<col; ++i )
        ;
    if( i<=terms && c[i][1] == col )       // already inserted.
        c[i][2] += val;
    else {                                 // a new entry should be
inserted at i.
        c[i][0] = row; c[i][1] = col; c[i][2] = val;
        c[0][2]++;
    }
}

void mmult( int a[][3], int b[][3], int c[][3] ) {
    /*
     * c = a*b;
     */
    int i, j, mn = M*N;
    int aterms = a[0][2], bterms = b[0][2];
    int rowsofb = b[0][0];
    int *t = (int *)malloc( rowsofb*sizeof(int) );
    int temp, temp2;
    int arow, acol, aval, brow, browstart, browend;

    c[0][0] = a[0][0];
    c[0][1] = b[0][1];

    // init c.
    for( i=0; i<=mn; ++i )
        c[i][2] = 0;

    // fill t[] : t[i] points to row of b where actual row i starts.
    // last+1 entry is also maintained for easing loops.
    for( i=0; i<=rowsofb; ++i )
        t[i] = 0;
    for( i=1; i<=bterms; ++i )
```

```
        t[b[i][0]]++;
    temp = t[0];
    t[0] = 1;
    for( i=1; i<=rowsofb; ++i )
        temp2 = t[i], t[i] = t[i-1]+temp, temp = temp2;

    // now start mult.
    for( i=1; i<=aterms; ++i ) {
        arow = a[i][0]; acol = a[i][1]; aval = a[i][2];
        brow = acol;
        browstart = t[brow]; browend = t[brow+1];
        for( j=browstart; j<browend; ++j )
                insert( c, arow, b[j][1], aval*b[j][2] );
    }
}

int main() {
    int a[][3] =     {
                                {4,2,3},
                                {0,1,2},
                                {1,0,3},
                                {3,1,4}
                        };
    int b[][3] =     {
                                {2,3,3},
                                {0,2,5},
                                {1,0,7},
                                {1,1,6}
                        };
    int c[M*N+1][3];

    printMatrix(a);
    printMatrix(b);
    mmult( a, b, c );
    printMatrix(c);

    return 0;
}
```

Explanation

1. A sparse matrix is represented by an array of triplets. Each triplet consists of a row, a column, and a value corresponding to one nonzero element in the sparse matrix. Thus, we maintain an array of size N × 3 for each sparse matrix, where N is the number of non-zero elements in the array. The first row of each sparse matrix contains the number of rows, columns, and nonzero elements in the matrix. An example follows.

I	A[I][0]	A[I][1]	A[I][2]
0	7	7	8
1	1	1	15
2	1	4	22
3	1	6	-15
4	2	2	11
5	2	3	3
6	3	4	-6
7	5	1	91
8	6	3	28

   `a[0][0]`=7 is the number of rows of the matrix `a`. `a[0][1]`=7 is the number of columns of the matrix `a`. `a[0][2]`=8 is the number of nonzero elements in the matrix `a`. The nonzero elements are saved in rows `i`=1 to `i`=8 where `a[i][0]` is the row, `a[i][1]` is the column, and `a[i][2]` is the value in the matrix `a`.

   Note that the elements are sorted by row, but inside a row they are sorted by columns.

2. The function `mmult(a, b, c)` takes each element of `a`, with row `arow` and column `acol`, and multiplies with each element of row `acol` of `b`. The product is added to the element in `c` with row `arow` and column equal to the column of the element in `b`. Note that since `a` and `b` are sorted by (row, column), the new entries generated for `c` are also in the same sorted order. Thus we can easily insert the new entries in `c` at the end of the currently stored entries.
3. The only problem now is how to reach the first entry in `b` with row `acol`. One way is to use a binary search to reach the row, as it is sorted by row. But we note that the number of rows of `b` is fixed. So we maintain an array of

indices pointing to the first entries for each row in b. Thus, if the matrix b is as shown previously, the indices maintained in a vector t are as shown here.

I	T[I]
0	0
1	1
2	4
3	6
4	0
5	7
6	8

t[i] = 0 signifies that row i of b does not contain any elements. Because of this, the original binary search of order O(log*n*) can now be done in O(1) time. But this needs extra O(Nb) time to construct t[] where Nb is the number of non-zero entries in b.

4. Example: Let

```
    |0 2|
a = |3 0|,  b = |0 0 5|
    |0 0|       |7 0 6|
    |1 4|
```

Here M = 4, P = 2, and N = 3. Then we should get

```
    |14 0 12|
c = | 0 0 15|
    | 0 0  0|
    |28 0 29|
```

The algorithm traverses each element of a, and for each element in a with row arow and column acol:

row acol of b is traversed

the element in a is multiplied by each element in the row-list of b with row acol and column bcol, and

the products are inserted in c as c[arow][bcol].

Thus in every step, an element in a contributes to c. If c[arow][bcol] exists, the product is added to the original value. The multiplication of a and b is given here for the previous example.

STEP	AROW	ACOL	BCOL	A[AROW][ACOL] *B[ACOL][BCOL]	C[AROW][BCOL]
1	0	1	0	14	14
2	0	1	2	12	12
3	1	0	2	15	15
4	3	0	2	5	5
5	3	1	0	28	28
6	3	1	2	24	29

Note that in step 6, the product is 24, while c[3][2] gets a value of 29. This happens because c[3][2] already contains a value 5 in step 4.

Points to Remember

1. By storing the sparse matrix as an array of triplets, we save a considerable amount of space required for pointers, in the case of a sparse matrix represented by horizontal and vertical lists.
2. Note how the matrix multiplication time reduces by storing the indices to the rows of b.
3. The complexity of `mmult()` is O(Na*Cb*Nc*Nc) where Na is the number of non-zero elements in a, Cb is the number of columns in b and Nc is the number of entries in c. The factor Nc*Nc can be reduced by maintaining the maximum (row, column) inserted in c and by performing a binary search while searching for an existing (row, column), or by maintaining an array similar to t[] for b. By storing such extra indices, insertion in c can be done in O(1) which will make mmult() be O(Na*Cb).

PROBLEM: MULTIPLICATION OF TWO SPARSE MATRICES (DIFFERENT VERSIONS)

Let a and b be two sparse matrices. Write a function sMatMul(a, b, c) to set up the structure for c=a*b.

Program

```
#include <stdio.h>

#define M 10
#define N 10
#define P 10
#define DEFAULTVAL 0
#define SUCCESS 0
#define ERROR -1

typedef int type;
typedef struct node node;

struct node {
    type data;
    int row;
    int col;
    node *hnext;
    node *vnext;
};

typedef struct {
    node *rows;
    node *cols;
} spmat;

void sInit( spmat *mat, int rows, int cols ) {
    /*
     * initialize the matrix.
     */
    int i;

    mat->rows = (node *)malloc( sizeof(node)*rows );
    mat->cols = (node *)malloc( sizeof(node)*cols );
```

```
        for( i=0; i<rows; ++i ) {
            mat->rows[i].hnext = mat->rows+i;
            mat->rows[i].row   = i;
        }
        for( i=0; i<cols; ++i ) {
            mat->cols[i].vnext = mat->cols+i;
            mat->cols[i].col   = i;
        }
    }

    int sAdd( spmat *mat, int row, int col, type data ) {
        /*
         * adds a new node to the sparse matrix.
         */
        node *ptr;

        if( data == DEFAULTVAL )
            return;
        ptr       = (node *)malloc( sizeof(node) );    // freed in
cColInsert() if reqd.
        ptr->data = data;
        ptr->row  = row;
        ptr->col  = col;
        cRowInsert( mat->rows+row, ptr );
        cColInsert( mat->cols+col, ptr );

        return SUCCESS;
    }

    int cRowInsert( node *head, node *dataptr ) {
        /*
         * inserts dataptr in appropriate row of sparse matrix.
         */
        node *ptr, *prev;
        for( prev=head, ptr=prev->hnext; ptr!=head && ptr->col<dataptr-
>col; prev=ptr, ptr=ptr->hnext )
            ;
        if( ptr!=head && ptr->col == dataptr->col ) { // data already
exists.
            ptr->data += dataptr->data; // this is for multiplication.
            return SUCCESS;
        }
```

```
    // dataptr should be added between prev and ptr.
    dataptr->hnext = ptr;
    prev->hnext = dataptr;

    return SUCCESS;
}

int cColInsert( node *head, node *dataptr ) {
    /*
     * inserts dataptr in appropriate col of sparse matrix.
     * Assume that cRowInsert() was called before.
     */
    node *ptr, *prev;
    for( prev=head, ptr=prev->vnext; ptr!=head && ptr->row<dataptr-
>row; prev=ptr, ptr=ptr->vnext )
        ;
    if( ptr!=head && ptr->row == dataptr->row )   { // data already
exists.
        free(dataptr);
        return SUCCESS;
    }
    // dataptr should be added between prev and ptr.
    dataptr->vnext = ptr;
    prev->vnext = dataptr;

    return SUCCESS;
}

void cRowPrint( node *head ) {
    /*
     * print a row.
     */
    node *ptr;

    printf( "%2d : ", head->row );
    for( ptr=head->hnext; ptr!=head; ptr=ptr->hnext )
        printf( "%d(%d,%d) ", ptr->data, ptr->row, ptr->col );
    printf( "\n" );
}

void cColPrint( node *head ) {
    /*
```

```
         * print a col.
         */
        node *ptr;

        printf( "%2d : ", head->col );
        for( ptr=head->vnext; ptr!=head; ptr=ptr->vnext )
            printf( "%2d(%d,%d) ", ptr->data, ptr->row, ptr->col );
        printf( "\n" );
    }

    void sHPrint( spmat *mat, int rows ) {
        /*
         * print sparse matrix by traversing it row-wise.
         */
        int i;
        for( i=0; i<rows; ++i )
            cRowPrint( mat->rows+i );
        printf( "\n" );
    }

    void sVPrint( spmat *mat, int cols ) {
        /*
         * print sparse matrix by traversing it col-wise.
         */
        int i;
        for( i=0; i<cols; ++i )
            cColPrint( mat->cols+i );
        printf( "\n" );
    }

    type sGetVal( spmat *a, int row, int col ) {
        /*
         * return a[row][col];
         */
        node *head = a->rows+row;
        node *ptr;

        for( ptr=head->hnext; ptr!=head; ptr=ptr->hnext )
            if( ptr->col == col )
                return ptr->data;
        return DEFAULTVAL;              // entry absent in matrix : default
value 0.
```

```
    }

    int sMatMulBad( spmat *a, spmat *b, spmat *c ) {
        /*
         * original inefficient implementation of matrix mult.
         */
        int i, j, k;

        for( i=0; i<M; ++i )
            for( j=0; j<N; ++j ) {
                    type data = 0;
                    for( k=0; k<P; ++k )
                            data += sGetVal(a,i,k)*sGetVal(b,k,j);
                    sAdd( c, i, j, data );
            }
        return SUCCESS;
    }

    int sMatMul( spmat *a, spmat *b, spmat *c ) {
        /*
         * matrix multiplication.
         */
        node *ptri, *ptrj;
        int i;

        for( i=0; i<M; ++i )
            for( ptri=a->rows[i].hnext; ptri!=a->rows+i; ptri=ptri->hnext )
{
                    int row = ptri->col;
                    for( ptrj=b->rows[row].hnext; ptrj!=b->rows+row;
ptrj=ptrj->hnext ) {
                            sAdd( c, i, ptrj->col, ptri->data*ptrj->data );
                    }
            }
        return SUCCESS;
    }

    int main() {
        spmat a, b, c;

        sInit(&a,M,P); sInit(&b,P,N); sInit(&c,M,N);
```

```
    sAdd( &a, 0,1, 2 );
    sAdd( &a, 1,0, 3 );
    sAdd( &a, 3,1, 4 );

    sAdd( &b, 0,2, 5 );
    sAdd( &b, 1,0, 7 );
    sAdd( &b, 1,1, 6 );

    sHPrint(&a,M);
    sHPrint(&b,P);

    sMatMul( &a, &b, &c );
    sHPrint(&c,M);
    sVPrint(&c,N);

    return 0;
}
```

Explanation

1. A sparse matrix is represented as two arrays of pointers: one for rows and the other for columns. Each row and each column is represented by horizontal and vertical circular lists. A nonzero entry `a[i][j]` is added as a node to the horizontal list of row `i` and the same node in the vertical list of column `j`. A node represents an entry. Thus it contains row, column, and values along with horizontal and vertical pointers in the lists.
2. Let the sizes of sparse matrices `a`, `b` be M×P, P×N. Thus the size of `c` is M×N. We describe the algorithm using an example. Let

$$a = \begin{vmatrix} 0 & 2 \\ 3 & 0 \\ 0 & 0 \\ 1 & 4 \end{vmatrix}, \quad b = \begin{vmatrix} 0 & 0 & 5 \\ 7 & 0 & 6 \end{vmatrix}$$

Here M = 4, P = 2, and N = 3. Then we should get

$$c = \begin{vmatrix} 14 & 0 & 12 \\ 0 & 0 & 15 \\ 0 & 0 & 0 \\ 28 & 0 & 29 \end{vmatrix}.$$

3. The algorithm traverses each row of a and checks its horizontal list corresponding to each row for any elements in it. For each element in a with row arow and column acol, row acol of b is traversed and the element in a is multiplied by each element in the row-list of b with row acol and column bcol. The products are inserted in c as c[arow][bcol]. Thus, in every step, an element in a contributes to c. If c[arow][bcol] exists, then the product is added to the original value. The multiplication of a and b is given here for the previous example.

STEP	AROW	ACOL	BCOL	A[AROW][ACOL]*B [ACOL][BCOL]	C[AROW][BCOL]
1	0	1	0	14	14
2	0	1	2	12	12
3	1	0	2	15	15
4	2				
5	3	0	2	5	5
6	3	1	0	28	28
7	3	1	2	24	29

Step 4 signifies that a[4] is taken for traversal but is not traversed as it is empty. Note that the product is 24 in step 7, while c[3][2] gets a value of 29. This happens because c[3][2] already contains a value of 5 in step 5.

4. Let na, nb be the number of nonzero entries in a and b. The basic step in this algorithm is adding the products of the elements in a and b to c. So even if the outer loop traverses from 0 to M–1, the sAdd() function gets invoked for each entry a[i][j]. This invoking is done for each nonzero element in row j of b. Thus at most, the number of multiplications for a[i][j] will be equal to N, number of columns of b. Thus the complexity of the matrix multiplication algorithm is O(na*N). However, if we add the complexity due to the outermost loop, the complexity is O(na*N+M*nb).

Points to Remember

1. A slight modification to the usual add(matrix,row,col,value) function resulted in an easing of the implementation of the matrix multiplication. In general, if a[i][j] is inserted and it already exists, then we either overwrite the value

or return an error. By adding the new value to the original value, we can add products of elements of a and b to c incrementally.

2. The sparse matrix, as it is represented by a complicated data structure, should be initialized properly.
3. An array-based matrix multiplication program has the complexity O(M*P*N). If we use a similar procedure in this representation as in sMatMulBad(), the complexity increases to O(M*P*N*P) as sGetVal is O(P).
4. Insertions were simplified by the representation of empty rows and columns by head nodes rather than NULL lists.

PROBLEM: IMPLEMENT *K*-WAY SORT-MERGE TO SORT A FILE CONTAINING RECORDS

Program

```
#include <stdio.h>
#include <malloc.h>

#define N 80
#define K 2                    // K-way merge.
#define DATAFILE "main.txt"
#define TEMPFILE "temp.txt"

typedef struct node node;
typedef enum {FALSE, TRUE} bool;

struct node {
    int val;
    char s[N];
};

node buf[K];            // buffer used for merging.
int  rec[K];            // record numbers of nodes in buffer.

int getNRecords( char *filename ) {
    /*
     * returns no of records in file filename using size of the file.
     */
    int off;
```

```
    FILE *fp = fopen( filename, "r" );
    fseek (fp, 0, SEEK_END );
    off = ftell(fp)/sizeof(node);   // no of records.
    fclose(fp);

    return off;
}

void writeToFile( char *filename, node *n ) {
    /*
     * writes record n to file filename.
     */
    FILE *fp = fopen( filename, "a" );
    fwrite( n, sizeof(node), 1, fp );
    fclose(fp);
}

void readFromFile( char *filename, node *n, int off ) {
    /*
     * reads a record at offset off from file filename into n.
     * off is number of records before the record in the file (NOT
bytes).
     * off starts from 0.
     */
    FILE *fp;
    //printf( "reading rec no %d...\n", off );
    if( off >= getNRecords(filename) ) {
        fprintf( stderr, "ERROR: reading beyond the file.\n" );
        return;
    }
    printf("total records are %d\n",getNRecords(filename));
    fp = fopen( filename, "r" );
    fseek( fp, off*sizeof(node), SEEK_CUR );
    fread( n, sizeof(node), 1, fp );
    fclose(fp);
}

void writeFun( char *filename ) {
    /*
     * writes some data to filename.
     */
    node data[10] = {
```

```
                                {5,"five"},
                                {3,"three"},
                                {4,"four"},
                                {8,"eight"},
                                {7,"seven"},
                                {6,"six"},
                                {9,"nine"},
                                {10,"ten"},
                                {1,"one"},
                                {2,"two"}
                     };
    int i;
    for( i=0; i<10; ++i )
        writeToFile( filename, data+i );
}

void readFun( char *filename ) {
    /*
     * reads filename and prints the data.
     */
    node n;
    int i, nrec = getNRecords(filename);

    for( i=0; i<nrec; ++i ) {
        readFromFile( filename, &n, i );
        printf( "%2d={%2d,%-5s}.\n", i, n.val, n.s );
    }
}

void copyrec( int *rec, int *rec2 ) {
     int i;
     *rec2 = *rec;

    for( i=0; i<K; ++i )
        rec2[i] = rec[i];
}

void fillbuf( int start, int l, int nrec, char *srcfile ) {
    /*
     * fills buf and rec with appropriate values.
     * l is length of each run.
     * start is rec no of first rec in first run.
     * data is in srcfile in nrec records.
```

```
    */
    int i;
    printf( "start=%d l=%d.\n", start, l );
    for( i=0; i<K; ++i ) {
       int startoff = start+l*i;
       if( startoff >= nrec )
             break;
       rec[i] = startoff;
       printf( "buf[%d]=%d.\n", i, startoff );
       readFromFile( srcfile, buf+i, startoff );
    }
    for( ; i<K; ++i )
       rec[i] = -1;
    getchar();
 }

 void updatebuf( node *buf, int *rec, int *rec2, int prevrec, int l,
char *srcfile, int nrec ) {
    /*
     * updates buf+rec2 as rec2[prevrec] was output.
     * read appropriate record from srcfile if necessary.
     * rec still contains the original rec nos which can be used for
     *  checking ends of runs.
     * l is runlength.
     */
    if( rec2[prevrec] < nrec-1 && rec2[prevrec] < rec[prevrec]+l-1 ) {
       // rec2[prevrec] was NOT the last rec of that run.
       rec2[prevrec]++;
       readFromFile( srcfile, buf+prevrec, rec2[prevrec] );
    }
    else {
       // rec2[prevrec] was the last rec of that run.
       rec2[prevrec] = -1;   // job of this run is over.
    }
 }

 int getMin( node *buf, int *rec2 ) {
    /*
     * returns index in buf of that record which has min sorting value.
     * rec2 is needed for checking whether a buf entry is valid.
     */
    int minval = 9999;
    int minindex = -1;
```

```
        int i;

        for( i=0; i<K; ++i )
            if( rec2[i] != -1 && buf[i].val < minval ) {
                    minval = buf[i].val;
                    minindex = i;
            }
        return minindex;
    }

    void merge( char *srcfile, char *dstfile, node *buf, int *rec2, int l,
int nrec ) {
        /*
         * rec2 contains record numbers being compared; global rec also
contains
         *  the same at this point.
         * buf contains their actual data.
         * l is runlength.
         * srcfile is needed for reading next data.
         * the data is appended to dstfile.
         * total no of records being written is min(l*k,nrec-rec[0]).
         */
        int totalrec = l*K;
        int i;
        int nrecremaining = nrec-rec[0];        // no of rec in srcfile yet
to be written to dstfile.

        if( nrecremaining < totalrec )
            totalrec = nrecremaining;
        printf( "totalrec=%d nrecremaining=%d.\n", totalrec, nrecremaining
);

        for( i=0; i<totalrec; ++i ) {
            int nextrec = getMin( buf, rec2 );  // here goes the comparison.
            printf( "after getMin: min=%d rec2=%d %d %d  buf=%d %d %d.\n",
nextrec, rec2[0], rec2[1], rec2[2], buf[0].val, buf[1].val, buf[2].val );
            if( nextrec == -1 ) {
                    fprintf( stderr, "ERROR: merge(): all rec2 are -1!\n" );
                    return;
            }
            //printf( "min=%d.\n", nextrec );
```

```
            // this is the index in rec2 of next record to be output.
            writeToFile( dstfile, buf+nextrec );
            // remove this written record. read new record from srcfile if
needed.
            updatebuf( buf, rec, rec2, nextrec, l, srcfile, nrec );
            //printf( "after updatebuf : rec2=%d %d %d.\n", rec2[0],
rec2[1], rec2[2] );
        }
    }

    void mergedriver( char *srcfile, char *dstfile ) {
        /*
         * sort+merge srcfile and store in dstfile.
         */
        int nrec = getNRecords(srcfile);
        int i, l;
        int rec2[K];
        char tempname[N];

        for( l=1; l<nrec; l*=K ) {
            // l is length of each run.
            // no of runs = ceil( nrec/l );
            // we need to consider only K runs at a time.

            for( i=0; i<nrec; i+=l*K ) {
                    // fill buf with appropriate values.
                    fillbuf( i, l, nrec, srcfile );
                    copyrec( rec, rec2 );
                    merge( srcfile, dstfile, buf, rec2, l, nrec );
            }
            unlink( srcfile );
            strcpy( tempname, srcfile );
            strcpy( srcfile, dstfile );
            strcpy( dstfile, tempname );
        }
        // sorted file is srcfile.
        printf( "\n\n\n" );
        readFun( srcfile );
    }

    int main() {
        char srcfile[N] = DATAFILE;
```

```
    char dstfile[N] = TEMPFILE;
    unlink( srcfile );
    unlink( dstfile );
    writeFun( srcfile );
    readFun( srcfile );
    printf( "nrec=%d.\n", getNRecords(srcfile) );
    mergedriver( srcfile, dstfile );

    return 0;
}
```

Explanation

1. The function `main()` creates a test file by using `writeFun()` and calls `mergedriver()`. `mergedriver()` calls the function `merge()`after reading (`fillbuf()`) K blocks of the file into memory.
2. The function `merge()` compares the blocks and sorts them on the predetermined key. The block with the minimum key (`getMin()`) is written to the file and its block is filled (`updatebuf()`) with the next record from the file. The sorting and merging thus proceeds simultaneously to successively sort the file.
3. The number of blocks being compared are called runs. Each run length increases with each iteration. It is 1 initially, then it becomes K, then K*K and so on. As it becomes greater than or equal to the number of records in the file (`getNRecords()`), the file is sorted because all the records in each run are sorted after each iteration.
4. **Example:** Assume the blocks are saved in a file as shown next and let K=3. Then the algorithm transforms the file as follows:

STEP	FILE	RUNLENGTH
1	5 3 4 8 7 6 9 10 2 1	1
2	3 4 5 6 7 8 2 9 10 1	3
3	2 3 4 5 6 7 8 9 10 1	9
4	1 2 3 4 5 6 7 8 9 10	27

Points to Remember

1. The complexity of merge-sort is $O(n \log n)$ where n is the number of records in the file. However, in general, the statements that are added to the complexity are comparisons or a nested assignment. But in case of files, one needs to consider reading and writing of records in the file, as the time required for execution of one such operation is much more than a comparison in memory or a simple assignment.
2. Sorting in files is called external sorting while sorting in main memory is termed internal sort.
3. By exchanging the names of source and destination files in `mergedriver()`, we avoided copying of files in each iteration.

PROBLEM: FIND A PLATEAU IN A MATRIX

Find a rectangular region in a matrix with the maximum sum of its elements. The elements may be negative.

Program

```
#include <stdio.h>

#define COLS 5
#define MININT -99999

int filter(int a[][COLS], int i, int j, int k, int l, int rows, int
cols) {
    /*
     * filter the matrix of size k*l starting from a[i][j].
     * size of the matrix is rows*cols.
     * k, l start with 1.
     */
    int iii, jjj;
    int sum = 0;

    if(i+k > rows || j+l > cols)    // the matrix was already
considered
                                                        // with
smaller k, l.
        return MININT;

    for(iii=0; iii<k && (i+iii)<rows; ++iii)
        for(jjj=0; jjj<l && (j+jjj)<cols; ++jjj) {
                sum += a[i+iii][j+jjj];
                if(sum < 0)
                        return sum;            // this is reqd: if all
vals -ve.
        }
    return sum;
}

void printMatrix(int a[][COLS], int i, int j, int k, int l, int rows,
int cols) {
    /*
     * print a rectangular region of a from a[i][j] of size k*l if
possible.
     * the matrix is bounded by rows*cols.
     */
    int iii, jjj;

    for(iii=0; iii<k && (i+iii)<rows; ++iii) {
```

```
            for(jjj=0; jjj<l && (j+jjj)<cols; ++jjj)
                    printf("%d ", a[i+iii][j+jjj]);
            printf("\n");
        }
        getchar();
    }

    void plateau(int a[][COLS], int rows, int cols) {
        /*
         * finds a rectangular region having max sum of elements in it.
         */
        int maxsum = a[0][0];
        int maxrow1=0, maxrow2=0, maxcol1=0, maxcol2=0;
        int i, j, k, l;
        int sum;

        for(i=0; i<rows; ++i)
            for(j=0; j<cols; ++j)
                    // generate k*l matrix using a[i][j].
                    for(k=1; k<=rows; ++k)
                            for(l=1; l<=cols; ++l) {
                                    sum=filter(a, i, j, k, l, rows, cols);
                                    if(sum > maxsum) {
                                            maxsum=sum, maxrow1=i, maxcol1=j,
maxrow2=i+k-1, maxcol2=j+l-1;
                                    }
                            }
        printf("sum=%d.\n", filter(a, maxrow1, maxcol1, maxrow2-maxrow1+1,
maxcol2-maxcol1+1, rows, cols));
        printMatrix(a, maxrow1, maxcol1, maxrow2-maxrow1+1, maxcol2-
maxcol1+1, rows, cols);

    }

    int main() {
        int a[][COLS] = {
                                    {5,-1,-2,-4,1},
                                    {-3,2,10,-6,3},
                                    {-1,9,-11,-7,9},
                                    {3,101,3,-2,-96},
                                    {-1,-2,0,3,-3},
                                    {-1,-2,-3,2,-3}
                            };
```

```
    plateau(a, sizeof(a)/COLS/sizeof(int), COLS);

    return 0;
}
```

Explanation

1. Considesr the following matrix. The rectangle of elements with the maximum sum is also given.

```
|5  -1  -2  -4  1  |          | 5  -1 |
|-3  2  10  -6  3  |          |-3   2 |
|-1  9 -11  -7  9  |          |-1   9 |  sum = 115.
|3 101   3  -2 -96 |          | 3 101 |
|-1 -2   0   3  -3 |
|-1 -2  -3   2  -3 |
```

An example matrix and its maximum-sum-plateau.

2. A straightforward algorithm to find a maximum sum is as follows.

```
for i=0 to rows-1
    for j=0 to columns-1 {
          maxsum = matrix[i][j].
          for height=1 to rows
   for width=1 to columns {
   let M = matrix of size height x width from matrix[i][j].
          let sum = sum of elements of M.
          if sum > maxsum
                  maxsum = sum.
   }
}
```

The values of i, j, width, and height at any point represent the matrix. Those values can be saved to print the matrix at the end of the algorithm.

The complexity of this procedure is O(m*m*m*n*n*n) where the size of the original matrix is m × n. The complexity of finding the sum of elements of M is O(m*n).

3. This procedure can be made more efficient by noting that not every internal matrix M needs to be generated. For an element `matrix[i][j]`, the maximum width of the matrix starting from the element can be columns-j and

maximum height can be rows–i. Thus, the number of iterations of loops over width and height can be decreased. Also, for an element e, if the sum of elements of a rectangular region of size height × width starting from e is negative, then no rectangular region of size more than height × width starting from e can have a sum higher than the final result. We show this in the example matrix.

Let `i=1` and `j=2`. `matrix[i][j] = 10`. Let height = 2 and width = 1. Thus the matrix contains two elements in it, {10, –11} and the sum is –1. Since the sum is negative, there is no point in increasing the size of the matrix and considering other elements because whatever their sum (say s), by adding this matrix to it, the total sum is definitely going to be lower than s. The matrix containing sum s may be a candidate for the result. So our matrix cannot have a sum greater than the final result. Thus, if we increase the height of the matrix further so that the elements are {10, –11, 3}, the sum becomes 2 which is less than a 1 × 1 matrix starting from 3. Instead, if we increase the width so that the elements are {10, –11, –6, –7}, the sum is –14 which is even lower.

The function `filter()` implements this strategy.

Points to Remember

1. By considering only those matrices that start from an element e, that is those increasing in width to the right of e and increasing in height downwards from e, we select all the possible rectangular regions without any repetition.
2. We need not increase the size of a submatrix once the sum of its elements becomes negative.
3. If all the values in the input matrix are negative, then the result is a 1×1 matrix with the only element having a minimum magnitude.
4. There can be multiple solutions to this problem.
5. The complexity of finding the sum of elements of a submatrix can be increased by noting the sum incrementally.

PROBLEM: IMPLEMENTATION OF A HASH SEARCH

Write a program that takes strings as inputs and stores them in a hash table. It should then ask the user for strings to be searched in the hash table. Use shift-folding as the hashing function and chaining for overflow handling.

Program

```
#include <stdio.h>
#include <string.h>
#include <malloc.h>

#define MAXLEN 80
#define HASHSIZE 23          // some prime val.
#define SHIFTBY 3            // each group size in hashing.

typedef struct node node;
typedef char *type;
typedef node *hashtable[HASHSIZE];

struct node {
    int val;
    char *key;
    node *next;
};

int hGetIndex(char *key) {
    /*
     * returns index into hashtable applying hash function.
     * uses shift-folding followed by mod function for hashing.
     */
    int i, n, finaln=0;
    char *keyptr;

    for(keyptr=key; *keyptr; finaln+=n)
        for(i=0, n=0; i<SHIFTBY && *keyptr; ++i, ++keyptr)
                n = n*10 + *keyptr;
    finaln %= HASHSIZE;

    return finaln;
}

void hInsert(hashtable h, char *key, int val) {
    /*
     * insert s in hashtable h.
     * use shift-folding followed by mod function for hashing.
     * does NOT check for duplicate insertion.
     */
    node *ptr = (node *)malloc(sizeof(node));
    int index = hGetIndex(key);
```

```
        ptr->key = strdup(key);
        ptr->val = val;
        ptr->next = h[index];

        h[index] = ptr;
        printf("h[%d] = %s.\n", index, key);
    }

    int hGetVal(hashtable h, char *key) {
        /*
         * returns val corresponding to key if present in h else -1.
         */
        node *ptr;

        for(ptr=h[hGetIndex(key)]; ptr && strcmp(ptr->key, key); ptr=ptr-
>next)
            ;
        if(ptr)
            return ptr->val;
        return -1;
    }

    void printHash(hashtable h) {
        /*
         * print the hashtable rowwise.
         */
        int i;
        node *ptr;

        for(i=0; i<HASHSIZE; ++i) {
            printf("%d: ", i);
            for(ptr=h[i]; ptr; ptr=ptr->next)
                    printf("%s=%d ", ptr->key, ptr->val);
            printf("\n");
        }
    }

    int main() {
        char s[MAXLEN];
        int i = 0;
        hashtable h = {"abc"};

        printf("Enter the string to be hashed: ");
```

```
    gets(s);

    while(*s) {
        hInsert(h, s, i++);
        printf("Enter the string to be hashed(enter to end): ");
        gets(s);
    }
    printf("Enter the string to be searched: ");
    gets(s);

    while(*s) {
        printf("%s was inserted at number %d.\n", s, hGetVal(h, s));
        printf("\nEnter the string to be searched(enter to end): ");
        gets(s);
    }
    //printHash(h);

    return 0;
}
```

Explanation

1. The hash table is maintained as an array of lists. Each list is either empty or contains nodes containing strings that map to the index in the hash table, after application of the hashing function. The string in the node is called the key. Each node also stores an integer that is the number at which the string was inserted in the hash table. Thus, each node contains a (key, value) pair. In a realistic situation, a value can be anything that has a key associated with it.
2. The program contains two loops. In the first, it asks the user to enter a series of strings and calls `hInsert()` to insert the strings in the hash table. The second loop asks the user to enter a string and returns the number at which it was inserted. An insertion number of –1 indicates that the string is not present in the hash table. This is done by using the function `hGetVal()`. Both these functions make use of the hashing function `hGetIndex()`, which, given a string, returns its hashing index. It folds the string into a pattern of m characters (perhaps except the last), and forms an integer out of each m characters. It then adds all these integers to get another number. This is then divided by the size of the hash table array to get the remainder as an index into the hash table. `hInsert()` adds this new string and its insertion sequence to a node and this node is added to the start of the list in the index.

hGetVal() searches the list at this index for the input string. If it finds such a string, its insertion sequence is returned, otherwise it returns –1.

3. Since the complexity of insertion of a node in the list is O(1), the complexity of hInsert() is the complexity of the hashing function. The complexity of the hashing function is $O(p)$ where p is the average length of the string. Thus the complexity of hInsert() is $O(p)$. The complexity of hGetVal() is $O(p+q)$ where q is the average number of nodes in each list. Chaining involves a linear search.

Points to Remember

1. The complexity of insertion in the hash table is decided by the hashing function if simple chaining is used for overflow handling.
2. The complexity of searching is decided by both the hashing function and the overflow handling technique.
3. An ideal hash function maps every input string to a different index and thus has zero collisions. Assuming that the complexity of a hash function is O(1), the insertions and searching into an ideal hash table are O(1).
4. Hash tables are used in compilers for symbol-table management. Hash tables have numerous other applications as well.
5. Different overflow handling techniques such as linear probing, quadratic probing, random probing, rehashing, etc., are in use depending on the application requirement.

PROBLEM: IMPLEMENTATION OF REHASHING

Write functions to insert and search in a hash table by using the rehashing technique. Use linear probing if the rehashing fails.

Program

```
#include <stdio.h>
#include <string.h>
#include <malloc.h>
#define MAXLEN 80
#define HASHSIZE 23          // some prime val.
```

```
typedef struct node node;
typedef char *type;
typedef node *hashtable[HASHSIZE];

struct node {
    int val;
    type key;
};

int hGetIndex1(type key) {
    /*
     * returns index into hashtable applying hash function.
     * uses sum of elements followed by mod function for hashing.
     */
    int n=0;
    char *keyptr;

    for(keyptr=key; *keyptr; ++keyptr)
       n += *keyptr;

    return n%HASHSIZE;
}

int hGetIndex2(type key) {
    /*
     * returns index into hashtable applying hash function.
     * sums the products of elements with their indices and then mod.
     */
    long n=0;
    int i;
    type keyptr;

    printf("Function 2:).\n");
    for(keyptr=key, i=1; *keyptr; ++keyptr, ++i)
       n += i**keyptr;

    return n%HASHSIZE;
}
int hGetEmptySlot(hashtable h, int index) {
    /*
     * search for an empty slot in h starting from index+1.
     */
```

```
    int i;

    for(i=index+1; i<HASHSIZE; ++i) // index+1 to end of hashtable.
        if(!h[i])
                return i;
    for(i=0; i<index; ++i)                    // starting from 0 to index-
1.
        if(!h[i])
                return i;
    return -1;
}

int hLinearProbe(hashtable h, type key, int index) {
    /*
     * search for node having key in h starting from index+1.
     */
    int i;

    for(i=index+1; i<HASHSIZE; ++i) // index to end of hashtable.
        if(h[i] && !strcmp(h[i]->key, key))
                return i;
        else if(!h[i])
                return -1;
    for(i=0; i<index; ++i)   // starting from 0 to index-1.
        if(h[i] && !strcmp(h[i]->key, key))
                return i;
        else if(!h[i])
                return -1;

    return -1;
}

void hInsert(hashtable h, type key, int val) {
    /*
     * insert s in hashtable h.
     * does NOT check for duplicate insertion.
     */
    node *ptr = (node *)malloc(sizeof(node));
    int index = hGetIndex1(key);

    if(h[index]) {
        index = hGetIndex2(key);
        if(h[index]) {
```

```
                index = hGetEmptySlot(h, index);
                if(index == -1) {
                        printf("ERROR: Hashtable full.\n");
                        return;
                }
        }
    }
    ptr->key = strdup(key);
    ptr->val = val;

    h[index] = ptr;
    printf("h[%d] = %s.\n", index, key);
}

int hGetVal(hashtable h, type key) {
    /*
     * returns val corresponding to key if present in h else -1.
     */
    int index = hGetIndex1(key);

    if(h[index] && strcmp(h[index]->key, key)) {
        index = hGetIndex2(key);
        if(h[index] && strcmp(h[index]->key, key)) {
                index = hLinearProbe(h, key, index);
                if(index == -1)
                        return -1;
        }
        else if(!h[index])
                return -1;
    }
    else if(!h[index])
        return -1;

    printf("index=%d ", index);
    return h[index]->val;
}
void printHash(hashtable h) {
    /*
     * print the hashtable.
     */
    int i;
```

```
    for(i=0; i<HASHSIZE; ++i)
        if(h[i]) {
                printf("%d: ", i);
                printf("%s=%d ", h[i]->key, h[i]->val);
                printf("\n");
        }
}

int main() {
    char s[MAXLEN];
    int i = 0;
    hashtable h = {"asd"};

    printf("Enter the string to be hashed: ");
    gets(s);

    while(*s) {
        hInsert(h, s, i++);
        printf("Enter the string to be hashed(enter to end): ");
        gets(s);
    }
    printf("Enter the string to be searched: ");
    gets(s);

    while(*s) {
        printf("%s was inserted at number %d.\n", s, hGetVal(h, s));
        printf("\nEnter the string to be searched(enter to end): ");
        gets(s);
    }
    printHash(h);

    return 0;
}
```

Explanation

1. Rehashing is one method of handling collisions. If a single application of the hash function results in a collision, and we detect that the space is already occupied by another element, we can use some other hash function to generate another index in the hash table. If this index is also already filled,

we can either use another hash function to generate another index or we can use linear probing or chaining. We use linear probing here to guarantee full utilization of the hash table in a finite amount of time.

2. We use simple `mod()` function as the first hash function (`hGetIndex1()`). We add the ASCII values of the input characters and apply `mod()` to get an index in the range of the hash table. We use a modification of this `mod()` function as the second hash function (`hGetIndex2()`). We add the products of the ASCII values of characters and their indices to get a final sum to which we apply the mod function to get the second index. If this slot is also filled, we use linear search (`hLinearProbe()`) over the hash table starting from index to the end of the hash table, and then starting from the start of the hash table to the index, to get an empty slot. If we do not find an empty slot, we output an error message. Otherwise the index of the empty slot is returned.
3. An identical procedure is used during searching (`hGetVal()`). It applies the first hash function to get the first index. If this is empty, –1 is returned. If it contains the node containing the search key, the value corresponding to that key is returned. If the key is different, we apply the second hash function to get another index. If this is empty, –1 is returned. If it contains the node containing the search key, the value corresponding to that key is returned. If the key is different, we apply linear probing (`hLinearProbe()`) to search for the key in the hash table. If we get such a node, the value corresponding to the key is returned. If such a node does not exist, an error message is printed.
4. Since linear probing (and not chaining) is used, the hash table is simply an array of pointers to nodes where each node contains only a key and value.
5. **Example:**

   Let `HASHSIZE = 23`.

   Let strings to be inserted be as follows: `'dj'`, `'na'`, `'id'`, `'q'`.

   `hGetIndex1('dj')` returns 22, so it is inserted in hash table[22].

   `hGetIndex1('na')` returns 0, so it is inserted in hash table[0].

   `hGetIndex1('id')` returns 21, so it is inserted in hash table[21].

   `hGetIndex1('q')` returns 21. Since the space is already filled by `'id'`, `hGetIndex2('q')` is called. It returns 22. But it is also filled by `'dj'`. So a linear search for an empty slot is done starting from index 22 (wrapping over the hash table). It finds that the next index 0 is filled with `'na'` and its next index 1 is empty. Hence `'q'` gets inserted in hash table[1].

Points to Remember

1. Since no deletions are taking place in the hash table, we can reduce the complexity of searching for an element. This is because during searching, if we get an empty slot, that means the search key cannot be present beyond that location. This follows from the predicate that a value is inserted into the first empty slot we get.
2. The hash table could have been an array of nodes instead of node pointers. But our implementation can save space if the `val` field is of a larger size and if many slots of the hash table are empty.
3. The complexities of both hash functions is O(n) where n is the average length of the string to be inserted. The complexity of linear probing is O(`HASHSIZE`) where `HASHSIZE` is the hashtable size.
4. The advantage of modular programming over `hash.c` is that the function `main()` remains the same even after changing the implementation of the hashing procedure.

24 Problems in Stacks and Queues

PROBLEM: CONVERT AN INFIX EXPRESSION TO PREFIX FORM

Program

```
#include <stdio.h>
#include <string.h>
#include <ctype.h>

#define N 80

typedef enum {FALSE, TRUE} bool;

#include “stack.h”
#include “queue.h”

#define NOPS 7

char operators [] = “()^/*+-”;
int  priorities[] = {4,4,3,2,2,1,1};
char associates[] = “  RLLLL”;

char t[N]; char *tptr = t;  // this is where prefix will be saved.

int getIndex( char op ) {
    /*
     * returns index of op in operators.
     */
    int i;
    for( i=0; i<NOPS; ++i )
        if( operators[i] == op )
               return i;
    return -1;
```

```
}

int getPriority( char op ) {
    /*
     * returns priority of op.
     */
    return priorities[ getIndex(op) ];
}

char getAssociativity( char op ) {
    /*
     * returns associativity of op.
     */
    return associates[ getIndex(op) ];
}
void processOp( char op, queue *q, stack *s ) {
    /*
     * performs processing of op.
     */
    switch(op) {
       case ')':
              printf( "\t S pushing )...\n" );
              sPush( s, op );
              break;
       case '(':
              while( !qEmpty(q) ) {
                     *tptr++ = qPop(q);
                     printf( "\tQ popping %c...\n", *(tptr-1) );
              }
              while( !sEmpty(s) ) {
                     char popop = sPop(s);
                     printf( "\tS popping %c...\n", popop );
                     if( popop == ')' )
                            break;
                     *tptr++ = popop;
              }
              break;
       default: {
              int priop;     // priority of op.
              char topop;    // operator on stack top.
              int pritop;    // priority of topop.
              char asstop;   // associativity of topop.
```

```
            while( !sEmpty(s) ) {
                priop  = getPriority(op);
                topop  = sTop(s);
                pritop = getPriority(topop);
                asstop = getAssociativity(topop);

                if( pritop < priop || (pritop == priop && asstop
== 'L') || topop == ')' )       // IMP.
                    break;
                while( !qEmpty(q) ) {
                    *tptr++ = qPop(q);
                    printf( "\tQ popping %c...\n", *(tptr-1) );
                }
                *tptr++ = sPop(s);
                printf( "\tS popping %c...\n", *(tptr-1) );
            }
            printf( "\tS pushing %c...\n", op );
            sPush( s, op );
            break;
        }
    }
}

bool isop( char op ) {
    /*
     * is op an operator?
     */
    return (getIndex(op) != -1);
}

char *in2pre( char *str ) {
    /*
     * returns valid infix expr in str to prefix.
     */
    char *sptr;
    queue q = {NULL};
    stack s = NULL;
    char *res = (char *)malloc( N*sizeof(char) );
    char *resptr = res;

    tptr = t;
    for( sptr=str+strlen(str)-1; sptr!=str-1; -sptr ) {
```

```
        printf( “processing %c tptr-t=%d...\n”, *sptr, tptr-t );
        if( isalpha(*sptr) ) // if operand.
            qPush( &q, *sptr );
        else if( isop(*sptr) )      // if valid operator.
            processOp( *sptr, &q, &s );
        else if( isspace(*sptr) )   // if whitespace.
            ;
        else {
            fprintf( stderr, “ERROR:invalid char %c.\n”, *sptr );
            return “”;
        }
    }
    while( !qEmpty(&q) ) {
        *tptr++ = qPop(&q);
        printf( “\tQ popping %c...\n”, *(tptr-1) );
    }
    while( !sEmpty(&s) ) {
        *tptr++ = sPop(&s);
        printf( “\tS popping %c...\n”, *(tptr-1) );
    }
    *tptr = 0;
    printf( “t=%s.\n”, t );
    for( –tptr; tptr!=t-1; –tptr ) {
        *resptr++ = *tptr;
    }
    *resptr = 0;

    return res;
}

int main() {
    char s[N];

    puts( “enter infix freespaces max 80.” );
    gets(s);
    while(*s) {
        puts( in2pre(s) );
        gets(s);
    }

    return 0;
}
```

Explanation

1. In an infix expression, a binary operator separates its operands (a unary operator precedes its operand). In a postfix expression, the operands of an operator precede the operator, while in a prefix expression the operator precedes its operands. Like postfix, a prefix expression is parenthesis-free, that is, any infix expression can be unambiguously written in its prefix equivalent without the need for parentheses.
2. When an infix expression is converted to reverse-prefix, it is scanned from right to left. A queue of operands is maintained, noting that the order of operands in infix and prefix remains the same. Thus, while scanning the infix expression, whenever an operand is encountered, it is pushed in a queue. If the scanned element is a right parenthesis (')'), it is pushed in a stack of operators. If the scanned element is a left parenthesis ('('), the queue of operands is emptied to the prefix output followed by popping of all the operators but excluding a right parenthesis in the operator stack.
3. If the scanned element is an arbitrary operator o, then the stack of operators is checked for operators with greater priority than the priority for o. Such operators are popped and written to the prefix output after emptying the operand queue. The operator o is finally pushed in the stack.
4. When the scanning of the infix expression is complete, first the operand queue and then the operator stack are emptied to the prefix output. Any whitespace in the infix input is ignored. Thus the prefix output we get can be reversed to get the required prefix expression of the infix input.
5. **Example:** If the infix expression is *a***b*+*c*/*d*, then different snapshots of the algorithm while scanning the expression from right to left are as follows:

step	remaining expression	scanned element	queue of operands	stack of operators	prefix output
0	a*b+c/d	nil	empty	empty	nil
1	a*b+c/	d	d	empty	nil
2	a*b+c	/	d	/	nil
3	a*b+	c	d c	/	nil
4	a*b	+	empty	+	dc/
5	a*	b	b	+	dc/

step	remaining expression	scanned element	queue of operands	stack of operators	prefix output
6	a	*	b	* +	dc/
7	nil	a	b a	* +	dc/
8	nil	nil	empty	empty	dc/ba*+

The final prefix output we get is *dc/ba**+, whose reverse is +**ab/cd*, which is indeed the prefix equivalent of the input infix expression *a*b+c*d*. Note that all the operands are simply pushed to the queue in steps 1, 3, 5, and 7. In step 2, the operator / is pushed to the empty stack of operators.

In step 4, the operator + is checked against the elements in the stack. Since / (division) has higher priority than + (addition), the queue is emptied to the prefix output (thus we get '*dc*' as the output) and then the operator / is written (thus we get '*dc/*' as the output). The operator + is then pushed to the stack. In step 6, the operator * is checked against the stack elements. Since * has a higher priority than +, * is pushed to the stack. Step 8 signifies that all of the infix expression has been scanned. Thus, the queue of operands is emptied to the prefix output (to get '*dc/ba*'), followed by emptying of the stack of operators (to get '*dc/ba**+').

Points to Remember

1. A prefix expression is parenthesis-free.
2. When an infix expression is converted to its postfix equivalent, it is scanned from right to left. The prefix expression we get is the reverse of the required prefix equivalent.
3. Conversion of infix to prefix requires a queue of operands and a stack, as in the conversion of an infix to postfix.
4. The order of operands in a prefix expression is the same as that in its infix equivalent.
5. If the scanned operator o1 and the operator o2 at the stack top have the same priority, then the associativity of o2 is checked. If o2 is right associative, it is popped from the stack.

PROBLEM: IMPLEMENTATION OF TWO STACKS USING AN ARRAY

Two stacks are represented in an array `twostacks[0...N-1]`, such that one stack starts from start of the array and grows towards its end while the other one starts from the end of the array and grows towards the start. Thus at any time, the maximum number of elements that the two stacks together can accommodate is N. Write functions `Push(stacki, data)` and `Delete(stacki)` to add element data and to delete an element from stack number `stacki`, 1 <= `stacki` <= 2. The functions should be able to add elements to the stacks as long as there are fewer than N elements in both stacks together.

Program

```
#include <stdio.h>

#define N 10                        // combined size of the two stacks.
#define EINDEXOUTOFBOUND -1         // error code on overflow in the
stacks.
#define SUCCESS 0                   // success code.

typedef int type;                       // type of each data item.

type twostacks[N];                  // stacks implemented using array.
int stop1 = -1;                         // pointer for stack 1.
int stop2 = N;                      // pointer for stack 2.

int sPush( int stacki, type data ) {
    /*
     * pushes data on top of stacki.
     * returns error on overflow.
     */
    if( stop2-stop1 == 1 )   // overflow.
        return EINDEXOUTOFBOUND;

    if( stacki == 1 ) {      // first stack.
        twostacks[ ++stop1 ] = data;
    }
    else {              // second stack.
        twostacks[ --stop2 ] = data;
    }
```

```
    return SUCCESS;
}

int sDelete( int stacki ) {
    /*
     * deletes element at top from stacki.
     */
    printf( “deleting from stack %d...\n”, stacki );

    if( stacki == 1 ) {      // first stack.
        if( stop1 >= 0 )
                –stop1;
        else
                return EINDEXOUTOFBOUND;
    }
    else {              // second stack.
        if( stop2 < N )
                ++stop2;
        else
                return EINDEXOUTOFBOUND;
    }
    return SUCCESS;
}
void sPrint() {
    /*
     * prints the two stacks.
     */
    int i;
    for( i=0; i<=stop1; ++i )
        printf( “%d “, twostacks[i] );
    printf( “: “ );
    for( i=stop2; i<N; ++i )
        printf( “%d “, twostacks[i] );
    printf( “\n” );
}

int main() {
    sPush(1,1);
    sPush(2,1);
    sPush(1,4);
    sPush(2,5);
    sPrint();
    sDelete(2);
```

```
    sDelete(2);
    sPrint();
    sDelete(2);
    sDelete(1);
    sDelete(1);
    sDelete(1);
    sPrint();
    sPush(2,2);
    sPush(1,9);
    sPrint();
    sDelete(1);
    sPrint();
    sDelete(2);
    sPrint();
    sPush(2,0);
    sPush(1,5);
    sPush(1,3);
    sPrint();
    sDelete(2);
    sPrint();

    return 0;
}
```

Explanation

1. The two stacks are characterized by their indices (stack pointers). Index 1 starts from –1 and goes up to N – 1 while index 2 starts from N and goes up to 0. When index 1 equals –1, it signifies that stack 1 is empty. When index 2 equals to N, that signifies that stack 2 is empty.
2. If the array is full, indices 1 and 2 point to consecutive elements. This situation is characterized as (index2 – index1 == 1).

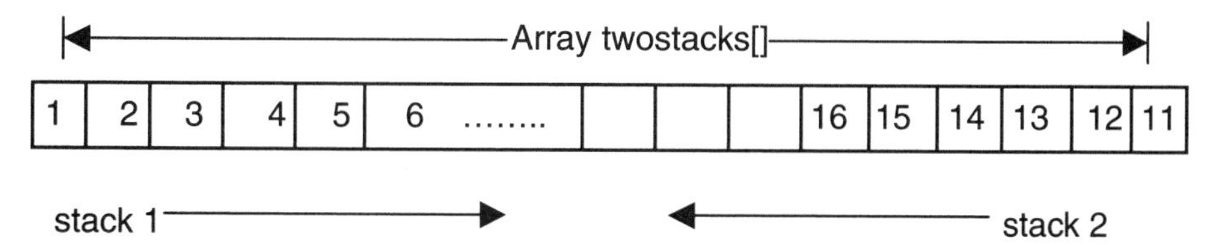

3. In `Push(stacki,data)`, this condition is checked to signify overflow. If space is available, insertion of the element proceeds. If `stacki` == 1 then index 1 is

incremented and the data is inserted at the index. If stacki == 2, index 2 is decremented and the data is inserted at the index.

4. In Delete(stacki), the condition for underflow is checked. If stacki == 1 then index 1 == –1 signifies underflow of stack 1. If stacki == 2 then index 2 == N signifies underflow of stack 2. If an element exists, then deletion of the element at the stack top proceeds. If stacki == 1 then index 1 is decremented, whereas if stacki == 2 then index 2 is incremented.

Points to Remember

1. To implement two stacks using an array, the two stacks should grow in opposite directions.
2. The maximum number of elements that the two stacks together can accommodate at any time is equal to the size of the array.
3. For an empty stack, making the stack pointer point to an element beyond the first element helps in an insertion, without needing to check for an extra condition. Thus, since we have a stack index/value of –1 during insertion, we need simply to increment it to point to the next empty slot. This is not possible with pointers as we need to explicitly check it for being NULL as in

```
if( stacktop == NULL )
   error("stack empty.");.
```

25 Problems in Linked Lists

PROBLEM: IMPLEMENTATION OF POLYNOMIALS USING LINKED LISTS

Write functions to add, subtract, and multiply two polynomials represented as lists. Also, write a function to evaluate the polynomial for a given value of its variable.

Program

```
/* compile with -lm option. */

#include <stdio.h>
#include <stdarg.h>
#include <math.h>

typedef struct node node;
typedef int type;
typedef node *polynomial;

struct node {
    type coeff;
    type power;
    node *next;
};

polynomial createPoly(int n, ...) {
    /*
     * create a list from the arguments.
     * n is the number of nodes which will be created.
     * thus after n there should be 2n arguments: (coeff, power) pairs.
```

```
     * it is assumed that their powers are decreasing.
     */
    va_list vl;
    polynomial p = NULL;
    int i;

    va_start(vl, n);

    for(i=0; i<n; ++i) {
        node *ptr = (node *)malloc(sizeof(node));
        ptr->coeff = va_arg(vl, int);
        ptr->power = va_arg(vl, int);
        ptr->next = p;
        p = ptr;
    }
    va_end(vl);

    return p;
}

void pPrint(polynomial p) {
    node *ptr;

    for(ptr=p; ptr; ptr=ptr->next)
        printf("%dx%d + ", ptr->coeff, ptr->power);
    printf("\n");
}

polynomial pSub(polynomial p1, polynomial p2) {
    /*
     * return p1-p2 recursively.
     */
    node *ptr = (node *)malloc(sizeof(node));

        if(p1 && p2 && p1->power == p2->power) {
                ptr->coeff = p1->coeff - p2->coeff;
                ptr->power = p1->power;
                ptr->next = pSub(p1->next, p2->next);
        }
        else if(p1 && ((p2 && p1->power < p2->power) || !p2)) {
                ptr->coeff = p1->coeff;
                ptr->power = p1->power;
```

```
            ptr->next = pSub(p1->next, p2);
        }
        else if(p2 && ((p1 && p1->power > p2->power) || !p1)) {
            ptr->coeff = -p2->coeff;
            ptr->power = p2->power;
            ptr->next = pSub(p1, p2->next);
        }
        else { // p1 == p2 == NULL.
            free(ptr);
            return NULL;
        }
    return ptr;
}

polynomial pAdd(polynomial p1, polynomial p2) {
    /*
     * return p1+p2 recursively.
     */
    node *ptr = (node *)malloc(sizeof(node));

        if(p1 && p2 && p1->power == p2->power) {
            ptr->coeff = p1->coeff + p2->coeff;
            ptr->power = p1->power;
            ptr->next = pAdd(p1->next, p2->next);
        }
        else if(p1 && ((p2 && p1->power < p2->power) || !p2)) {
            ptr->coeff = p1->coeff;
            ptr->power = p1->power;
            ptr->next = pAdd(p1->next, p2);
        }
        else if(p2 && ((p1 && p1->power > p2->power) || !p1)) {
            ptr->coeff = p2->coeff;
            ptr->power = p2->power;
            ptr->next = pAdd(p1, p2->next);
        }
        else { // p1 == p2 == NULL.
            free(ptr);
            return NULL;
        }
    return ptr;
}
```

```
void pInsertOrAdd(polynomial p, node *ptr) {
    /*
     * p->next anytime contains a partial product.
     * add ptr at appropriate place in it keeping powers in order.
     */
    node *curr, *prev;

    for(prev=p, curr=prev->next; curr && curr->power < ptr->power;
prev=curr, curr=curr->next)
        ;
    // prev will always be NON-NULL :).
    if(curr && curr->power == ptr->power)
        curr->coeff += ptr->coeff, free(ptr);
    else
        prev->next = ptr, ptr->next = curr;
}

polynomial pMult(polynomial p1, polynomial p2) {
    /*
     * return p1*p2.
     */
    node p3, *ptr1, *ptr2, *ptr;

    p3.next = NULL;

    for(ptr1=p1; ptr1; ptr1=ptr1->next)
        for(ptr2=p2; ptr2; ptr2=ptr2->next) {
                ptr = (node *)malloc(sizeof(node));
                ptr->coeff = ptr1->coeff * ptr2->coeff;
                ptr->power = ptr1->power + ptr2->power;
                pInsertOrAdd(&p3, ptr);
        }
    return p3.next;
}

int pEval(polynomial p1, int x) {
    /*
     * evaluate p1 at x.
     */
    node *ptr;
    int result = 0;
```

```
    for(ptr=p1; ptr; ptr=ptr->next)
       result += ptr->coeff * pow(x, ptr->power);

    return result;
}

int main() {
    polynomial p1 = createPoly(3, 3, 5, -1, 3, -10, 0),
               p2 = createPoly(3, -2, 3, 0, 2, 20, 0);

    pPrint(p1);
    pPrint(p2);

    pPrint(pAdd(p1, p2));
    pPrint(pSub(p1, p2));
    pPrint(pMult(p1, p2));
    printf("value of p1 at x=%d is %d.\n", 2, pEval(p1, 2));

    return 0;
}
```

Explanation

1. A polynomial, such as *ax3+bx+c*, is represented by using a linked list of nodes. Each node contains a coefficient and a power of the variable x. Thus this above expression is stored as follows:

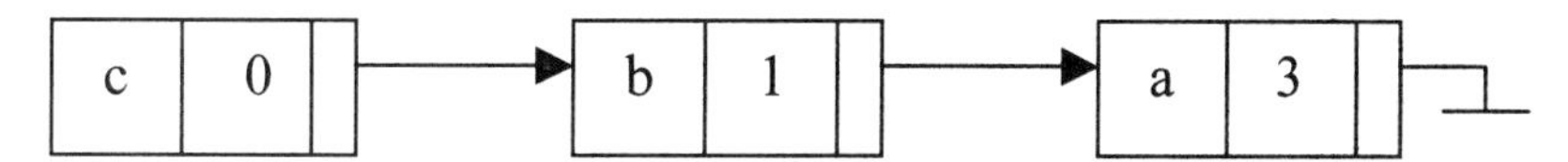

2. The function `createPoly()` creates a linked list of the given coefficients and powers sent to the function, using the variable number of arguments technique.
3. `pAdd(p1, p2)` adds polynomials p1 and p2 and returns the sum. It adds coefficients of the nodes of lists p1 and p2 containing the same power. The nodes in p1 and p2 for which there is no node in the other list with the same power is copied to the result as it is. The traversal of p1 and p2 is done such that the resulting list is also sorted based on power in ascending order. This traversal is done using recursion. The function `pSub()` for subtraction is identical. For example,

$(3 \times 5 + 4 \times 3 + 9) + (5 \times 3 - 4x) = (3 \times 5 + 9 \times 3 - 4x + 9)$

and

$(3 \times 5 + 4x\ 3 + 9) - (5 \times 3 - 4x) = (3 \times 5 - 1 \times 3 + 4x + 9)$

4. The function `pMult(p1, p2)` traverses each node n2 of list p2 for each node n1 of list p1, and prepares a new node whose coefficient is the product of the coefficients in the two nodes n1 and n2 and whose power is the sum of the powers of n1 and n2. It is possible to get the same resulting power for two multiplications. So the procedure `pInsertOrAdd()` traverses the resulting list p3 for occurrence of the node with the same power. If such a node n3 exists, then the new coefficient is added to the old coefficient of n3; otherwise a new node with the new coefficient is inserted in p3. For example,

   $(3 \times 5 + 4 \times 3 + 9) \times (5 \times 3 - 4x)$

   $= (15 \times 8 - 12 \times 6) + (20 \times 6 - 16 \times 4) + (45 \times 3 - 36x)$

   $= (15 \times 8 + 8 \times 6 - 16 \times 4 + 45 \times 3 - 36x)$.

5. The function pEval(p1,x) evaluates the expression in p1 at point x and returns the value. This is done by traversing the list p1 once and adding the values coefficient *x^ power for each node. For example, the value of the polynomial (3x5+4x3+9) at x=2 is (3*2^5+4*2^3+9) = (96+32+9) = 137.

6. The complexity of `pAdd()` and `pSub()` is O(m1+m2), where m1 and m2 are the lengths of the input lists. The complexity of `pInsertOrAdd()` is O(m1+m2). Thus the complexity of `pMult()` is O(m1*m2*(m1+m2)). The complexity of `pEval()` is O(m) where m is the length of the input list.

Points to Remember

1. Polynomials can be represented using linked lists. This has the advantage of reduced space if many of the coefficients in the list are zero. Also, the procedures operating on polynomials represented by arrays and lists are not different as far as complexity is concerned.
2. To avoid sending a variable number of arguments to `createPoly()`, an array can be passed as a parameter.
3. The program should be compiled with the –lm option in order to link `libm` library for the function `pow()`.

PROBLEM: IMPLEMENTATION OF CIRCULAR LISTS BY USING ARRAYS

A linear list is maintained circularly in an array clist[0...N-1] with rear and front set up as for circular queues. Write functions to delete the k-th element in the list and to insert an element e immediately after the k-th element.

Program

```
#include <stdio.h>

#define N 10                  // size of the list.
#define FIRSTINDEX 0 // index of first element in the list.
#define ILLEGALINDEX -1       // illegal index – for special cases.
#define EINDEXOUTOFBOUND -1 // error code on overflow in the list.
#define SUCCESS 0             // success code.

typedef int type;              // type of each data item.

type clist[N];          // list implemented using array.
int front = ILLEGALINDEX;    // points to first element in the list.
int rear;                              // points to last element in the
list.

int lPush( type data ) {
    /*
     * appends 'data' to the end of the list if space is available.
     * otherwise returns error.
     */
    if( front == ILLEGALINDEX ) {   // list empty.
        front = rear = FIRSTINDEX;
    }
    else if( (rear+1)%N == front ) {       // list overflow.
        return EINDEXOUTOFBOUND;
    }
    else        // normal case.
        rear = (rear+1)%N;    // %N for wrapping around of index.
```

```
        clist[rear] = data;
        return SUCCESS;
    }

    void lPrint() {
        /*
         * prints elements in the list from front to rear.
         */
        int i;
        int nelem = lGetNElements();

        for( i=0; i<nelem; ++i )
            printf( "%d ", clist[ (front+i)%N ] );
        printf( "\n" );
    }

    int lGetNElements() {
        /*
         * returns no of elements in the list.
         */
        if( front == ILLEGALINDEX )      // empty list.
            return 0;
        else if( front <= rear ) // no wrapping of rear.
            return (rear-front+1);
        else                                            // wrapping of rear.
            return (N-front+rear+1);
    }

    int lDeleteK( int k ) {
        /*
         * deletes k'th element in the list if present.
         * otherwise returns error.
         * k starts from 1.
         * this procedure may be improved by checking for number of
elements
         * to be shifted after deleting k'th element. thus we can shift
either
         * k+1..N elements or 1..K-1 elements.
         */
        int index, i;
```

```
    int nelem = lGetNElements();
    printf( “deleting %d’th element...\n”, k );
    if( k > nelem || k < 1 )
        return EINDEXOUTOFBOUND;

    index = (front+k-1)%N;   // index of the element to be deleted.

    for( i=k+1; i<=nelem; ++i )
        clist[ (front+i-2)%N ] = clist[ (front+i-1)%N ];

    if( nelem == 1 )  // list is empty now.
        front = ILLEGALINDEX;
    else if( k == 1 )
        front = (front+1)%N;
    else
        rear = (rear-1+N)%N;

    return SUCCESS;
}

int lInsertAfterK( type data, int k ) {
    /*
     * inserts ‘data’ after k’th element in the list.
     * if list is full or k is out of bounds, error is returned.
     * k starts from 0.
     */
    int i, index;

    int nelem = lGetNElements();
    printf( “inserting %d after %d’th element...\n”, data, k );
    if( k > nelem || k < 0 || nelem == N )
        return EINDEXOUTOFBOUND;
    if( nelem == 0 )  // list empty.
        front = rear = FIRSTINDEX;
    else
        rear = (rear+1)%N;

    index = (front+k)%N;     // index at which data should be inserted.

    for( i=nelem; i>k; --i )
        clist[ (front+i)%N ] = clist[ (front+i-1)%N ];
```

```
    clist[ (front+k)%N ] = data;
}

int main() {
    lInsertAfterK(100,0);
    lPrint();
    lPush(0);
    lPush(4);
    lPush(7);
    lPush(1);
    lPush(13);
    lPush(2);
    lPush(5);
    lPrint();
    lInsertAfterK(2,1);
    lPrint();
    lDeleteK(4);
    lPrint();
    lPush(6);
    lPush(3);
    lPush(23);
    lPrint();
    lDeleteK(9);
    lPrint();
    lInsertAfterK(20,9);
    lPrint();
    lDeleteK(10);
    lPrint();
    lInsertAfterK(20,0);
    lPrint();

    return 0;
}
```

Explanation

1. `clist[0...N-1]` is a global queue of integers. The `rear` and `front` are its two pointers (indices) maintained for insertion and deletion of elements. The `rear` points to the last element inserted in the queue while `front` points to the next element to be removed from the queue.
2. An empty queue is represented by front = –1. The `front` and `rear` are updated on insertions and deletions. Since the queue is circular and the array size is

fixed, element e is inserted when rear = N – 1 goes to `clist[0]`, if it is empty. Also, after removal of an element from the queue when front = N – 1, front points to `clist[0]`, if it exists. Thus the indices `front` and `rear` wrap around in the range 0...N – 1.

3. The number of elements in the queue at any time (`lGetNElements()`) can be found using the following formula:

   Number of elements = 0 if front == –1.
   = rear – front + 1 if front <= rear.
   = N – front + rear + 1 otherwise.

4. To delete the k-th element (k > 0) (`lDeleteK()`), it is first checked to see if the number of elements in the queue is less than or equal to k. If there is a k-th element, its index in the array can be found using the formula (front+k–1)%N. After deletion of the element, all the elements after it are shifted back by one position and `rear` is updated. If k == 1, or if the deleted element was the last element in the queue, then `front` needs to be updated accordingly. For example, let the queue look like this:

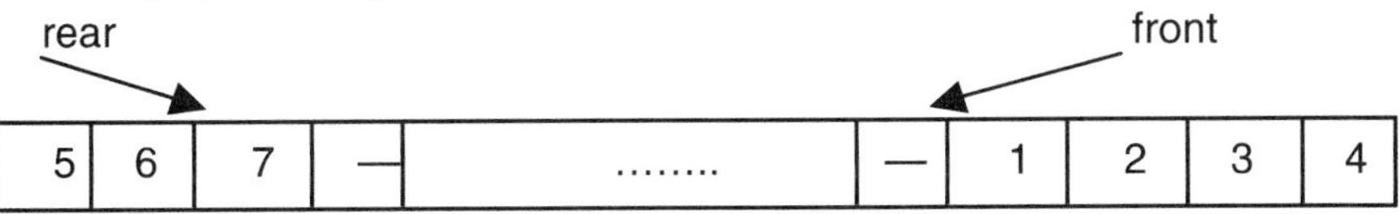

   If k == 5, then after `lDeleteK()` runs, the queue looks like this:

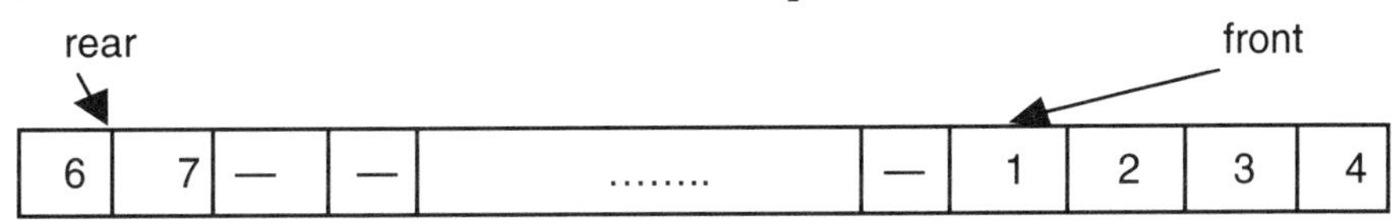

   Note that elements 6 and 7 have been shifted back by one position.

5. To insert an element after the k-th element (k >= 0) (`lInsertAfterK()`), again k is checked against the number of elements in the queue. The index for the new element would be `(front+k)%N`. To create space for the new element, all the elements after k-th element need to be shifted foward by one position. The new element can then be inserted and the indices `front` and `rear` are updated accordingly. For example, say the queue looks like the following:

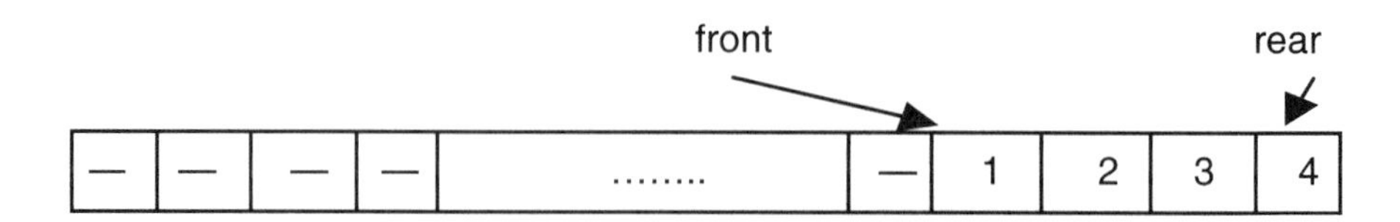

Then, after lInsertAfterK() is run with k == 4, the queue becomes

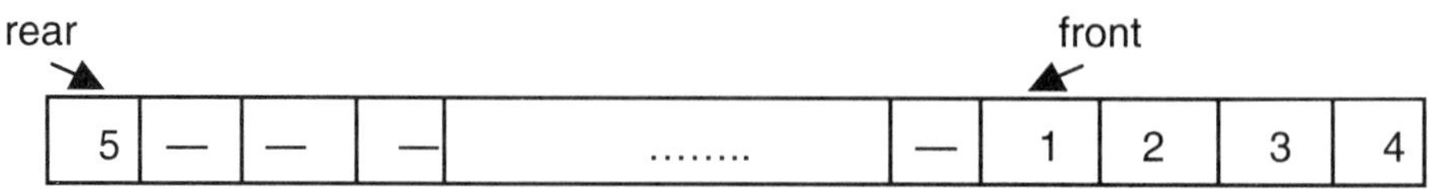

Note how the rear is wrapped around.

6. Because of the shifting required in insertion and deletion, the time complexity of both the functions is $O(n)$. We note that if insertions and deletions take place at rear and front, respectively, the complexity remains O(1). The constant of proportionality of the complexity can be improved by checking the number of elements before and after the k-th element and then shifting the smaller number of the two.

Points to Remember

1. In an array-based circular queue implementation, the indices `front` and `rear` wrap around the N elements.
2. The index of a k-th element is calculated as `(front+k-1)%N`.
3. The number of elements in the queue is calculated as 0, `(rear-front+1)`, or (N–`front`+`rear`+1) depending on whether the queue is empty (`front` == –1). There is no wrapping of indices (`front` <= `rear`) and `rear` is wrapped around (`front` > `rear`).
4. To insert an element after the k-th element, all the elements after it should be first shifted forward by one position. To delete an element at the k-th position, all the elements after it should be shifted back by one position.
5. Corner cases such as queue full and queue empty should be handled properly.

PROBLEM: REVERSING LINKS IN THE CASE OF CIRCULAR LIST

Write a function for a singly linked circular list that reverses the direction of the links.

Program

```
#include <stdio.h>

#define SUCCESS 0
#define ERROR   -1

typedef int type;
typedef struct node node;

struct node {
    type data;
    node *next;
};

node *head = NULL;

int lInsert( type data ) {
    /*
     * inserts a new node containing data at start of the list.
     */
    node *ptr = (node *)malloc( sizeof(node) );
    ptr->data = data;
    if( head == NULL ) {     // this is the first element in the list.
        ptr->next = ptr;
        head  = ptr;
    }
    else {
        ptr->next = head->next;
        head->next = ptr;
    }
    return SUCCESS;
}

void lPrint() {
    node *ptr;

    if( head == NULL )
        return;
    else
        printf( "%d ", head->data );
```

```
    for( ptr=head->next; ptr!=head; ptr=ptr->next )
        printf( "%d ", ptr->data );
    printf( "\n" );
}

int lReverse() {
    /*
     * insitu reverses the list.
     */
    node *curr, *prev, *next;

    printf( "reversing list...\n" );
    if( head == NULL )
        return SUCCESS;
    for( prev=head, curr=prev->next, next=curr->next; curr!=head;
prev=curr, curr=next, next=next->next ) {
        curr->next = prev;
    }
    head->next = prev;

    return SUCCESS;
}

int main() {
    lPrint();
    lInsert(1);
    lPrint();
    lInsert(2);
    lInsert(3);
    lInsert(4);
    lInsert(5);
    lInsert(6);
    lPrint();
    lReverse();
    lPrint();
    return 0;
}
```

Explanation

1. `main()` creates a linked list of integers and then calls `lReverse()` to reverse the list. Note that the list is circular.

2. The function `lReverse()` maintains three pointers, `prev`, `curr`, and `next`, while traversing the list. They point to consecutive nodes in the list. If the list is non-empty, then `prev` points to the head node while `curr` and `next` are accordingly assigned to the next elements in the list. The circular list is traversed until `curr` = `head` of the list. At every step, `curr->next` points to `prev`. At the end of the loop, all the nodes in the list except the head node point to their original previous nodes. Thus at the end of the loop, `head->next` is assigned `prev` which is the last node in the list.
3. **Example:**

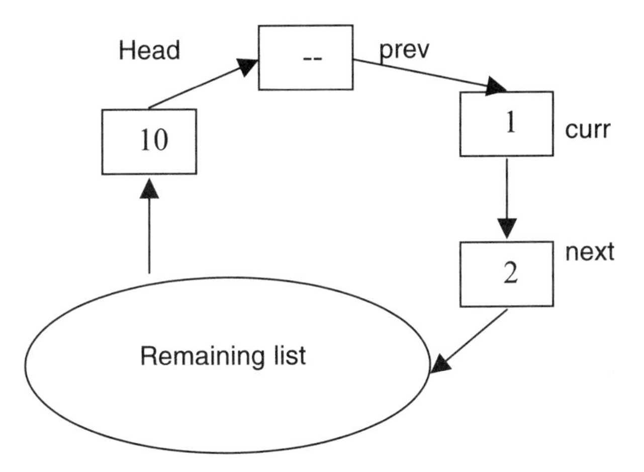

After reversal this list appears as shown here:

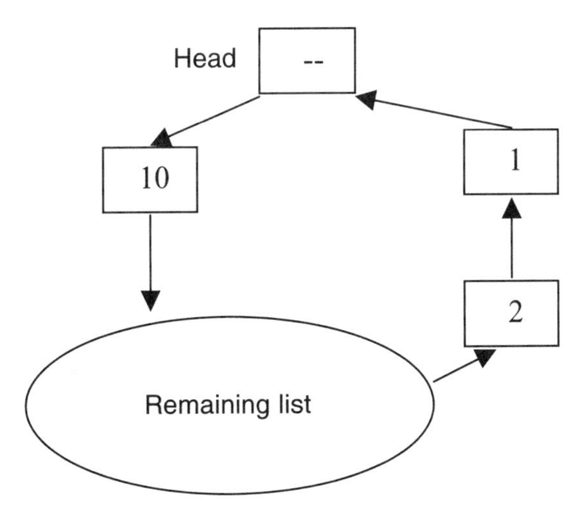

Points to Remember

1. Note that the loop advances curr as curr=next rather than curr=curr->next, because curr->next gets changed in the loop.
2. The complexity of the reversal procedure is $O(n)$.
3. If the head is maintained as a fixed node, that is, an empty list is denoted by a single node rather than NULL, then the functions operating on the list get simplified.

PROBLEM: MEMORY MANAGEMENT USING LISTS

Design a storage management scheme where all requests for memory are of the same size, say K. Write functions to free and allocate storage in this scheme.

Program

```
#include <stdio.h>

#define N 90
#define K 10

#define SUCCESS 0
#define ERROR   -1

typedef struct node node;

struct node {
   void *ptr; // points to free block of size K.
   node *next;
};

struct head {
   //int nnodes;
   node *next;
   char *bytes;              // mem will be allocated from this pool.
}freelist;

void init() {
```

```
    /*
     * initialize the memory space.
     */
    int i;
    void memfree( void * );

    //freelist.nnodes = 0;
    freelist.next   = NULL;
    freelist.bytes  = (char *)malloc(N);

    for( i=N/K-1; i>=0; –i ) {
        memfree( freelist.bytes+K*i );
    }
}

void *memalloc() {           // assume request to be of size K.
    /*
     * returns a void * pointer to area from freelist.
     */
    void *ptr;
    node *nodeptr;

    if( freelist.next == NULL )
        return (void *)NULL;
    nodeptr = freelist.next;
    ptr = nodeptr->ptr;
    freelist.next = freelist.next->next;
    free(nodeptr);           // this is standard free().

    return ptr;
}

void memfree( void *ptr ) {
    /*
     * adds ptr to freelist.
     */
    node *nodeptr;

    if( ptr == NULL )
        return;
    nodeptr      = (node *)malloc( sizeof(node) );
```

```
        nodeptr->ptr  = ptr;
        nodeptr->next = freelist.next;
        freelist.next = nodeptr;
}

void print() {
        node *ptr;

        for( ptr=freelist.next; ptr!=NULL; ptr=ptr->next ) {
            printf( "%u ", ptr->ptr );
        }
        printf( "\n\n" );
}

int main() {
        void *p1, *p2, *p3, *p4;
        init();
        printf( "after init...\n" );
        print();
        p1 = memalloc();
        printf( "after memalloc(p1)...\n" );
        p2 = memalloc(); p3 = memalloc();
        print();
        memfree(p1); memfree(p2); memfree(p3);
        printf( "after memfree(p1)...\n" );
        print();

        return 0;
}
```

Explanation

1. Since all requests are of the same size, theoretically only one bit is required with each block of size K to tag it as free or allocated. However, we do not know the number of blocks (the size of the memory pool). So we need a list of such status bits. An array-based list would have been sufficient but the size of a memory pool can be very large. So we prefer a linked list of status bits.
2. The functions on the memory pool are `memfree()` and `memalloc()`, which need pointers to the memory areas to be exchanged between the memory manager and the user program. So instead of bits, we store only the pointers to the memory areas. We further shorten this list by storing only pointers to

the free blocks in the lists. Thus we maintain a free-list of free pointers.

3. We take advantage of all the requests of the same size to make memfree() and memalloc() functions O(1). memalloc() returns the first free pointer in the list and memfree() adds the free pointer to the head of the free-list. The only loop in the program is required only once, for initializing the free-list at the start of the program. A global memory pool of size N acts as the free pool. It is divided into blocks of size K and pointers to the start of each block are inserted into the free-list.

Points to Remember

1. If all the requests are of the same size, then allocation and free operations can be done in O(1) time.
2. Instead of a separate free-list, each free block of size K can contain a pointer to the next free block of memory.

PROBLEM: MEMORY MANAGEMENT USING VARIOUS SCHEMES

Implement a memory allocation scheme by using the algorithms *first-fit*, *next-fit*, and *best-fit*.

Program

```
#include <stdio.h>

#define N 100

typedef struct node node;
typedef enum {FALSE, TRUE} bool;

struct node {
    char *ptr; // start addr.
    int size;         // size of the free block.
    node *next;       // next free block.
};

char mem[N];          // total memory pool.
node freelist;        /*
               * the freelist should be sorted on start addr.
               * this will ease coalescing adjacent blocks.
```

```
                */
void init() {
    /*
     * init freelist to contain the whole mem.
     */
    node *ptr = (node *)malloc( sizeof(node) );
    ptr->ptr  = mem;
    ptr->size  = N;
    ptr->next = NULL;
    freelist.next = ptr;
}

void removenode( node *ptr, node *prev ) {
    /*
     * remove a node ptr from the list whose previous node is prev.
     */
    prev->next = ptr->next;
    free(ptr);
}

char *firstfit( int size ) {
    /*
     * returns ptr to free pool of size size from freelist.
     */
    node *ptr, *prev;
    char *memptr;

    for( prev=&freelist, ptr=prev->next; ptr; prev=ptr, ptr=ptr->next )
        if( ptr->size > size ) {
                memptr = ptr->ptr;
                ptr->size -= size;
                ptr->ptr += size;
                return memptr;
        }
        else if( ptr->size == size ) {
                memptr = ptr->ptr;
                removenode( ptr, prev );
                return memptr;
        }
    return NULL;
}
```

```
char *nextfit( int size ) {
    /*
     * returns ptr to free pool of size size from freelist.
     * the free pool is second allocatable block instead of first.
     * if no second block then first is returned.
     */
    bool isSecond = FALSE;
    node *prev, *ptr;
    node *firstprev, *firstptr;

    for( prev=&freelist, ptr=prev->next; ptr; prev=ptr, ptr=ptr->next )
        if( ptr->size >= size && isSecond == FALSE ) {
                isSecond  = TRUE;
                firstprev = prev;
                firstptr  = ptr;
        }
        else if( ptr->size > size && isSecond == TRUE ) {
                char *memptr = ptr->ptr;
                ptr->size -= size;
                ptr->ptr += size;
                return memptr;
        }
        else if( ptr->size == size && isSecond == TRUE ) {
                char *memptr = ptr->ptr;
                removenode( ptr, prev );
                return memptr;
        }
    // ptr is NULL.
    ptr = firstptr;
    prev = firstprev;

    if( ptr->size > size && isSecond == TRUE ) {
        char *memptr = ptr->ptr;
        ptr->size -= size;
        ptr->ptr += size;
        return memptr;
    }
    else if( ptr->size == size && isSecond == TRUE ) {
        char *memptr = ptr->ptr;
        removenode( ptr, prev );
```

```
            return memptr;
        }
        else        // isSecond == FALSE
            return NULL;
    }

    char *bestfit( int size ) {
        /*
         * returns ptr to free pool of size size from freelist.
         * the allocated block's original size - size is min in the
freelist.
         */
        node *ptr, *prev;
        char *memptr;
        int minwaste = N+1;
        node *minptr = NULL, *minprev;

        for( prev=&freelist, ptr=prev->next; ptr; prev=ptr, ptr=ptr->next )
            if( ptr->size >= size && ptr->size-size < minwaste ) {
                    minwaste = ptr->size-size;
                    minptr = ptr;
                    minprev = prev;
            }
        if( minptr == NULL )     // could NOT get any allocatable mem.
            return NULL;
        ptr = minptr;
        prev = minprev;
        if( ptr->size > size ) {
                    memptr = ptr->ptr;
                    ptr->size -= size;
                    ptr->ptr += size;
                    return memptr;
        }
        else if( ptr->size == size ) {
                    memptr = ptr->ptr;
                    removenode( ptr, prev );
                    return memptr;
        }
        return NULL;
    }

    void addtofreelist( char *memptr, int size ) {
```

```
        /*
         * add memptr of size to freelist.
         * remember that block ptrs are sorted on mem addr.
         */
        node *prev, *ptr, *newptr;

        for( prev=&freelist, ptr=prev->next; ptr && ptr->ptr<memptr;
prev=ptr, ptr=ptr->next )
            ;
        // memptr is to be added between prev and ptr.
        newptr = (node *)malloc( sizeof(node) );
        newptr->ptr = memptr;
        newptr->size = size;
        newptr->next = ptr;
        prev->next = newptr;
    }

    void coalesce() {
        /*
         * combine adj blocks of list if necessary.
         */
        node *prev, *ptr;

        for( prev=&freelist, ptr=prev->next; ptr; prev=ptr, ptr=ptr->next )
            // check for prev mem addr and size against ptr->ptr.
            if( prev != &freelist && prev->ptr+prev->size == ptr->ptr ) {//
    prev->size += ptr->size;            //
                prev->next  = ptr->next;
                free(ptr);
                ptr = prev;             // ).
            }
    }

    char *memalloc( int size ) {
        /*
         * return ptr to pool of mem of the size.
         * return NULL if NOT available.
         * ptr-sizeof(int) contains size of the pool allocated, like
malloc.
         */
        char *ptr = bestfit( size+sizeof(int) );      // change this to
```

```
firstfit() or nextfit().
        printf( "allocating %d using bestfit...\n", size );
        if( ptr == NULL )
            return NULL;
        *(int *)ptr = size;

        return ptr+sizeof(int);
    }

    void memfree( char *ptr ) {
        /*
         * adds ptr to freelist and combine adj blocks if necessary.
         * size of the mem being freed is at ptr-sizeof(int).
         */
        int size = *(int *)(ptr-sizeof(int));
        printf( "freeing %d...\n", size );
        addtofreelist( ptr-sizeof(int), size+sizeof(int) );
        coalesce();         // combine adjacent blocks.
    }

    void printfreelist() {
        node *ptr;
        printf( "\t" );
        for( ptr=freelist.next; ptr; ptr=ptr->next )
            printf( "{%u %d} ", ptr->ptr, ptr->size );
        printf( "\n" );
    }

    int main() {
        char *p1, *p2, *p3, *p4, *p5;
        init();
        printfreelist();

        p1 = memalloc(10);
        printfreelist();
        p2 = memalloc(15);
        printfreelist();
        p3 = memalloc(23);
        printfreelist();
        p4 = memalloc(3);
        printfreelist();
        p5 = memalloc(8);
```

```
    printfreelist();
    memfree(p1);
    printfreelist();
    memfree(p5);
    printfreelist();
    memfree(p3);
    printfreelist();
    p1 = memalloc(23);
    printfreelist();
    p1 = memalloc(23);
    printfreelist();
    memfree(p2);
    printfreelist();
    p1 = memalloc(3);
    printfreelist();
    memfree(p4);
    printfreelist();
    p2 = memalloc(1);
    printfreelist();
    memfree(p1);
    printfreelist();
    memfree(p2);
    printfreelist();

    return 0;
}
```

Explanation

1. A memory manager provides a pool of memory when requested (`memalloc()`) and frees a pool of memory (`memfree()`) to be used for the next allocation request. It maintains a free-list of pointers to memory blocks along with their sizes. Whenever there is a request for a free pool of memory having the size `size`, this free-list is searched for the appropriate block depending on the algorithm. If such a block is found, it is removed from the free-list and a pointer to it is returned. Whenever a pool of memory is freed using `memfree(p)`, the pointer `p` is added to the free-list. If possible, the adjacent blocks are combined using `coalesce()` to get a bigger free pool.
2. In a first-fit algorithm, the first free pool of memory is granted if it has sufficient size to satisfy the request. Thus, if sizes of free pools in the free-list are {10 9 20 34 43 12 22}, and the request is for size 21, then the pointer to the pool pointed to by the node having size 34 is returned.

3. In a next-fit algorithm, the second free pool of memory is granted if it has sufficient size to satisfy the request. If no such second free pool is available, then the first such free pool is granted. Thus, if the free-list is as shown earlier, then a request for size 21 is fulfilled using the block having size 43.
4. In a best-fit algorithm, that free pool of memory is granted which retains the minimum amount of space after allocation. Thus, if the free-list is as shown earlier, then a request for size 21 is fulfilled by the block having size 22, as its residual memory is 22 – 21 = 1.
5. Note that the free-list is sorted on the address each node saves and not on size. Sorting on size can help improve the performance of `memalloc()` by a constant factor. The advantage of sorting on addresses is realized during coalescing of adjacent blocks. This happens when two adjacent blocks that are free are stored in the free-list in adjacent nodes. This may keep a request unsatisfied, even if a free pool of the requested size existed. So, we check for the start address of a block added to its size with the start address of the next block. If they match, we combine the two nodes. This procedure is followed until this condition is violated.
6. The complexity of each of the three algorithms is $O(n)$ where n is the number of nodes in the free-list. It has been seen using experiments that first- and best-fit have nearly similar performances and somewhat better than the next-fit algorithm. The complexity of `memalloc()` is the same as the algorithm it implements. The complexity of `memfree()` is $O(n)$, as it needs to insert the free pointer in a sorted list.

Points to Remember

1. In first-fit, the first free pool satisfying the request is returned. In next-fit, the second such free pool is returned. In best-fit, the free pool which leaves a free pool of minimum size after allocation is returned.
2. If the application requires two pointers to point to the same area, then the problem of a dangling pointer may arise in the system, where a pointer may point to an area that is not allocated. This happens when another pointer pointing to the same memory area frees the pool.
3. Garbage collection is the process of collecting the memory that was previously allocated but is no longer being used by the application. Garbage is generated as a result of bad programming practice, wherein we allocate memory as required but do not free it after its use. Algorithms using reference counts and the mark-and-sweep algorithm are generally used for garbage collection.

PROBLEM: GARBAGE COLLECTION—THE FIRST METHOD

Implement a mark() procedure used in garbage collection to mark all the nodes traversible from a head node by using a stack.

Program

```
/************************ mark1.s.c *************************/
#include <stdio.h>

typedef struct snode snode;
typedef list stype;
typedef struct snode *stack;

struct snode {
    stype op;
    snode *next;
};

bool sEmpty( stack *s ) {
    return (*s == NULL);
}

void sPush( stack *s, stype op ) {
    /*
     * pushes op in stack s.
     */
    snode *ptr = (snode *)malloc( sizeof(snode) );
    ptr->op = op;
    ptr->next = *s;
    *s = ptr;
}

stype sTop( stack *s ) {
    /*
     * returns top op from stack s without popping.
     */
    if( sEmpty(s) )
       return NULL;
    return (*s)->op;
}
```

```
stype sPop( stack *s ) {
    /*
     * pops op from top of stack s.
     */
    snode *ptr = *s;
    stype op;

    if( sEmpty(s) )
       return NULL;
    *s = (*s)->next;
    op = ptr->op;
    free(ptr);

    return op;
}

/************************ mark1.c ***************************/
#include <stdio.h>

typedef struct node node;
typedef int type;
typedef enum {FALSE, TRUE} bool;
typedef node *list;

#include "mark1.s.c"

struct node {
    bool mark;
    type val;
    node *horiz;
    node *vert;
};

list newNode() {
    /*
     * return a new node.
     */
    return (list)calloc(1, sizeof(node));
}

list createList() {
    /*
     * return a dummy list created.
```

```
 */
list ptr;
snode *s = NULL;
list sixptr;
list nineptr;

ptr = newNode();
sPush(&s, ptr);
ptr->val = 1;
ptr->horiz = newNode();
ptr = ptr->horiz;
sPush(&s, ptr);
ptr->val = 2;
ptr->vert = newNode();
ptr = ptr->vert;
sPush(&s, ptr);
ptr->val = 3;
ptr->vert = newNode();
ptr = ptr->vert;
ptr->val = 4;
ptr = sPop(&s);
ptr->horiz = newNode();
ptr = ptr->horiz;
ptr->val = 5;
ptr->horiz = newNode();
ptr = ptr->horiz;
sPush(&s, ptr);
sixptr = ptr;
ptr->val = 6;
ptr->vert = newNode();
ptr = ptr->vert;
ptr->val = 7;
ptr = sPop(&s);
ptr->horiz = newNode();
ptr = ptr->horiz;
ptr->val = 8;
ptr = sPop(&s);
ptr->horiz = newNode();
ptr = ptr->horiz;
sPush(&s, ptr);
nineptr = ptr;
ptr->val = 9;
ptr->vert = newNode();
```

```
    ptr = ptr->vert;
    ptr->val = 10;
    ptr->horiz = newNode();
    ptr = ptr->horiz;
    ptr->val = 11;
    ptr = sPop(&s);
    ptr->horiz = newNode();
    ptr = ptr->horiz;
    ptr->vert = nineptr;     // an internal link.
    ptr->val = 12;
    ptr->horiz = newNode();
    ptr = ptr->horiz;
    sPush(&s, ptr);
    ptr->val = 13;
    ptr->vert = newNode();
    ptr = ptr->vert;
    ptr->val = 14;
    ptr->horiz = sixptr;     // an internal link.
    ptr = sPop(&s);

    return sPop(&s);
}

void markList(list ptr) {
    /*
     * print the horiz and vert lists iteratively using a stack.
     */
    snode *s = NULL;
    list horiz;

    if(!ptr || ptr->mark)
        return;

    ptr->mark = TRUE;
    printf("marked=%d.\n", ptr->val);
    sPush(&s, ptr);

    while(!sEmpty(&s)) {
        ptr = sPop(&s);
        do {
                    horiz = ptr->horiz;
                    if(horiz && horiz->mark == FALSE) {
```

```
                    horiz->mark = TRUE;
                    printf("marked=%d.\n", horiz->val);
                    sPush(&s, horiz);
                }
                ptr = ptr->vert;
                if(!ptr || ptr->mark)
                    break;
                ptr->mark = TRUE;
                printf("marked=%d.\n", ptr->val);
        } while(TRUE);
    }
}

void markListRec(list ptr) {
    /*
     * mark the list pointed to by ptr recursively.
     */
    for(; ptr && !ptr->mark; ptr=ptr->horiz) {
        ptr->mark = TRUE;
        printf("marked=%d.\n", ptr->val);
        markListRec(ptr->vert);
    }
}

int main() {
    list head = createList();
    markListRec(head);

    return 0;
}
```

Explanation

1. An algorithm called *mark-and-sweep* is used for garbage collection. In this, from every variable, memory is traversed and marked as used. Then a traversal over the whole memory is done to add all the unmarked nodes to the free-list, thus collecting the garbage. The former is called `mark` while the latter is called `sweep`.
2. We implement the marking procedure here. We assume the node structure as follows:

```
struct node {
    bool mark;
    type val;
    node *horiz;
    node *vert;
};
```

Thus the lists are generalized and can grow in both horizontal and vertical directions.

3. The recursive marking algorithm is as shown here:

```
markListRec(ptr) {
    for(; ptr && !ptr->mark; ptr=ptr->horiz) {
        ptr->mark = TRUE;
        markListRec(ptr->vert);
    }
}
```

   This procedure uses the system stack for recursion (internally).

4. The iterative procedure `markList(ptr)` uses an explicit stack to store the nodes. It contains a nested loop in which the list is traversed horizontally and vertically. The next node on the horizontal list is marked and pushed on stack and the whole vertical list is then traversed. The same strategy is used for each node in the vertical list. Whenever a node is visited, it is checked to see if it was marked. If yes, then it is not processed; otherwise, it is marked.

5. **Example:** The function `createlist()` creates an arbitrary list structure as shown here.

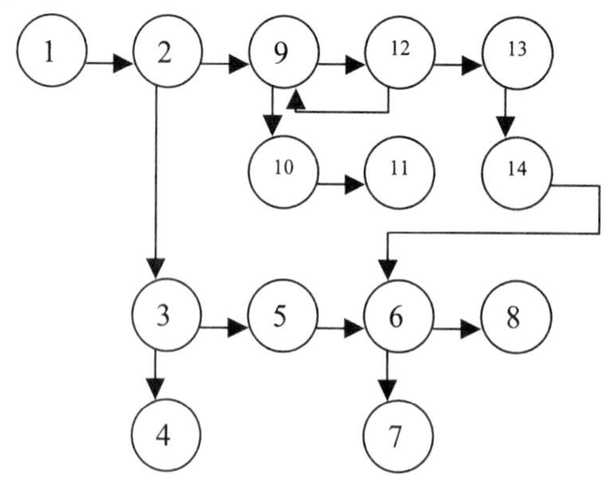

   Different steps of the algorithm:

STEP	PTR	PTR->MARK	HORIZ	HORIZ->MARK	STACK	OUTPUT
0	1	F	—	—	1	1
1	1	T	2	F	2	1 2
2	2	T	9	F	9	1 2 9
3	3	F	9	T	9	1 2 9 3
4	3	T	5	F	9 5	1 2 9 3 5
5	4	F	5	T	9 5	1 2 9 3 5 4
6	5	T	6	F	9 6	1 2 9 3 5 4 6
7	6	T	8	F	9 8	1 2 9 3 5 4 6 8
8	7	F	8	T	9 8	1 2 9 3 5 4 6 8 7
9	8	T	nil	—	9	1 2 9 3 5 4 6 8 7
10	9	T	12	F	12	1 2 9 3 5 4 6 8 7 12
11	10	F	12	F	12	1 2 9 3 5 4 6 8 7 12 10
12	10	T	11	F	12 11	1 2 9 3 5 4 6 8 7 12 10 11
13	11	T	11	T	12	1 2 9 3 5 4 6 8 7 12 10 11
14	12	T	13	F	13	1 2 9 3 5 4 6 8 7 12 10 11 13
15	9	T	13	T	13	1 2 9 3 5 4 6 8 7 12 10 11 13
16	13	T	nil	—	empty	1 2 9 3 5 4 6 8 7 12 10 11 13
17	14	F	6	T	empty	1 2 9 3 5 4 6 8 7 12 10 11 13 14
18	14	T	6	T	empty	1 2 9 3 5 4 6 8 7 12 10 11 13 14

Thus in each step, the horizontal node is pushed as the return point and the whole vertical list is traversed. The return point is then popped and the procedure is continued until the stack becomes empty.

In step 9, since 8 is already marked, nothing is output. In step 14, `ptr` points to 12, which is marked, and `horiz` points to 13, which is unmarked. Thus it marks 13 and pushes it on stack. In the next step, the vertical list is traversed and `ptr` points to 9 while `horiz` is still 13. Since 9 is already marked, it is left and 13 is popped. `horiz` is now nil. Now the vertical list of 13 is traversed and `ptr` points to 14, which is unmarked, and so is marked. `horiz` now points to 6, which is marked, so nothing is done to it. A vertical pointer of 14 is traversed and it is found to be nil, so the inner loop quits. The stack is empty at this point, so the outer loop also quits and the procedure terminates.

Points to Remember

1. A simple recursive procedure such as `markListRec()` can be more readable and compact than its iterative version `markList()`.
2. The complexity of the marking algorithm is $O(e)$, where e is equal to the number of links in the graph.
3. To mark all the memory traversible from program variables, `markList()` should be called for each of the variables.
4. There is a serious caveat in using this algorithm for garbage collection. Usually, garbage is created when the user runs out of memory. This algorithm, however, requires a stack to work, so it will not proceed when there is not enough memory. So an algorithm that does not require more than a constant amount of extra memory should be used for garbage collection.

PROBLEM: GARBAGE COLLECTION— THE SECOND METHOD

Implement the marking procedure used in garbage collection without using more than a constant amount of memory.

Program

```
/************************ mark2.s.c *************************/
#include <stdio.h>

typedef struct snode snode;
typedef list stype;
typedef struct snode *stack;

struct snode {
    stype op;
    snode *next;
};
bool sEmpty( stack *s ) {
    return (*s == NULL);
}

void sPush( stack *s, stype op ) {
    /*
```

```
     * pushes op in stack s.
     */
    snode *ptr = (snode *)malloc( sizeof(snode) );
    ptr->op = op;
    ptr->next = *s;
    *s = ptr;
}

stype sTop( stack *s ) {
    /*
     * returns top op from stack s without popping.
     */
    if( sEmpty(s) )
        return NULL;
    return (*s)->op;
}

stype sPop( stack *s ) {
    /*
     * pops op from top of stack s.
     */
    snode *ptr = *s;
    stype op;

    if( sEmpty(s) )
        return NULL;
    *s = (*s)->next;
    op = ptr->op;
    free(ptr);

    return op;
}

/************************ mark2.c *************************/
#include <stdio.h>

typedef struct node node;
typedef int type;
typedef enum {FALSE, TRUE} bool;
typedef node *list;

struct node {
    bool mark;
```

```
    bool tag;
    type val;
    node *horiz;
    node *vert;
};

#include "mark2.s.c"

list newNode() {
    /*
     * return a new node.
     */
    list ptr = (list)calloc(1, sizeof(node));
    return ptr;
}

list createList() {
    /*
     * return a dummy list created.
     */
    list ptr;
    snode *s = NULL;
    list sixptr;
    list nineptr;

    ptr = newNode();
    sPush(&s, ptr);
    ptr->val = 1;
    ptr->horiz = newNode();
    ptr = ptr->horiz;
    sPush(&s, ptr);
    ptr->val = 2;
    ptr->vert = newNode();
    ptr = ptr->vert;
    sPush(&s, ptr);
    ptr->val = 3;
    ptr->vert = newNode();
    ptr = ptr->vert;
    ptr->val = 4;
    ptr = sPop(&s);
    ptr->horiz = newNode();
    ptr = ptr->horiz;
    ptr->val = 5;
```

```
    ptr->horiz = newNode();
    ptr = ptr->horiz;
    sPush(&s, ptr);
    sixptr = ptr;
    ptr->val = 6;
    ptr->vert = newNode();
    ptr = ptr->vert;
    ptr->val = 7;
    ptr = sPop(&s);
    ptr->horiz = newNode();
    ptr = ptr->horiz;
    ptr->val = 8;
    ptr = sPop(&s);
    ptr->horiz = newNode();
    ptr = ptr->horiz;
    sPush(&s, ptr);
    nineptr = ptr;
    ptr->val = 9;
    ptr->vert = newNode();
    ptr = ptr->vert;
    ptr->val = 10;
    ptr->horiz = newNode();
    ptr = ptr->horiz;
    ptr->val = 11;
    ptr = sPop(&s);
    ptr->horiz = newNode();
    ptr = ptr->horiz;
    ptr->vert = nineptr;     // an internal link.
    ptr->val = 12;
    ptr->horiz = newNode();
    ptr = ptr->horiz;
    sPush(&s, ptr);
    ptr->val = 13;
    ptr->vert = newNode();
    ptr = ptr->vert;
    ptr->val = 14;
    ptr->horiz = sixptr;     // an internal link.
    ptr = sPop(&s);

    return sPop(&s);
}
```

```
void markList(list ptr) {
    /*
     * print the horiz and vert lists iteratively without using a
stack.
     */
    list p, q, t;

    p = ptr; t = NULL;
    do {
        printf("p->val=%d.\n", p->val);
        q = p->vert;
        if(q)
                if(q->mark == FALSE && q->tag == FALSE) {
                        q->mark = TRUE; p->tag = TRUE;
                        p->vert = t; t = p;
                        p = q;
                }
                else {
                        q->mark = TRUE;
label:
                        q = p->horiz;
                        if(q)
                                if(q->mark == FALSE && q->tag == FALSE) {
                                        q->mark = TRUE; p->horiz = t;
                                        t = p; p = q;
                                }
                                else {
                                        q->mark = TRUE;
label2:
                                        while(t) {
                                                q = t;
                                                if(q->tag) {
                                        t = q->vert; q->vert = p;
                                        q->tag = FALSE; p = q;
                                                goto label;
                                                }
                                        t = q->horiz; q->horiz = p;
                                                p = q;
                                        }
                                }
                        else
```

```
                        goto label2;
            }
        else
            goto label;
    } while(t);
}
int main() {
    list head = createList();
    markList(head);

    return 0;
}
```

Explanation

1. The node structure is assumed to be as follows.

```
struct node {
        bool mark;
        bool tag;
        type val;
        node *horiz;
        node *vert;
};
```

   Thus, using horiz and vert, generalized lists can be created containing data in val. mark is used for marking the nodes as visited or not. An additional Boolean tag is needed since we are not using an extra amount of memory in the algorithm, except for a few pointers. The mark and tag fields are initially FALSE for each node.

   The explicit stack is used for creating the list (createList()), not for marking the nodes.

2. This algorithm modifies some of the links in the list. However, by the time it finishes its task, the list structure is restored. Starting from ptr, markList() traces all possible paths made up of horiz and vert. Whenever a choice is to be made, the vert direction is explored first. Instead of maintaining a stack of return points, we now maintain the path taken from ptr to the node p that is currently being examined. This path is maintained by changing some of the pointers along the path from ptr to p.

3. **Example:** Consider this example list. We omit the val field as it is not important here.

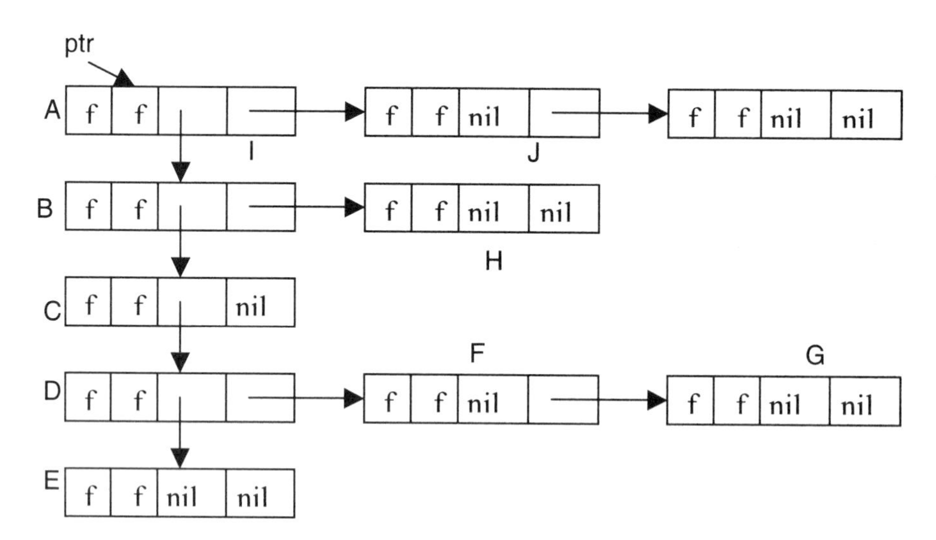

Initial list.

Initially, all nodes except node A are unmarked. From A we can move down to B or right to I. This algorithm always moves down when faced with such an alternative. We use p to point to the node currently being examined and t to point to the node preceding p in the path from ptr to p. The path t to ptr will be examined as a chain composed of the nodes on this t - ptr (read as "t to ptr") path. If we advance from node p to node q, then either q=p->horiz or q=p->vert, and q will become the node currently being examined. The node preceding q on the ptr-q path is p, and so the path list must be updated to represent the path from p to ptr. This is done simply by adding the node p to the t-ptr path that has been already constructed. Nodes will be linked onto this path through either their vert or horiz field. When node p is added to the path chain, p is linked to t via its vert field, if q=p->vert. When q=p->horiz, p is linked to t via its horiz field. In order to be able to determine whether a node on the t-ptr path list is linked through either the vert or horiz field, we make use of the tag field. When vert is used for linking, this tag will be TRUE. Thus, for nodes on t-ptr path we have:

tag = FALSE if the node is linked via horiz field.
= TRUE if the node is linked via vert field.

The tag will be reset to FALSE when the node gets off the t-ptr path list. Different snapshots of the list are shown here:

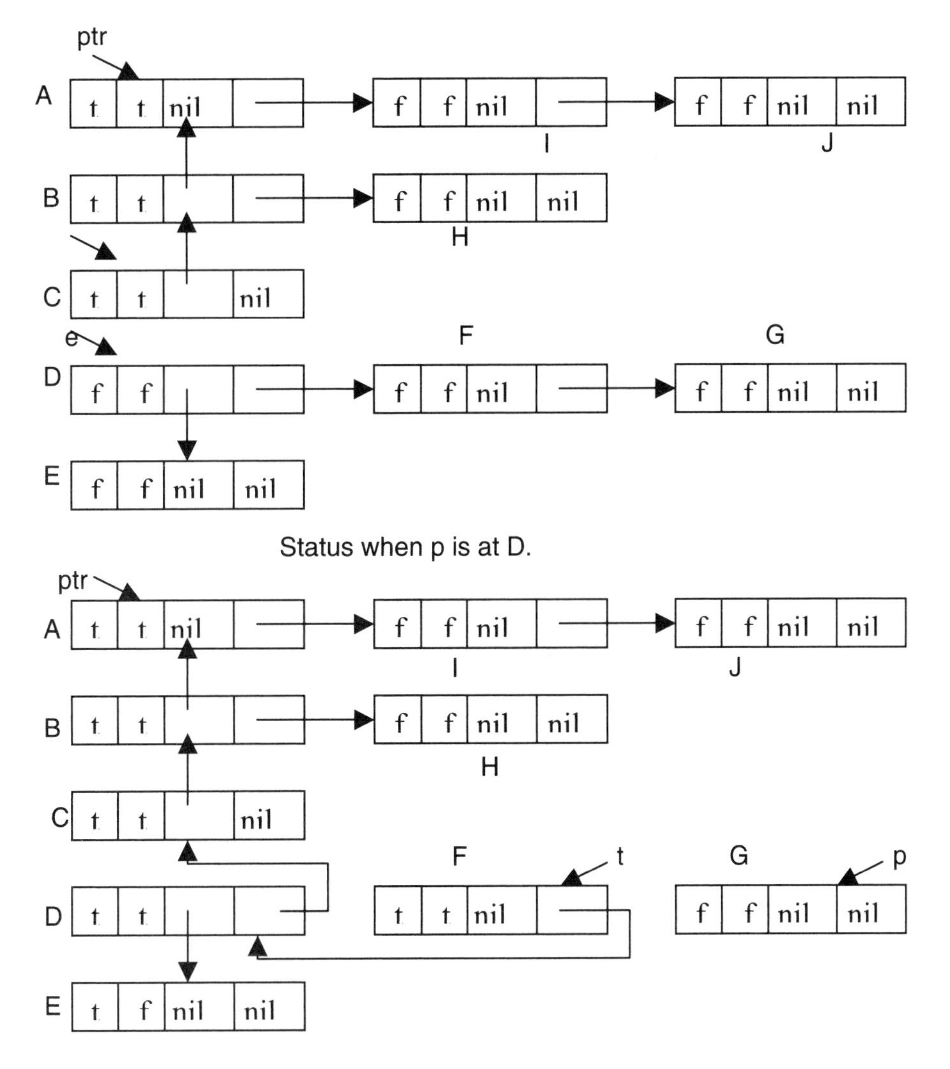

Status when p is at D.

Status when p is at G.

At the end of the algorithm, all the nodes are marked and all tag fields are restored to FALSE. Thus the list looks like this.

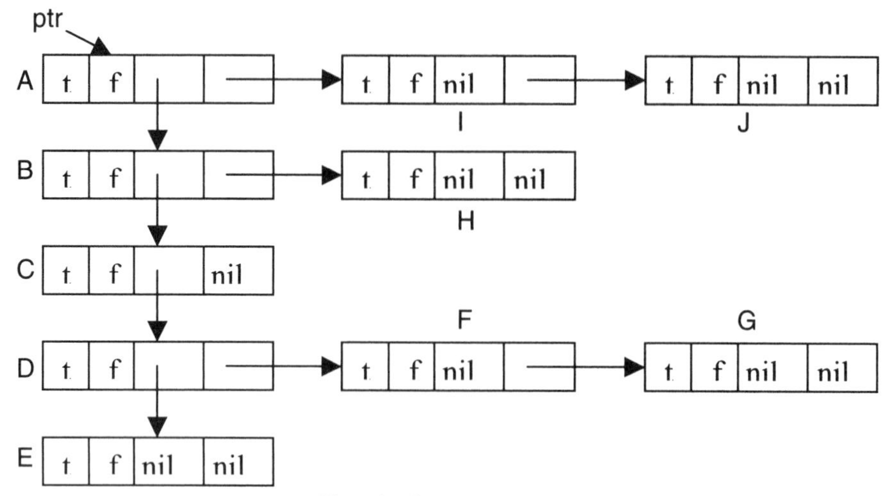

Terminal status.

Points to Remember

1. We require an extra tag field if the marking is to be done without using an extra amount of memory.
2. The computing time of markList() is O(m) where m is the number of newly marked nodes.
3. By properly maintaining the predicate for t-ptr list, the code can be easily understood.
4. It is said that gotos should be avoided as much as possible. However, judicious use of gotos at proper places can make an otherwise tiring code easier to understand.

PROBLEM: COMPUTE N EQUIVALENCE CLASSES

Write a program to produce N equivalent classes as linked lists from a linked list of integers, applying the mod function.

Program

```
#include <stdio.h>
typedef struct node node;
```

```
struct node {
    int val;
    node *next;
};

node *getList(int a[], int n) {
    /*
     * form a list of n integers in a[].
     */
    int i;
    node *list = NULL;

    for( i=0; i<n; ++i ) {
        node *ptr = (node *)malloc( sizeof(node) );
        ptr->val  = a[i];
        ptr->next = list;
        list      = ptr;
    }
    return list;
}

node **applyMod(node *list, int n) {
    /*
     * apply (mod n) on every element of list and store that element in
     * the list modlists[(mod n)].
     * thus we form array[n] of lists.
     * each list is one equivalence class.
     */
    node **modlists;
    node *ptr, *next;

    modlists = (node **)calloc( n, sizeof(node *) );
    for(ptr=list, next=(ptr?ptr->next:NULL); ptr; ptr=next,
next=(next?next->next:NULL))
        ptr->next = modlists[ptr->val%n], modlists[ptr->val%n] = ptr;
    return modlists;
}

void printModlists(node **modlists, int n) {
    /*
     * prints the equivalence classes.
```

```
    */
    node *ptr;
    int i;

    for(i=0; i<n; ++i) {
        printf("%d: ", i);
        for(ptr=modlists[i]; ptr; ptr=ptr->next)
                printf("%d ", ptr->val);
        printf("\n");
    }
    printf("\n");
}

main() {
    int a[] = {10,2,3,44,432,35,6576,34,12,5456,23423,234,23};
    node *list = getList(a, sizeof(a)/sizeof(int));
    node **modlists;
    int n;

    printf( "Enter number of equivalence classes: " );
    scanf( "%d", &n );
    modlists = applyMod(list, n);
    printModlists(modlists, n);

    return 0;
}
```

Explanation

1. A relation that is symmetric, reflexive, and transitive is termed an *equivalence relation*. Each equivalence relation divides its elements into partitions called *equivalence classes*.

   A relation R is symmetric if aRb => bRa for all a, b in R and a!=b.

   A relation R is reflexive if aRa for all a in R.

   A relation R is transitive if aRb and bRc => aRc for all a, b, c in R.

2. `mod()`, (the modulus function) is an equivalence relation. It partitions its set of elements into N equivalence classes where N is the number used for division.

3. The program creates a linked list of integers (`getList()`). It then asks for the value of `N` for applying the `mod()` function. The function `applyMod()`

takes a list of integers and the value of `N` as arguments and applies `mod(N)` over each element e of the list to get a value v as the remainder. The element e is then put in `modlists[v]`, where `modlists[N]` is an array of list of integers. Each list `modlists[i]` represents one partition. Each list is finally printed.

4. The complexity of `applyMod()` is O(m) where m is the number of elements in the original list. For each element, `mod()` is applied, which is assumed to be an O(1) operation, and the insertion of the element in the list is O(1). So `applyMod()` has linear complexity.

Points to Remember

1. A relation which is symmetric, reflexive, and transitive is called an equivalence relation.
2. An equivalence relation partitions its elements into disjointed sets.
3. `mod()` is an equivalence relation.
4. We save space in `applyMod()` by reusing the nodes from the original list into the array of lists.

26 Problems in Strings

PROBLEM: MAXIMIZE A COMBINATION UNDER CONSTRAINTS

Given n arbitrary strings S1, S2,...,Sn, maximize the function

f(Si, Si+1, ..., Sj) = length(Si) + length(Si+1) + ... + length(Sj)

under the given constraints c(Sp,...,Sq), which means strings Sp,...,Sq cannot be taken together.

Program

```
#include <stdio.h>

#define NANSWERS 3    // max no of strings in the answer.
#define MAXCONS 5     // max length of any constraint.

typedef enum {FALSE, TRUE} bool;

bool isAbsent(int strnum, int *answer, int nans) {
    /*
     * returns TRUE if answer[nans] does NOT contain strnum.
     */
    int i;

    for(i=0; i<nans; ++i)
       if(answer[i] == strnum)
              return FALSE;
    return TRUE;
}

   bool satisfies(int *answer, int nans, int constraints[][MAXCONS+1], int
ncons) {
    /*
```

```
        * returns TRUE if nans answers in answer satisfy ncons
constraints.
        * note that each constraint ends with -1.
        */
       int i, j;

       for(i=0; i<ncons; ++i) {
          for(j=0; constraints[i][j] != -1; ++j)
                if(isAbsent(constraints[i][j], answer, nans))
                      break;
          if(constraints[i][j] == -1)
                return FALSE;
       }
       return TRUE;
   }

   void findMaxComb(int *lengths, int nstr, int constraints[][MAXCONS+1],
int ncons, int *answer, int nans, int *maxsum, int startstr, int startans,
int currsum ) {
       /*
        * find the max sum of lengths of nans strings out of nstr strings
of
        * lengths lengths[] satisfying ncons constraints constraints[].
        * save the max sum in *maxsum and the string indices in answer[].
        */
       int i;

       if(startans < nans) {
          for(i=startstr; i<nstr; ++i) {
                answer[startans] = i;
                findMaxComb(lengths, nstr, constraints, ncons, answer,
nans, maxsum, i+1, startans+1, currsum+lengths[i]);
          }
       }
       else if(currsum > *maxsum) {
          if(satisfies(answer, nans, constraints, ncons))
                *maxsum = currsum;
       }
   }
```

```
int main() {
    int lengths[] = {9, 8, 6, 5, 4, 3};    // lengths[i] =
length(string[i]).
    int answer[NANSWERS];             // indices of strings.
    int constraints[][MAXCONS+1] = {       // index of string starts
with 1 so that
                        {1,3,-1},           // end of constraint
can be signified
                        {2,-1},        // by 0.
                        {1,5,-1},           // we decrement every
index so that
                        {3,4,-1},           // each constraint
ends with -1.
                        {0,4,5,-1},
                        {0,4,-1},
                        {0,3,5,-1}
                    };
    int ncons = sizeof(constraints)/sizeof(int)/(MAXCONS+1);    // no
of constraints.
    int nstr  = sizeof(lengths)/sizeof(int);      // no of strings.
    int i, j;
    int maxsum = 0;

    for(i=1; i<=NANSWERS; ++i) {
        findMaxComb(lengths, nstr, constraints, ncons, answer, i,
&maxsum, 0, 0, 0);
        printf("After %d strings: maxsum=%d.\n", i, maxsum);
        //maxsum = 0; // this will keep all length combinations
separate.
    }

    return 0;
}
```

Explanation

1. We represent the given strings' lengths in an integer array. Each constraint consists of an array of indices of strings that cannot appear together in the final answer. Since each constraint can contain different number of indices, we end each constraint by –1. If the final answer we require consists of at most NANSWERS strings, we find the combination of i strings, the sum of whose lengths will be maximum, where 1<=i<=NANSWERS. The function

findMaxComb() finds such a combination for given value of i, if it exists. The variable maxsum signifies the current maximum sum of the combination of strings found so far. In order to find such a combination for every length, one needs to set maxsum=0 in the loop in main(). Also, if one wants to find a combination consisting of only NANSWERS strings, then the loop in main() can be changed as for(i=NANSWERS; i<=NANSWERS; ++i). In our implementation, the loop traverses from 1 to NANSWERS and maxsum is not set to 0 inside the loop. Thus we find a combination with any number of strings, up to NANSWERS.

2. The function findMaxSum() recursively finds a combination of maximum length and successively fills the answer[] array, which contains the final answer. The variable currsum contains the sum of the current strings selected in answer[]. The number of entries in answer[] to be filled is sent as a parameter from main() in the variable nans. If the nans fields in the array answer[] are filled, we check whether the current sum is greater than the current maximum sum found. If it is, then we check whether the new combination satisfies all constraints. If it satisfies all constraints, we have found a new combination that becomes the current maximum. So maxsum is updated.
3. Whether a string combination satisfies all constraints is checked in the function satisfies(). If there exists a constraint, all of whose string indices are present in the answer[] array of string indices, then that means the string combination does not satisfy that constraint, because each constraint c(Sp,...,Sq) says that the strings Sp...Sq cannot be combined. Thus, if we find that the string combination in answer[] satisfies all the constraints, then the function satisifes() returns TRUE, otherwise, it returns FALSE.
4. **Example:** Let the strings have lengths 9, 8, 6, 5, 4, and 3. The string indices are from 0 to 5. Let the constraints be { {1,3}, {2}, {1,5}, {3,4}, {0,4,5}, {0,4}, and {0,3,5} }, which means strings 1 and 3 cannot be included together, string 2 cannot be taken, strings 1 and 5 cannot appear together, strings 0, 4, and 5 cannot appear together, strings 0 and 4 cannot appear together, and strings 0, 3, and 5 cannot appear together. Note that the constraint {0,4,5} is included in the constraint {0,4}. Let NANSWERS = 3. The different steps of the algorithm are presented in the following table.

I	string-combination	currsum	maxsum	currsum > maxsum	satisfies-all-constraints?	maxsum
1	0	9	0	yes	yes	9
	1 to 5	<9	9	no	—	9
2	0, 1	17	9	yes	yes	17
other combinations		<17	17	no	—	17
3	0, 1, 2	23	17	yes	no	17
	0, 1, 3	22	17	yes	no	17
	0, 1, 4	21	17	yes	no	17
	0, 1, 5	20	17	yes	no	17
	0, 2, 3	20	17	yes	no	17
	0, 2, 4	19	17	yes	no	17
	0, 2, 5	18	17	yes	no	17
	0, 3, 4	18	17	yes	no	17
	0, 3, 5	17	17	no	—	17
	0, 4, 5	16	17	no	—	17
	1, 2, 3	19	17	yes	no	17
	1, 2, 4	18	17	yes	no	17
	1, 2, 5	17	17	no	—	17
	2, 3, 4	15	17	no	—	17
other combinations		<17	17	no	—	17

Points to Remember

1. The complexity of the algorithm is exponential over the number of strings.
2. Some speed-up can be achieved by sorting the lengths array first, which will result in avoiding some combinations. In that case A* algorithm can be used to find the answer.
3. If each constraint consists of a fixed number of strings, then this problem can be solved in polynomial time by reducing this problem to the longest path problem.

PROBLEM: MAXIMIZE A COMBINATION OF STRINGS—THE SECOND METHOD

Given n arbitrary strings S1, S2, ..., Sn, maximize the function

f(Si, Si+1, ..., Sj) = length(Si) + length(Si+1) + ... + length(Sj)

under the given constraints c(Si, Sj) which means Si and Sj cannot be taken together.

Program

```
#include <stdio.h>
#define MININT -1000
#define MAXVERTICES 10
#define MAXPATHVERT 3

void printCosts( int a[][MAXVERTICES], int nvert, int
pathvert[][MAXVERTICES] ) {
      /*
       * prints min cost matrix a.
       */
      int i, j;

      for( i=0; i<nvert; ++i ) {
         for( j=0; j<nvert; ++j )
               if( a[i][j] <= MININT )
                     printf( "%4c(%d) ", 'M', pathvert[i][j] );
               else
                     printf( "%4d(%d) ", a[i][j], pathvert[i][j] );
         printf( "\n" );
      }
      printf( "\n" );
}

int getMaxSum(int a[][MAXVERTICES], int b[][MAXVERTICES][MAXVERTICES],
int i, int j, int k, int *h1, int *h2) {
      /*
       * find such h1 and h2 that b[h1][i][k]+b[h2][k][j] is max and >
a[i][j];
       * and h1+h2-1 < MAXPATHVERT;
```

```
         * return the sum.
         */
        int p, q;
        int maxsum = 0;
        *h1 = *h2 = -1;

        for(p=2; p<MAXPATHVERT; ++p)    // 0 and 1 NOT necessary.
           for(q=p; q<=MAXPATHVERT; ++q) {
                 if(p+q-1 <= MAXPATHVERT && b[p][i][k]>0 && b[q][k][j]>0
&& b[p][i][k]+b[q][k][j] > maxsum)
                       maxsum=b[p][i][k]+b[q][k][j], *h1=p, *h2=q;
           }
        if(maxsum > a[i][j])
           return maxsum;
        *h1 = *h2 = -1;
        return -1;
    }

    void allCosts(int cost[][MAXVERTICES], int a[][MAXVERTICES], int nvert)
{
        int i, j, k;
        int pathvert[MAXVERTICES][MAXVERTICES] ;
        int b[MAXPATHVERT+1][MAXVERTICES][MAXVERTICES] ;
        int sum, h1, h2;
        int l;

        for( i=0; i<nvert; ++i )
           for( j=0; j<nvert; ++j )
                 a[i][j] = cost[i][j], pathvert[i][j] = 2, b[2][i][j] =
cost[i][j];

        printCosts(a, nvert, pathvert);

        for(l=2; l<=MAXPATHVERT; ++l)
          for(k=0; k<nvert; ++k) {
           for(i=0; i<nvert; ++i)
                 for(j=0; j<nvert; ++j) {
                       sum = getMaxSum(a, b, i, j, k, &h1, &h2);
                       if(sum != -1 && h1 != -1 && a[i][j]>=0) {
                             //printf("a[i][j]=%d, h1=%d, h2=%d.\n",
sum, h1, h2);
                             a[i][j] = sum, b[h1+h2-1][i][j] = sum,
pathvert[i][j]=h1+h2-1;
                       }
```

```
            }
        printCosts(a, nvert, pathvert);
        }
}

int main() {
    int cost[MAXVERTICES][MAXVERTICES] =
                                                    {
{0,50,10,MININT,45,MININT},

{MININT,0,15,MININT,10,MININT},

{20,MININT,0,15,MININT,MININT},

{MININT,20,MININT,0,35,MININT},

{MININT,MININT,MININT,30,0,MININT},

{MININT,MININT,MININT,3,MININT,0} };

                                                    /*  {20,30,40,50},
                                                      {MININT,40,50,60},

{MININT,MININT,60,70},

{MININT,MININT,MININT,80} ,
                                                      {2,3,4,5},
                                                      {3,4,5,6},
                                                      {4,5,6,MININT},
                                                      {5,6,MININT,8} ,

                                                      {0,1,2,MININT},
                                                      {3,0,MININT,4},

{MININT,MININT,0,6},
                                                      {MININT,5,MININT,0}

                                                    };  */
    int a[MAXVERTICES][MAXVERTICES] ;
    int nvert = 6;            // no of vertices.

    /*print( cost, nvert ); */
    allCosts( cost, a, nvert );
    //printCosts( a, nvert );
    return 0;
}
```

Explanation

1. Consider a graph represented by a cost adjacency matrix cost[n][n], where n is the number of vertices in the graph cost[i][i] == 0. cost[i][j] represents the length of the edge from vertex i to vertex j.
2. We used this cost adjacency matrix to solve the problem of finding the shortest path between any pair of vertices in the graph in O(n^3). We set cost[i][j] = infinity whenever there is no edge from vertex i to vertex j.

   We define a similar problem of finding the longest path between any pair of vertices in the graph, called the longest path problem. If we allow inclusion of a vertex more than once, there is a possibility of getting into an infinite loop and the procedure may not terminate! So we put a limit on the number of vertices that can be included in the longest path.
3. We reduce the given problem to the longest path problem. Each string Si is mapped to a vertex Vi. In the cost adjacency matrix, an entry cost[i][j] contains length(Si)+length(Sj). Thus the matrix is symmetric. Also, cost[i][i] is no longer zero. It contains the value length(Si)+length(Si). After this, we put in the constraints given by c(Si, Sj) and we mark such entries cost[i][j] as -infinity (MININT). Thus we get the cost adjacency matrix to represent the given constraints.
4. The function allCosts() finds the cost of the longest path between any pair of vertices. It contains a matrix a[][], which contains the current maximum path between any pair of vertices at any given time. It is initialized with cost[][]. In the loop, we find longest paths of lengths 2, 3, ... up to the limit defined by MAXPATHVERT. It starts with length == 2 because cost[][] contains the path containing 2 vertices, that is, an edge. Thus the original three loops of the shortest path problem are enclosed in another loop that goes from 2 to MAXPATHVERT. We also maintain a 3-dimensional array b[][][]. An entry b[h][i][j] contains the length of the longest path from vertex i to vertex j of length h. We also maintain a matrix pathvert[][] in which pathvert[i][j] contains the length of the longest path from i to j. Inside the innermost loop, the function getMaxSum() is called. In this function, for a[i][j], all combinations of paths from vertex i to vertex k and from vertex k to vertex j are checked and the one with the maximum cost is returned. a[i][j], then contains this new value if it is greater than the earlier one. The program finally calculates the cost of the longest path between any pair of vertices in the graph.

5. Let m=MAXPATHVERT. Then the complexity of getMaxSum() is O(m^2). The complexity of allCosts() is then O($m^{3}n$^3) where n is the number of strings in the given problem. The complexity of getMaxSum() may be reduced to O($m$log$m$) using a procedure similar to merge sort.

Points to Remember

1. Reducing one problem to another known problem can help in reusing the code from the earlier problem, as well as in the complexity analysis.
2. The complexity of the problem stated is O($m^{3}n$^3), where n is the number of strings and m is the number of maximum strings allowed.
3. If we allow the constraint function c() to contain an arbitrary number of strings, then the complexity becomes exponential O(n^n).

PROBLEM: CLOSURE OF SETS

Write a program to find closure of a set of characters input as a string.

Program

```
#include <stdio.h>

#define MAXLEN 80

void init(char *answer, int slen) {
    /*
     * initialize first slen entries in answer[] to 0.
     */
    int i;

    for(i=0; i<slen; ++i)
        *answer++ = 0;
    *answer = 0;    // eos.
}

void printComb(char *s, int slen, char *answer) {
    /*
     * fixes a character of s and then calls printComb() recursively
```

```
     * to get all combinations of the remaining chars.
     */
    int i;
    static int count = 0;

    if(*s == 0) {
        count++;
        printf("%2d: %s.\n", count, answer);
        return;
    }
    for(i=0; i<slen; ++i)
        if(answer[i] == 0) {
            answer[i] = *s;
            printComb(s+1, slen, answer);
            answer[i] = 0;
        }
}

void fillBitwise(int i, char *str, char *s, int slen) {
    /*
     * the pattern in i is the characteristic function of each char in
s.
     * whenever this bit is 1, fill str.
     */
    int j;

    for(j=0; j<slen; ++j)
        if((i & (1<<j)) != 0)
            *str++ = s[j];
    *str = 0;
}

void findClosure(char *s, int slen) {
    /*
     * finds closure of chars in s.
     * closure includes all substrings of sizes <= slen.
     * this function also prints their combinations using the
     * function printComb().
     */
    int i;
    char answer[MAXLEN];
    char str[MAXLEN];
```

```
    for(i=(1<<slen)-1; i>=0; -i) {
        // i represents the bit pattern.
        // 1 in the bit pattern means char should be displayed.
        fillBitwise(i, str, s, slen);
        init(answer, strlen(str));
        printf("printComb(%s).\n", str);
        printComb(str, strlen(str), answer);
        getchar();
    }
}

int main() {
    char s[MAXLEN];

    printf("Enter characters for closure: ");
    gets(s);

    while(*s) {
        //init(answer, strlen(s));
        //printComb(s, strlen(s), answer);
        findClosure(s, strlen(s));
        printf("Enter characters for closure(press enter to end): ");
        gets(s);
    }
    return 0;
}
```

Explanation

1. Closure of string 'abc' is a set of strings that are combinations of all 3-character strings plus combinations of all 2-character strings plus combinations of all 1-character strings plus combinations of all strings of length zero.

   Thus, closure(abc) = comb(abc) /* strings of length 3. */
   + comb(ab) + comb(bc) + comb(ac) /* strings of length 2. */
   + comb(a) + comb(b) + comb(c) /* strings of length 1. */
   + comb('') /* empty string. */

2. Note that the number of times comb() gets called is 2^*n* where *n* is the length of the input string. We can visualize this using bits. For a 3-character string,

invocation of `comb()` with different input parameters can be represented by 3-bit strings as follows:

abc	ó	111
ab	ó	110
ac	ó	101
a	ó	100
bc	ó	011
b	ó	010
c	ó	001
‘’	ó	000

Thus, we generate bit patterns as shown to represent different strings and then call `comb()` with these input parameters. Note further that these bit patterns are nothing but numbers 0 to 2^n-1.

3. The function `findClosure()` contains a loop which goes from 2^n-1 to 0 generating 2^n bit patterns representing 2^n strings, as just shown. According to these bit patterns, the function `fillBitwise()` builds the input string. This input string is then given to function `printComb()`, which finds all combinations of the input string.
4. The complexity of `printComb()` is O(n!) and it is called 2^n times in the loop. So the complexity of `findClosure()` is $O(n!2^n)$.

Points to Remember

1. We use bitwise operators to check whether a bit is on (1) or off (0). This speeds the processing.
2. Note carefully the mapping between the bit pattern of an integer and the character string, and how this mapping helped us build the algorithm.
3. Note the reuse of the procedure `printComb()`.
4. The complexity of `findClosure()` is $O(n!2^n)$.

PROBLEM: DISTANCE BETWEEN TWO STRINGS

Write a program to find the edit distance between two character strings.

Program

```
#include <stdio.h>

#define MAXLEN 80

int findMin(int d1, int d2, int d3) {
    /*
     * return min of d1, d2 and d3.
     */
    if(d1 < d2 && d1 < d3)
       return d1;
    else if(d1 < d3)
       return d2;
    else if(d2 < d3)
       return d2;
    else
       return d3;
}

int findEditDistance(char *s1, char *s2) {
    /*
     * returns edit distance between s1 and s2.
     */
    int d1, d2, d3;

    if(*s1 == 0)
       return strlen(s2);
    if(*s2 == 0)
       return strlen(s1);

    if(*s1 == *s2)
       d1 = findEditDistance(s1+1, s2+1);
    else
       d1 = 1 + findEditDistance(s1+1, s2+1);     // update.
    d2 = 1+findEditDistance(s1, s2+1);                    // insert.
```

```
    d3 = 1+findEditDistance(s1+1, s2);                    // delete.

    return findMin(d1, d2, d3);
}

int main() {
    char s1[MAXLEN], s2[MAXLEN];

    printf("Enter string 1: ");
    gets(s1);

    while(*s1) {
        printf("Enter string 2: ");
        gets(s2);
        printf("Edit distance(%s, %s) = %d.\n", s1, s2,
findEditDistance(s1, s2));
        printf("Enter string 1(enter to end): ");
        gets(s1);
    }

    return 0;
}
```

Explanation

1. The edit distance between two strings is the minimum number of characters in one string to be updated, inserted, or deleted to get the second string.

   Example: The edit distance between 'abc' and 'abd' is 1, as one character in 'abc' needs to be updated to get 'abd'.

   The edit distance between 'abc' and 'bd' is 2, as one character 'a' in 'abc' needs to be deleted and one character 'c' should be updated to 'd' to get 'bd'.

   The edit distance between 'abc' and 'abc' is 0.

2. We solve this problem elegantly using recursion. The function `findEditDistance(s1, s2)` checks whether any of its input strings, s1 and s2, are empty. If so, then it returns the length of the other string as the edit distance. If not, it checks whether the first characters of the two strings match. If they do, then a count d1 is obtained by calling `findEditDistance()` recursively with inputs s1+1 and s2+1. If the first two characters of s1 and s2 do not match, then it is assumed to be one updation and the count d1 is

obtained by adding one (for updation) to the edit distance between strings s1+1 and s2+1. The function then calls itself recursively again to get a count d2 by adding one to the edit distance between strings s1 and s2+1, to account for the deletion of one character from s1 to get s2. A symmetrical thing is done for s2 to get the count d3. After finding d1, d2, and d3, the minimum of the three counts is the edit distance between s1 and s2. The function `findMin()` does this job.

3. Since for each character in s1, the function calls itself recursively 3 times, the complexity can be calculated using the following recurrence relation:

   $T(n) = 3*T(n-1)$

   where n is the minimum of the two lengths of the strings. Solving this recurrence relation gives us the complexity O(3^*n*).

Points to Remember

1. Edit distance between two strings is the minimum number of insertions, deletions, or updations required in one string to get the other string.
2. Messages over a noisy channel can be compared with some approximation using the edit distance technique. This technique is also useful in voice and image recognition.
3. The complexity of `findEditDistance()` is O(3^*n*) where *n* is the minimum of the lengths of the two strings. The factor of 3 comes into the picture because for approximately each character, the function calls itself 3 times. To see how enormously this exponential complexity grows, try inputs to this program in increasing order of lengths.

PROBLEM: FINDING THE MAXIMUM MATCHING PATTERN IN THE STRING

Find the maximum matching pattern in an input string. Note that the matching pattern may be separated by some other patterns.

Program

```
#include <stdio.h>
#define MAXLEN 80
```

```
void findMaxPat(char *s, char *pat, int *maxpat) {
    int i;

    for(i=0; *s && *pat; ++s, ++i)
        if(*s == *pat)
            *maxpat++=i, pat++;
    if(!*pat)
        printf("whole pat found.\n");
    else
        printf("whole pat NOT found.\n");
    *maxpat = -1;   // end of maxpat.
}

void printMaxPat(char *s, int *maxpat) {
    char *sptr = s;

    puts(s);
    for(; *sptr && *maxpat != -1; ++sptr) {
        if(sptr-s == *maxpat) {
            printf("^");
            maxpat++;
        }
        else
            printf("%c", ' ');
    }
    printf("\n");
}

int main() {
    char s[MAXLEN];
    char pat[MAXLEN];
    int maxpat[MAXLEN];

    printf("Enter main string: ");
    gets(s);
    while(*s) {
        printf("Enter pattern to be searched: ");
        gets(pat);
        findMaxPat(s, pat, maxpat);
        printMaxPat(s, maxpat);
        printf("Enter main string: ");
```

```
        gets(s);
    }
    return 0;
}
```

Explanation

1. This program finds the maximum match of a pattern in a string. The characters matching the pattern in the string may be separated by other characters which are of no interest to us. Thus the job of this program is like a noise disposal parser that parses a valid syntactic entity separated by noise. We restrict ourselves to one string and one pattern.
2. **Example:** Let the input string be 'hello world' and the matching pattern be 'lord'. Then the program searches for each character in the pattern in the input string and marks each matching character as follows:

   ```
   h e l l o  w o r l d
       ^ ^      ^  ^.
   ```
3. `main()` iterates and asks for the input string and pattern until the input string is empty. It then calls `findMaxPat()` to find indices of characters in the input string that match characters in the input pattern. This array of indices is then passed to `printMaxPat()`, which marks the indexed characters.
4. The complexity of `findMaxPat()` is $O(n)$ where n is the length of the input string.

Points to Remember

1. Noise disposal parsing is useful in parsing languages such as English. It is also useful in filtering of data (signal).
2. The procedure `findMaxPat()` can be useful in approximate pattern-matching algorithms.
3. The complexity of `findMaxPat()` is $O(n)$ where n is the length of the input string.

PROBLEM: IMPLEMENTATION OF THE SOUNDEX FUNCTION

Write a function to compare two strings using the soundex method.

Program

```
#include <stdio.h>

#define MAXLEN 80
#define NALPHA 26

char *soundexGroups[] = {
                "aeiouhyw",
                            "kcgjqsxz",
                            "td",
                            "bpfv",
                            "l",
                            "mn",
                            "r"
                        };
int soundexCodes[NALPHA];

void soundexInit() {
    /*
     * build an inverted index from the global table soundexGroups[].
     * the inverted index is stored in global soundexCodes[].
     */
    int i;
    char *sptr;

    for(i=sizeof(soundexGroups)/sizeof(char *)-1; i>=0; --i)
        for(sptr=soundexGroups[i]; *sptr; ++sptr)
            soundexCodes[*sptr-'a'] = i;
}

int compareCodes(int *soundex1, int *soundex2) {
    int *ptr1, *ptr2;

    for(ptr1=soundex1, ptr2=soundex2; *ptr1!=-1 && *ptr2!=-1 &&
*ptr1==*ptr2; ++
ptr1, ++ptr2)
        ;
    return *ptr1 == *ptr2;
}

void findSoundex(char *s, int *soundex, char lastchar) {
    /*
```

```
         * find the soundex code for s and save in soundex.
         * the stored value is the index in the array of soundex codes.
         * function is recursive.
         * start by changing multiple occurrences of chars in consecutive
positions
         * by single occurrences.
         * end soundex by -1.
         * lastchar == -1 implies this is the first call to this function.
         */
        if(!*s)
            *soundex = -1, lastchar = 0;
        else if(*s == lastchar)
            findSoundex(s+1, soundex, lastchar);
        else if(lastchar == -1) // *s is the first char.
            *soundex=soundexCodes[*s-'a'], findSoundex(s+1, soundex+1, *s);
        else if(soundexCodes[*s-'a'] == 0)  // vowel group.
            findSoundex(s+1, soundex, *s);
        else
            *soundex=soundexCodes[*s-'a'], findSoundex(s+1, soundex+1, *s);
    }

    int compareSoundex(char *s1, char *s2) {
        /*
         * find soundex codes for s1 and s2.
         * return 1 if codes are equal else 0.
         */
        int soundex1[MAXLEN], soundex2[MAXLEN];

        findSoundex(s1, soundex1, -1);
        findSoundex(s2, soundex2, -1);

        return compareCodes(soundex1, soundex2);
    }

    int main() {
        char s1[MAXLEN];
        char s2[MAXLEN];
```

```
    soundexInit();

    printf("Enter string 1: ");
    gets(s1);

    while(*s1) {
        printf("Enter string 2: ");
        gets(s2);
        printf("(%s == %s) = %d.\n", s1, s2, compareSoundex(s1, s2));
        printf("Enter string 1(enter to end): ");
        gets(s1);
    }
    return 0;
}
```

Explanation

1. Soundex is a technique in phonetics used to compare various phonetic elements. This technique is useful to compare voices. The soundex scheme can also help in correcting an incorrect phonetic element against its dictionary.
2. The soundex scheme groups similar-sounding characters in one group. When applied to the English alphabet, the groups are as follows.

Group Number	Characters
0	aeiouhwy
1	bpfv
2	cgjkqsxz
3	dt
4	l
5	mn
6	r

Thus, the soundex code for the word 'cross' is '26022'. By soundex, the words 'think' and 'thing' will be the same as they both have the same soundex code, '30052'. Note that they both sound similar.

3. The results become more interesting if we do the following:

- The leading character is retained.
- The consecutive duplicate characters are changed to a single character.
- The vowels' group is dropped.

   Thus, the new soundex code for 'cross' is '262' and the one for 'alpha' is '041'.

4. The function soundexInit() builds an inverted index soundexCodes[] from the soundex groups in soundexGroups[]. The function compareSoundex(s1, s2) compares strings s1 and s2 using the soundex scheme. It uses functions findSoundex() to find the soundex code for a string and compareCodes() to compare the two soundex codes found. A soundex code of a string is simply an array of integers, so the function compareCodes() is straightforward. The function compareSoundex(s1, s2) returns TRUE if s1 and s2 are equal by soundex, otherwise it returns FALSE.
5. The function findSoundex() finds the soundex code for its input string. The code is terminated by –1. The function is recursive and goes characterwise. The input variable lastchar is used to remove multiple consecutive occurrences of characters. If the soundex group is 0, it is not added to the code.

Points to Remember

1. The soundex scheme is used to compare phonetically equal strings. This is useful in voice recognition.
2. Building an inverted index such as soundCodes[] proves to be more efficient than using soundexGroups[] directly.
3. There are additional rules depending on language, dialect, and accents.

27 Problems in Trees

PROBLEM: WRITE A NON-RECURSIVE VERSION OF PREORDER

Program

```
/**************** stack.c *********************/
#include <stdio.h>

#define SUCCESS 0
#define ERROR    -1

struct tnode;

typedef struct tnode *stype;
typedef struct snode snode;

struct snode {
    stype data;
    snode *next;
};

snode *stop = NULL;  // signifies empty stack.

int sPush( stype data ) {
    /*
     * push data at stop.
     */
    snode *ptr = (snode *)malloc( sizeof(snode) );
    ptr->data = data;
    ptr->next = stop;
    stop = ptr;

    return SUCCESS;
}
```

```
stype sPop() {
    /*
     * returns data at stop.
     */
    stype data;
    snode *ptr;

    if( sEmpty() ) {
        fprintf( stderr, "ERROR : popping from empty stack." );
        return (stype)NULL;
    }
    data = stop->data;
    ptr = stop;
    stop = stop->next;
    free(ptr);

    return data;
}

int sEmpty() {
    return ( stop == NULL );
}

/*************** queue.c *********************/
#include <stdio.h>

#define SUCCESS 0
#define ERROR    -1

struct tnode;

typedef struct tnode *qtype;
typedef struct qnode qnode;

struct qnode {
    qtype data;
    qnode *next;
};

qnode *front = NULL, // signifies empty queue.
      *rear  = NULL;
```

```
int qInsert( qtype data ) {
    /*
     * inserts data at rear.
     */
    qnode *ptr = (qnode *)malloc( sizeof(qnode) );
    ptr->data = data;
    ptr->next = NULL;
    if( qEmpty() )
        front = ptr;
    else
        rear->next = ptr;
    rear = ptr;

    return SUCCESS;
}

qtype qRetrieve() {
    /*
     * retrieve data from front and remove it from queue.
     */
    qtype data;
    if( qEmpty() ) {
        fprintf( stderr, "ERROR : retrieving from empty queue." );
        return (qtype)NULL;
    }
    data = front->data;
    front = front->next;
    if( qEmpty() )    // last node removed.
        rear = NULL;
    return data;
}

int qEmpty() {
    return ( front == NULL );
}

/**************** main.c *********************/
#include <stdio.h>
#include <malloc.h>
#include "queue.c"
#include "stack1.c"
typedef int ttype;
```

```
typedef struct tnode tnode;

struct tnode {
    ttype data;
    tnode *left;
    tnode *right;
};

tnode *tree = NULL;

int tInsert( ttype data ) {
    /*
     * insert data into global tree.
     */
    tnode *ptr  = (tnode *)malloc( sizeof(tnode) );
    ptr->data  = data;
    ptr->left  = NULL;
    ptr->right = NULL;
    tInsertPtr( &tree, ptr );
}

int tInsertPtr( tnode **tree, tnode *ptr ) {
    /*
     * inserts ptr into tree recursively.
     */
    if( *tree != NULL ) {
       if( ptr->data < (*tree)->data )
              tInsertPtr( &((*tree)->left), ptr );
       else
              tInsertPtr( &((*tree)->right), ptr );
    }
    else
       *tree = ptr;
    return SUCCESS;
}

void tPrint( tnode *tree ) {
    /*
     * prints tree in inorder recursively.
     */
    if( tree != NULL ) {
       printf( "going left of %d...\n", tree->data );
```

```
        tPrint( tree->left );
        printf( "%d ", tree->data );
        printf( "going right of %d...\n", tree->data );
        tPrint( tree->right );
        printf( "tree of %d is over.\n", tree->data );
    }
}

void tIterBFS( tnode *tree ) {
    /*
     * prints tree in breadth-first manner iteratively using a queue.
     */
    if( tree != NULL ) {
        qInsert( tree );

        while( !qEmpty() ) {
                tree = qRetrieve();
                printf( "[[ %d ]]\n", tree->data );
                if( tree->left != NULL )
                        qInsert( tree->left );
                if( tree->right != NULL )
                        qInsert( tree->right );
        }
    }
}

void tIterPreorder( tnode *tree ) {
    /*
     * prints tree in preorder iteratively using a stack.
     */
    if( tree != NULL ) {
        sPush( tree );

        while( !sEmpty() ) {
                tree = sPop();
                printf( "[[ %d ]]\n", tree->data );
                if( tree->right != NULL )
                        sPush( tree->right );
                if( tree->left != NULL )
                        sPush( tree->left );
        }
    }
```

```
}

int main() {
    tInsert(4);
    tInsert(2);
    tInsert(6);
    tInsert(1);
    tInsert(3);
    tInsert(5);
    tInsert(7);
    tPrint(tree);
    tIterPreorder(tree);
}
```

Explanation

1. In a preorder traversal, a node is processed first, followed by its left and right children nodes. This is easy with recursion. But while doing it iteratively, the right child node needs to be saved so that after the processing of the whole left subtree, the saved node can be taken for execution. This procedure needs to be followed for every node. Therefore, we need to use a stack.
2. We process the root node, push its right child (if it exists) in a stack, and take its left child (if it exists) for processing. This is done in a loop which results in the processing of all the nodes in the left subtree of the root node. We then pop the right child node and start processing it the same way. Thus we get a preorder traversal of the tree.
3. **Example:**

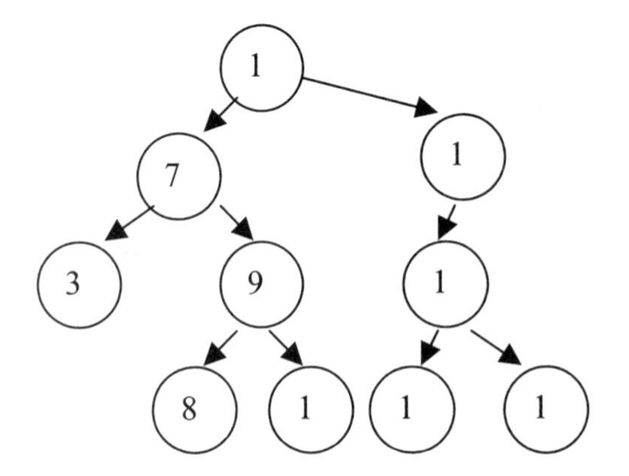

The preorder traversal is 12, 7, 3, 9, 8, 11, 18, 14, 13, 15.

The stepwise run of the algorithm is as follows.

step	node	stack	output
0	nil	empty	nil
1	12	18 7	12
2	7	18 9 3	12 7
3	3	18 9	12 7 3
4	9	18 11 8	12 7 3 9
5	8	18 11	12 7 3 9 8
6	11	18	12 7 3 9 8 11
7	18	14	12 7 3 9 8 11 18
8	14	15 13	12 7 3 9 8 11 18 14
9	13	15	12 7 3 9 8 11 18 14 13
10	15	empty	12 7 3 9 8 11 18 14 13 15

Points to Remember

1. The preorder traversal needs a stack.
2. Keeping stack processing and tree processing separate helps us make the program modular. This can help us in using the same implementation of stack/queue for some other application, by making changes in only the data type contained in the stack/queue.
3. In the preorder traversal, the right child should be pushed in the stack first, followed by the left child.
4. The recursive version is more readable than the iterative version. This is because the tree data structure itself is inherently recursive.

PROBLEM: WRITE A NON-RECURSIVE VERSION OF POSTORDER

Program

```
/******************** stack.c *********************/
#include <stdio.h>

#define SUCCESS 0
#define ERROR   -1
```

```
struct tnode;

typedef struct tnode *stype;
typedef struct snode snode;
typedef struct snode *stack;

struct snode {
    stype data;
    snode *next;
};

void sInit( stack *s ) {
    *s = NULL;
}

int sPush( stack *s, stype data ) {
    /*
     * push data in stack s.
     */
    snode *ptr = (snode *)malloc( sizeof(snode) );
    ptr->data = data;
    ptr->next = *s;
    *s = ptr;

    return SUCCESS;
}

stype sPop( stack *s ) {
    /*
     * returns data at top of stack s.
     */
    stype data;
    snode *ptr;

    if( sEmpty(*s) ) {
        fprintf( stderr, "ERROR : popping from empty stack.\n" );
        return (stype)NULL;
    }
    data = (*s)->data;
    ptr = *s;
    (*s) = (*s)->next;
    free(ptr);
```

```
    return data;
}

int sEmpty( stack s ) {
    return ( s == NULL );
}

/******************** main.c ********************/
#include <stdio.h>
#include <malloc.h>
#include "stack.c"


typedef int ttype;
typedef struct tnode tnode;

struct tnode {
    ttype data;
    tnode *left;
    tnode *right;
};

tnode *tree = NULL;

int tInsert( ttype data ) {
    /*
     * insert data into global tree.
     */
    tnode *ptr  = (tnode *)malloc( sizeof(tnode) );
    ptr->data  = data;
    ptr->left  = NULL;
    ptr->right = NULL;
    tInsertPtr( &tree, ptr );
}

int tInsertPtr( tnode **tree, tnode *ptr ) {
    /*
     * inserts ptr into tree recursively.
     */
    if( *tree != NULL ) {
        if( ptr->data < (*tree)->data )
            tInsertPtr( &((*tree)->left), ptr );
```

```
        else
            tInsertPtr( &((*tree)->right), ptr );
    }
    else
        *tree = ptr;
    return SUCCESS;
}

void tPrint( tnode *tree ) {
    /*
     * prints tree in inorder recursively.
     */
    if( tree != NULL ) {
        printf( "going left of %d...\n", tree->data );
        tPrint( tree->left );
        printf( "%d ", tree->data );
        printf( "going right of %d...\n", tree->data );
        tPrint( tree->right );
        printf( "tree of %d is over.\n", tree->data );
    }
}

void tIterPostorder( tnode *tree ) {
    /*
     * prints tree in postorder iteratively using 2 stacks.
     */
    stack s1, s2;
    sInit(&s1); sInit(&s2);

    if( tree != NULL ) {
        sPush( &s1, tree );

        while( !sEmpty(s1) ) {
            tnode *t = sPop(&s1);
            sPush( &s2, t );
            if( t->left != NULL )
                sPush( &s1, t->left );
            if( t->right != NULL )
                sPush( &s1, t->right );
        }
        while( !sEmpty(s2) )
            printf( "%d\n", sPop(&s2)->data );
```

```
    }
}

int main() {
    tInsert(4);
    tInsert(2);
    tInsert(5);
    tInsert(1);
    tInsert(3);
    tInsert(6);
    //tInsert(7);
    tPrint(tree);
    tIterPostorder(tree);
}
```

Explanation

1. In a postorder traversal, a node's left subtree is first output, followed by the right subtree, and finally the node is output. Thus, we need a stack in which we push the right child, followed by the left child. But we also need the node itself in the output. So, we will have to push it again in the stack before pushing its children. But then how will we keep track of whether that node is already processed? If we don't, we will again push it and its children in the stack, forming an infinite loop!
2. One way is to tag it as 'processed.' This can be done by adding a field to each node that will specify its status. Another way is to keep a list of processed nodes. We use this latter approach, maintaining a stack of processed nodes. The reason for using a stack is that when all the nodes in the tree are processed, the nodes in the stack give postorder traversal in reverse. Thus, by popping the nodes and outputting one by one, we get the required postorder traversal.

3. **Example:**

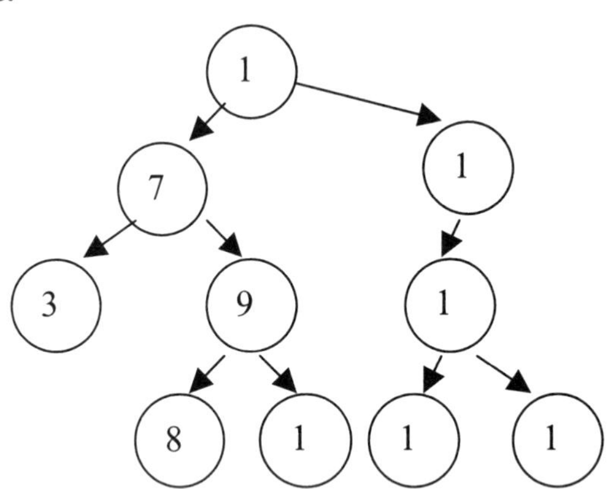

The postorder traversal of this tree is 3, 8, 11, 9, 7, 13, 15, 14, 18, 12. The stepwise run of the algorithm is shown next:

step	node	stack 1	stack 2
0	nil	empty	empty
1	12	7 18	12
2	18	7 14	12 18
3	14	7 13 15	12 18 14
4	15	7 13	12 18 14 15
5	13	7	12 18 14 15 13
6	7	3 9	12 18 14 15 13 7
7	9	3 8 11	12 18 14 15 13 7 9
8	11	3 8	12 18 14 15 13 7 9 11
9	8	3	12 18 14 15 13 7 9 11 8
10	3	empty	12 18 14 15 13 7 9 11 8 3

Stack 2 can now be output to get the required postorder traversal of the tree.

Points to Remember

1. Iterative postorder requires a stack.

2. Keeping stack processing and tree processing separate helps us to make the program modular. This can help us in using the same implementation of the stack for some other application, by making changes in only the data type contained in the stack.
3. The left child of a node is first pushed to the stack, followed by the right child.
4. The recursive version is more readable than the iterative version. This is because the tree data structure itself is inherently recursive.

PROBLEM: PREORDER TRAVERSAL OF A THREADED BINARY TREE

Write a function to traverse an inorder threaded binary tree in preorder.

Program

```
#include <stdio.h>

#define SUCCESS 0
#define ERROR    -1

typedef int type;
typedef struct node node;
typedef enum {FALSE, TRUE} bool;

struct node {
    type data;
    node *lchild, *rchild;
    bool lthread, rthread;
};

/* NOTE: since this is a threaded binary tree, there wont be any
condition */
/*       of type (ptr == NULL).                                  */

node tree;

void tInit() {
```

```
    tree.lchild  = tree.rchild = &tree;
    tree.lthread = TRUE;
    tree.rthread = FALSE;
    tree.data = 99999999;
}

node *insucc( node *t ) {
    /*
     * find inorder successor of t.
     */
    node *temp = t->rchild;
    if( t->rthread == FALSE )
       while( temp->lthread == FALSE )
              temp = temp->lchild;
    return temp;
}

node *inpred( node *t ) {
    /*
     * find inorder predecessor of t.
     */
    node *temp = t->lchild;
    if( t->lthread == FALSE )
       while( temp->rthread == FALSE )
              temp = temp->rchild;
    return temp;
}

int tInsertRight( node *s, node *t ) {
    /*
     * insert t as right child of s.
     */
    node *temp;

    t->rchild = s->rchild;
    t->rthread = s->rthread;
    t->lchild = s;
    t->lthread = TRUE;
    s->rchild = t;
    s->rthread = FALSE;
    if( t->rthread == FALSE ) {
       temp = insucc(t);
       temp->lchild = t;
```

```
    }
    return SUCCESS;
}

int tInsertLeft( node *s, node *t ) {
    /*
     * insert t as left child of s.
     */
    node *temp;

    t->lchild = s->lchild;
    t->lthread = s->lthread;
    t->rchild = s;
    t->rthread = TRUE;
    s->lchild = t;
    s->lthread = FALSE;

    if( t->lthread == FALSE ) {
        temp = inpred(t);
        temp->rchild = t;
    }
    return SUCCESS;
}

node *tGetNewNode( type data ) {
    /*
     * returns a new node containing the data.
     */
    node *ptr = (node *)malloc( sizeof(node) );
    ptr->data = data;
    ptr->lchild = ptr->rchild = NULL;
    ptr->lthread = ptr->rthread = FALSE;
    return ptr;
}

int tInsert( node *t, type data ) {
    /*
     * insert data in t recursively.
     */
    if( data < t->data )
        if( t->lthread == TRUE )
                tInsertLeft( t, tGetNewNode(data) );
```

```
        else
                tInsert( t->lchild, data );
    else
        if( t->rthread == TRUE )
                tInsertRight( t, tGetNewNode(data) );
        else
                tInsert( t->rchild, data );
    return SUCCESS;
}

void tPrint( node *t ) {
    /*
     * prints t inorder recursively without using threads.
     */
    if( t != &tree ) {
        if( t->lthread == FALSE )
                tPrint( t->lchild );
        printf( "%d\n", t->data );
        if( t->rthread == FALSE )
                tPrint( t->rchild );
    }
}

void tPrintPreorder( node *t ) {
    /*
     * prints tree preorder (no use of threads).
     */
    if( t != &tree ) {
        printf( "%d\n", t->data );
        if( t->lthread == FALSE )
                tPrintPreorder( t->lchild );
        if( t->rthread == FALSE )
                tPrintPreorder( t->rchild );
    }
}

void tPrintInorder( node *tree ) {
    /*
     * prints tree inorder using threads.
     */
    node *temp = tree;
    do {
        temp = insucc(temp);
```

```
        if( temp != tree )
                printf( "%d\n", temp->data );
    } while( temp != tree );
}

int main() {
    tInit();
    tInsert( &tree, 4 );
    tInsert( &tree, 2 );
    tInsert( &tree, 1 );
    tInsert( &tree, 3 );
    tInsert( &tree, 6 );
    tInsert( &tree, 5 );
    tInsert( &tree, 7 );
    tPrint( tree.lchild );
    printf( "\n" );
    tPrintPreorder( tree.lchild );

    return 0;
}
```

Explanation

1. In a threaded binary tree, the NULL links of leaf nodes are replaced by pointers (called threads) to other nodes in the tree. If p→rightchild is normally equal to NULL, it is replaced by a pointer to the node which would be printed after p when traversing the tree in inorder. A NULL leftchild link at node p is replaced by a pointer to the node that immediately precedes node p in inorder. The left link of the first node and the right link of the last node printed in the inorder traversal point to a dummy head node of the tree, and all the nodes appear in the left subtree of this head node. For example, in this representation, a tree such as the one shown next, where solid pointers are normal links and the dotted pointers are threads, t means TRUE and f means FALSE.

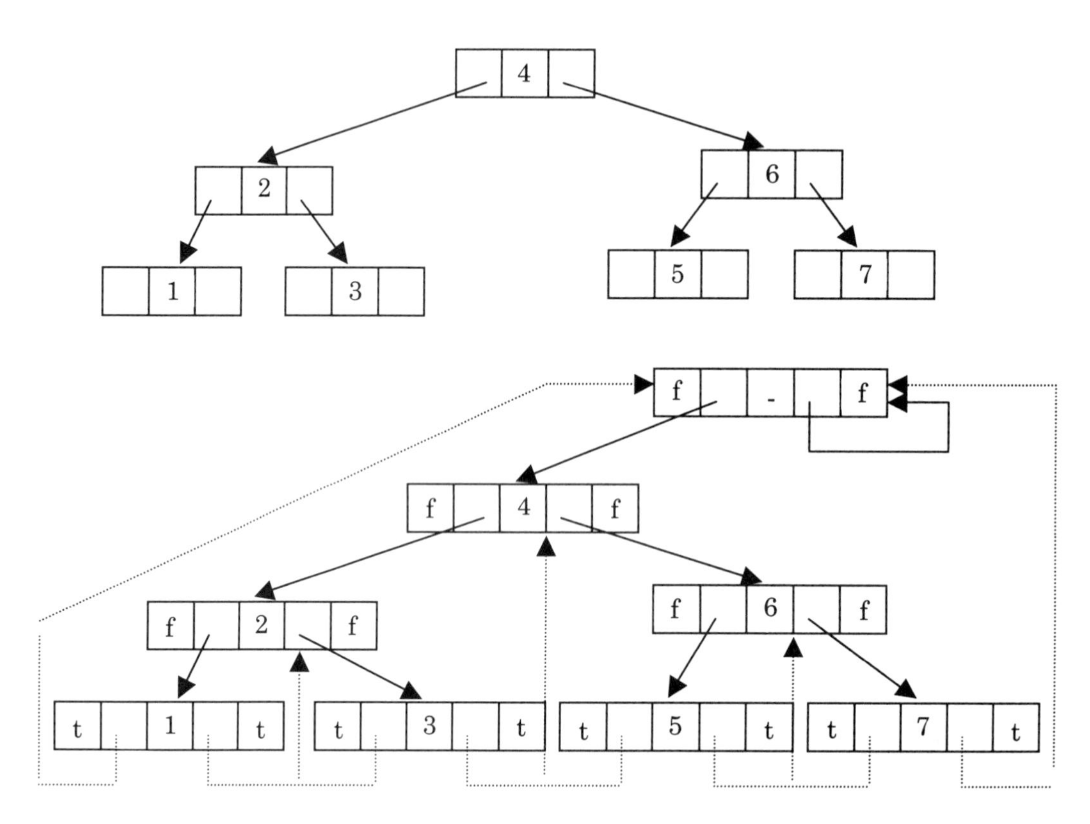

2. In the memory representation, we should be able to distinguish between threads and normal pointers. This is done by adding two Boolean fields to the structure: `leftthread` and `rightthread`. If `tree->leftthread==TRUE`, then `tree->leftchild` contains a thread; otherwise, it contains a pointer to the `leftchild`. Similarly, if `tree->rightthread == TRUE`, then `tree->rightchild` contains a thread. Otherwise it contains a pointer to the `rightchild`.
3. The function to traverse a threaded binary tree remains as simple as that for the normal binary tree. One simply needs to check for a link not that is not a thread, and traverse it. So the recursive function `tPrintPreorder()` is self-explanatory. For example, the preorder traversal of the tree above is 4, 2, 1, 3, 6, 5, 7.

Points to Remember

1. Some traversing algorithms are simplified by making a tree threaded.

2. Making a tree threaded makes insertions and deletions clumsy. Also, the node size increases. So this increased complexity should be taken into consideration when using a threaded binary tree for an application.
3. Keeping a dummy head node helps in easy insertions, deletions, and traversals.

PROBLEM: IMPLEMENTATION OF A SET USING A BINARY TREE

Implement union() and find() operations on sets represented as trees.

Program

```
#include <stdio.h>

#define N 100 // max no of elements together in all sets.

typedef int type;
typedef struct node node;

struct node {
    type val;           // this is value of member.
    int parent;                 // this is index of parent in the array.
};

node sets[N];           // all sets are contained in it.
int setsindex = 0;      // total no of elements in sets.

int insertRoot( type val ) {
    /*
     * insert val in sets as a root of a new tree.
     */
    sets[setsindex].val = val;
    sets[setsindex].parent = -1;
    setsindex++;

    return setsindex-1;
}
```

```
void insertElement( int rootindex, type val ) {
    /*
     * insert element val in set whose root is indexed at rootindex.
     */
    sets[setsindex].val = val;
    sets[setsindex].parent = rootindex;
    setsindex++;
}

int buildSet( type a[], int n ) {
    /*
     * repeated calls to this fun with diff arrays will insert diff set
in sets.
     * forms a tree representation of elements in a.
     * n is number of elements in the set.
     * empty set(n==0) cannot be represented here.
     * returns index of root.
     */
    int i, rootindex;
    if( n <= 0 ) {
        fprintf( stderr, "n should be > 0.\n" );
        return -1;
    }
    // check whether there is enough space for n elements.
    if( setsindex+n > N ) {
        fprintf( stderr, "ERROR: set overflow.\n" );
        return -1;
    }
    // a[0] becomes the root.
    rootindex = insertRoot( a[0] );
    for( i=1; i<n; ++i )
        insertElement( rootindex, a[i] );

    return rootindex;
}

void printSets() {
    int i;
    printf( "\n" );
    for( i=0; i<setsindex; ++i )
        printf( "%d %d.\n", sets[i].val, sets[i].parent );
    printf( "\n" );
}
```

```
int unionSets( int rindex1, int rindex2 ) {
    /*
     * makes a union of sets whose root indices are rindex1 and
rindex2.
     */
    sets[rindex2].parent = rindex1;
    // or the reverse.
    return rindex1;          // root of the union.
}

int findSet( int valindex ) {
    /*
     * given a val at index valindex in the array, finds index of its
root.
     */
    for( ; sets[valindex].parent!=-1; valindex=sets[valindex].parent )
        ;
    return valindex;
}

int getIndex( type val ) {
    /*
     * dummy procedure to return index in array of val.
     */
    int i;
    for( i=0; i<setsindex; ++i )
        if( sets[i].val == val )
                return i;
    return -1;
}

int main() {
    type s1[] = {1,7,8,9};
    type s2[] = {5,2,10};
    type s3[] = {3,4,6};
    int i1 = buildSet( s1, 4 );
    int i2 = buildSet( s2, 3 );
    int i3 = buildSet( s3, 3 );

    //printSets();

    i1 = unionSets( i1, i2 );
```

```
    printf( "%d %d.\n", 3, sets[findSet( getIndex(3) )].val );
    printf( "%d %d.\n", 5, sets[findSet( getIndex(5) )].val );
    printf( "%d %d.\n", 2, sets[findSet( getIndex(2) )].val );

    i3 = unionSets( i3, i1 );
    printf( "%d %d.\n", 3, sets[findSet( getIndex(3) )].val );
    printf( "%d %d.\n", 5, sets[findSet( getIndex(5) )].val );
    printf( "%d %d.\n", 7, sets[findSet( getIndex(2) )].val );

    printSets();
    return 0;
}
```

Explanation

1. We represent sets as trees in which different elements are stored as nodes in a tree. Since the order of elements is immaterial in sets, the order of nodes also does not matter in the tree. However, the links of the tree are reversed, that is, the child nodes have links to the parent instead of the parent pointing to its children. The reason for this representation is discussed next.

   Example:

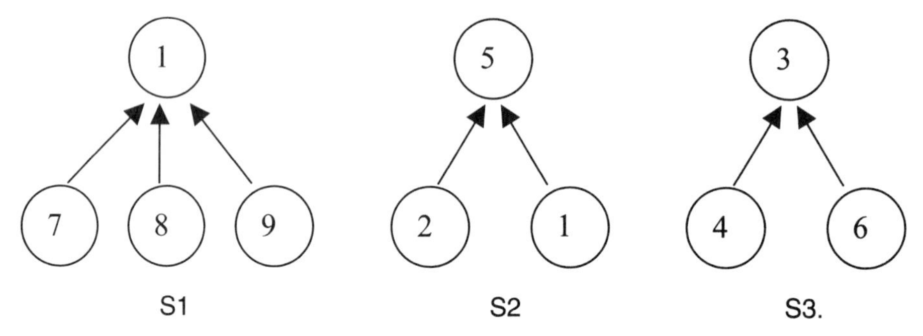

2. Union of two disjointed sets S1 and S2 (`unionSets()`) under the tree representation can be carried out simply by making any node of S1 the parent node of the root node of S2, or vice-versa. This operation can be carried out in a constant amount of time.

Example:

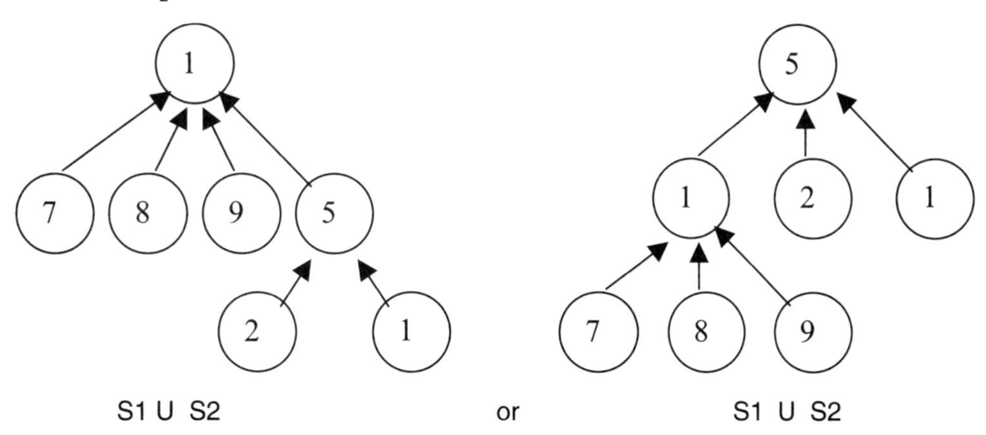

3. The find operation (`findSet()`) searches for the root of an element in the tree. It travels from that element upwards in the hierarchy until it reaches a node that has no parent. Thus, its root is determined. The complexity of the `find()` operation is $O(h)$ where h is the height of the tree. If the height remains below $O(\log n)$, the operation is faster. However, the worst-case time is linear. This $O(n)$ time appears when each node (except the root) has only one predecessor.

   Example: The root of vertex 8 in S1 in the last example is 1. However, the root of 8 in (S1 U S2) is 1 or 5 depending on how the union is done.

4. The program stores all the sets in a global array of nodes. Each node contains the value of the element and the index of its parent element. To make `find()` efficient, each node is connected directly to the root node. The function `unionSets(i1, i2)` take indices of root elements and makes `i1` the parent of the root node pointed to by `i2`. The function `findSets(index)` travels backwards from the node pointed to by the index until it reaches any of the root nodes. The root node has a parent index equal to -1.

Points to Remember

1. The function `union()` does not work if the sets are not disjointed.
2. Operations `union()` and `find()` were devised from specific applications that use symbol tables. This asks for a specific representation of data in the form of a tree with reversed links. This shows how closely algorithms and data structures are related.

3. `union()` has a complexily of O(1), while `find()` has the worst-case complexity of O(h), where h is the height of the element in the tree.

PROBLEM: HUFFMAN CODING

Write a program to find Huffman codes for a set of characters, given the frequency of occurrence of each of these characters.

Program

```
/********************** huffman.q.h ***********************/
#include <stdio.h>

typedef struct qnode qnode;
typedef tree qtype;
typedef struct {
    qnode *front;
    qnode *rear;
} queue;

struct qnode {
    qtype op;
    qnode *next;
};

bool qEmpty( queue *q ) {
    return (q->front == NULL);
}

void qPushBack( queue *q, qtype op ) {
    /*
     * pushes op at q.rear.
     */
    qnode *ptr = (qnode *)malloc( sizeof(qnode) );
    ptr->op = op;
    ptr->next = NULL;
    if( qEmpty(q) )   // first element in q.
        q->front = ptr;
    else
```

```
        q->rear->next = ptr;
    q->rear = ptr;
}

void qInsertSorted(queue *q, qtype op) {
    /*
     * inserts val in sorted q and keeps new q sorted.
     */
    qnode *ptr = (qnode *)malloc(sizeof(qnode));
    qnode *curr, *prev;
    ptr->op = op;

    for(prev=NULL, curr=q->front; curr && curr->op->w < op->w;
prev=curr, curr=curr->next)
        ;
    if(!curr && !prev)       // q empty.
        ptr->next = NULL, q->rear = q->front = ptr;
    else if(!curr)    // op is the max value.
        ptr->next = NULL, prev->next = q->rear = ptr;
    else if(!prev)    // op is the min value.
        ptr->next = curr, q->front = ptr;
    else {     // if prev and ptr both exist.
        ptr->next = curr;
        prev->next = ptr;
    }
}

qtype qPopFront( queue *q ) {
    /*
     * pops op from q->front.
     */
    qnode *ptr = q->front;
    qtype op;

    if( qEmpty(q) )
        return (qtype)NULL;
    q->front = q->front->next;
    if( qEmpty(q) )
        q->rear = NULL;

    op = ptr->op;
    free(ptr);
```

```
    return op;
}

/******************** huffman.c ************************/
#include <stdio.h>

#define MAXLEN 80

typedef struct node node;
typedef char type;
typedef node *tree;
typedef enum {FALSE, TRUE} bool;

struct node {
    int w;
    type val;
    node *left, *right;
};

#include "huffman.q.h"

int compare(const void *e1, const void *e2) {
    /*
     * compare the two elements in e1 and e2.
     * each element is a vector of two elements.
     */
    return ((int *)e1)[1] > ((int *)e2)[1];
}

void printTree(tree t, char *outputstr) {
    /*
     * print the huffman codes for each element of t.
     * outputstr contains huffman code for t (NOT parent of t).
     * assumes t!=NULL.
     */
    char str[2] = "1";

    if(t->right) {
        strcat(outputstr, str);
        printTree(t->right, outputstr);
        outputstr[strlen(outputstr)-1] = 0; // restore.
    }
```

```
        if(t->left) {
            str[0] = '0';
            strcat(outputstr, str);
            printTree(t->left, outputstr);
            outputstr[strlen(outputstr)-1] = 0; // restore.
        }
        else if(!t->right)
            printf("%c=%d=%s.\n", t->val, t->w, outputstr);
    }

    tree buildTree(int a[][2], int n) {
        /*
         * build a huffman tree using frequency in a[i][1] where a[0][j]
indicates
         * the character
         * for that sort a on frequency.
         * n is the size of a.
         */
        int i;
        tree t = NULL;
        queue sortedq = {NULL};

        // sort a on frequency.
        qsort(a, n, sizeof(a[0]), compare);

        // insert each element in tree.
        for(i=0; i<n; ++i) {
            tree temp = (tree)calloc(1, sizeof(node));
            temp->w = a[i][1];
            temp->val = (type)a[i][0];
            qPushBack(&sortedq, temp);
        }
        // assume n>0.
        while(TRUE) {
            tree t2 = NULL,
                     t1 = qPopFront(&sortedq);
            if(!qEmpty(&sortedq))
                    t2 = qPopFront(&sortedq);
            else {
                    t = t1;
                    break;
            }
```

```
        t = (tree)malloc(sizeof(node));
        t->w = t1->w + t2->w;
        t->left = t1;
        t->right = t2;
         {
              qnode *ptr;
              for(ptr=sortedq.front; ptr; ptr=ptr->next)
                    printf("%d ", ptr->op->w);
              printf("\n");
        }
        printf("insertsorted=%d.\n", t->w);
        qInsertSorted(&sortedq, t);
    }
    return t;
}

int main() {
    int a[][2] = {    {'a', 20},
                            {'b', 23},
                            {'c', 14},
                            {'d', 56},
                            {'e', 35},
                            {'f', 29},
                            {'g', 5 }
                      };
    char outputstr[MAXLEN] = "";
    tree t = buildTree(a, sizeof(a)/2/sizeof(int));

    if(t)
       printTree(t, outputstr);

    return 0;
}
```

Explanation

1. Many applications prefer that frequently occurring messages have smaller lengths during coding, to make efficient use of the available bandwidth. This can be done by finding a binary tree with minimum weighted external path length. An external path length of a binary tree is the sum of all external nodes of the lengths of the paths, from the root to those nodes. The `null` nodes in a tree are called external nodes. Weighted external path length can

be obtained by multiplying the external path length of each node by the frequency contained in the node, and than adding all of these values.

Example: External path lengths of nodes A, B, C, D are 2, 2, 2, 2 and the weighted external path length of the tree is

2*2 + 2*4 + 2*5 + 2*15 = 52.

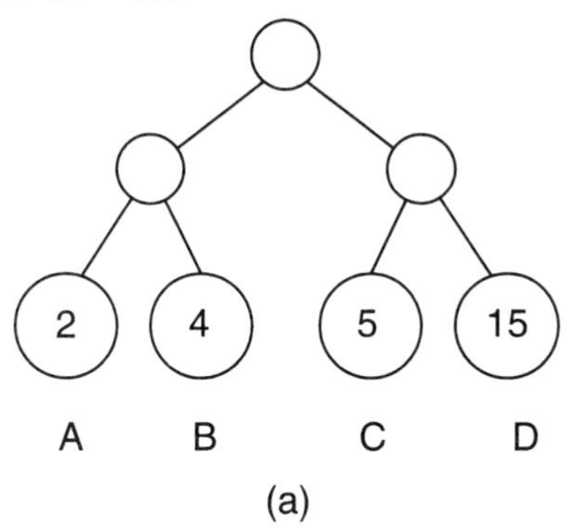

(a)

Note that this is not the minimum weighted path length. The minimum weighted path length can be obtained by restructuring the tree as shown here.

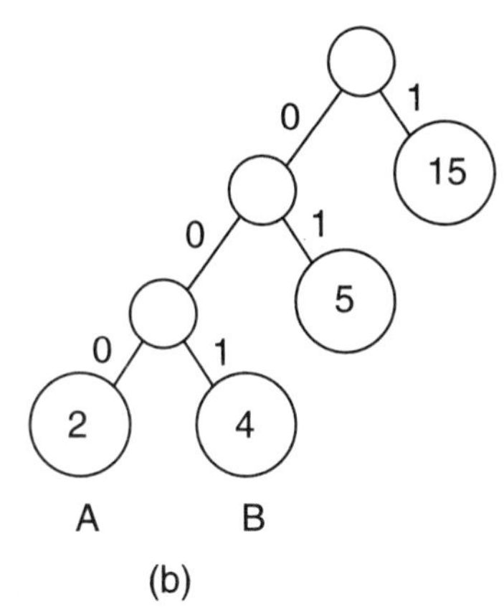

(b)

The path length of this tree is 1*15 + 2*5 + 3*2 + 3*4 = 43, which is minimum. The numbers 0 and 1 of the edges will be clear in the following description.

2. An easy and nice solution to the problem of finding a binary tree with minimum weighted external path length is given by D. Huffman. We implement the algorithm in the function `buildTree()`. This algorithm first prepares a list of nodes out of the input characters and their frequencies. This list is sorted on the frequency in ascending order. Thus every retrieval retrieves the node with minimum frequency count from the list. The function

then contains a loop that runs until the list becomes empty. In each iteration, two nodes (with minimum frequencies f1 and f2) are retrieved from the list, and another node with the frequency count f1+f2 is added to the list. This node also becomes the parent node of the two retrieved nodes. Thus every iteration reduces the length of the list by 1. In the end, when the list contains only one node, the node is returned, as the root of the Huffman tree is built.

3. If we assume that the codes contain only two symbols, 0 and 1, then the edge to a left child can be named 0 while that to a right child can be named 1, or vice versa, as given in (b) in the last example. Thus, the codes for A, B, C, and D are 000, 001, 01, and 1.

Points to Remember

1. Huffman codes are useful to encode a message using minimum number of symbols.
2. Another big advantage of Huffman codes is that given a message string built from Huffman codes, we can uniquely divide the message into the patterns of individual characters. For example, the pattern 01001101 uniquely identifies the character sequence CBDC.
3. The complexity of `buildTree()` is $O(n\log n+n+n^2)$ for `qsort()`, `for` loop, and `while` loop. Note that the complexity of `qPushBack()` is O(1) while that of `qInsertSorted()` is O(n). Thus the overall complexity of `buildTree()` is O(n^2). The complexity of `qInsertSorted()` may be improved by using a heap instead of a sorted list.

PROBLEM: IMPLEMENTATION OF A B-TREE

Write procedures for insertion and deletion in a B-tree.

Program

```
/************************** b.q.c ******************************/
#include <malloc.h>
#include <stdlib.h>

typedef struct qnode qnode;
typedef btree qtype;
```

```
typedef struct {
    qnode *front;
    qnode *rear;
} queue;
struct qnode {
    qtype op;
    qnode *next;
};

bool qEmpty( queue *q ) {
    return (q->front == NULL);
}

void qPush( queue *q, qtype op ) {
    /*
     * pushes op at q.rear.
     */
    qnode *ptr = (qnode *)malloc( sizeof(qnode) );
    ptr->op = op;
    ptr->next = NULL;
    if( qEmpty(q) )   // first element in q.
        q->front = ptr;
    else
        q->rear->next = ptr;
    q->rear = ptr;
}

qtype qPop( queue *q ) {
    /*
     * pops op from q->front.
     */
    qnode *ptr = q->front;
    qtype op;

    if( qEmpty(q) )
        return (qtype)-1;
    q->front = q->front->next;
    if( qEmpty(q) )
        q->rear = NULL;

    op = ptr->op;
```

```
    free(ptr);

    return op;
}

/*************************** b.c ****************************/
#include <stdio.h>

#define K 5

typedef struct node node;
typedef int type;
typedef node *btree;
typedef enum {FALSE, TRUE} bool;
#include "b.q.c"

struct node {
    type val[K];       // data in the node.
    // actually first K-1 vals are valid. Kth entry is used
              // during breaking of the node.
    btree ptr[K+1];   // pointers to other nodes in the node.
    // one extra ptr to be used in some loops.
    int nptr;          // no of pointers in the node = non-null ptrs.
    // this can also be used to see whether a node is a leaf.
    int nval;          // no of values in the node.
    // this is reqd even when nptr is present: for leaf nodes.
              // for non-leaf nodes: nptr = nval+1.
              // for leaf nodes: nptr = 0.
};

btree bNew() {
    /*
     * returns a new initialized node.
     */
    btree tree = (btree)malloc(sizeof(node));
    tree->nval = tree->nptr = 0;
    return tree;
}

void insertVal(btree tree, type val, btree ptr) {
    /*
     * insert (val, ptr) in node pointed to by tree without any checks.
     */
```

```
        int i, j;

        for(i=0; i<tree->nval && tree->val[i]<val; ++i)      // since K is
usually < 10, dont use binsrch.
            ;
        // the val should be inserted at tree->val[i].
        // shift-next the next values by one position.
        for(j=tree->nval-1; j>=i; —j) {
            tree->val[j+1] = tree->val[j];
            tree->ptr[j+2] = tree->ptr[j+1];
        }
        // insert val now at i.
        tree->val[i] = val;
        tree->nval++;
        tree->ptr[i+1] = ptr;
        if(ptr != NULL)   // tree is NOT a leaf.
            tree->nptr++;
        printf("\tval %d inserted at position %d, tree->nptr=%d tree-
>nval=%d.\n", val, i, tree->nptr, tree->nval);
    }

    btree getSplitNode(btree tree) {
        /*
         * returns a new node containing vals and pointers from tree.
         */
        int i, j;
        btree ptr = bNew();

        // copy vals.
        for(i=(K-1)/2+1, j=0; i<K; ++i, ++j)
            ptr->val[j] = tree->val[i];
        ptr->nval = K/2;
        tree->nval = K-K/2-1;    // temporarily this node contains an extra
val.

        // copy ptrs.
        for(i=(K-1)/2+1, j=0; i<=K; ++i, ++j)
            ptr->ptr[j] = tree->ptr[i];
        if(tree->nptr > 0) {     // non-leaf nodes.
            ptr->nptr = K/2+1;
            tree->nptr = K-K/2;
        }
        else                // leaf nodes.
```

```
        tree->nptr = ptr->nptr = 0;
    return ptr;
}

btree bMakeChanges(btree tree, btree ptr, btree parent) {
    /*
     * last val of tree contains an extra val which should be
     * inserted in parent.
     * if parent == NULL, then tree was root.
     * tree = parent->ptr[i] if parent is NOT NULL.
     */
    // extract the last value from tree.
    type val = tree->val[tree->nval];
    printf("in bMakeChanges().\n");

    if(parent == NULL) {
        parent = bNew();
        parent->ptr[0] = tree;
        parent->nptr = 1;
    }
    if(parent->nval < K-1) { // parent has space.
        insertVal(parent, val, ptr);
    }
    else {                                          // parent full.
        printf("parent is full.\n");
        insertVal(parent, val, ptr);
        return getSplitNode(parent);
    }
    return parent;
}

btree bInsert(btree tree, type val) {
    /*
     * calls insert to insert val in tree.
     * if the return node is diff from tree, that means a new node has
     * been created. Thus calls bMakeChanges().
     */
    btree insert(btree tree, type val);

    btree ptr = insert(tree, val);
    if(ptr != tree)   // node was split.
        return bMakeChanges(tree, ptr, NULL);
    return tree;
```

```
}

btree insert(btree tree, type val) {
    /*
     * inserts val in tree.
     * returns tree if there is no change.
     * if there is creation of a new node then the new node is
returned.
     */
    int i;
    btree ptr;

    if(tree->nptr > 0) {     // non-leaf.
        for(i=0; i<tree->nval && tree->val[i]<val; ++i)   // since K is
usually < 10, dont use binsrch.
                ;
        // the val should be in a tree pointed to by tree->ptr[i].
        printf("\tval should be inserted in tree->ptr[%d].\n", i);
        ptr = insert(tree->ptr[i], val);
        if(ptr != tree->ptr[i])
                return bMakeChanges(tree->ptr[i], ptr, tree);
        return tree;
    }
    else {      // tree is a leaf.
        if(tree->nval < K-1) {       // space is available in the leaf.
                insertVal(tree, val, NULL);
                return tree;
        }
        else {          // leaf full!
                printf("\tleaf is full.\n");
                insertVal(tree, val, NULL);
                // now break the leaf node.
                return getSplitNode(tree);
        }
    }
}

btree getAdjTree(btree tree, int offset, btree parent, int treeindex) {
    /*
     * returns parent[treeindex+offset] if exists else NULL.
     */
    int newindex = treeindex+offset;
```

```
    if(newindex >= 0 && newindex < parent->nptr)
        return parent->ptr[newindex];
    return NULL;
}

void combineNodes(btree left, btree right) {
    /*
     * left += right.
     */
    int i;
    int nptrleft = left->nptr;

    for(i=0; i<right->nval; ++i)
        left->val[left->nval++] = right->val[i];

    for(i=0; i<right->nptr; ++i)
        left->ptr[nptrleft+i] = right->ptr[i];
    if(left->nptr > 0)        // non-leaf.
        left->nptr += i;
}

type getNextVal(btree ptr) {
    /*
     * return the first value in the first leaf accessible from ptr.
     */
    if(ptr->nptr > 0) // still this is a non-leaf.
        return getNextVal(ptr->ptr[0]);
    return ptr->val[0];       // got it!
}

btree deleteVal(btree tree, int i) {
    /*
     * remove (val, ptr) at position i from tree without any checks.
     * and return tree->ptr[i+1].
     */
    btree rightptr = tree->ptr[i+1];        // ptr being removed along
with val.

    if(i == -1)       // special case for a dummy call to this function.
        return;

    for(++i; i<tree->nval; ++i) {
        tree->val[i-1] = tree->val[i];
```

```
            tree->ptr[i] = tree->ptr[i+1];
        }
        tree->nval–;

        if(tree->nptr > 0)        // if it is a non-leaf.
            tree->nptr–;

        return rightptr;
    }

    btree bApplyChanges(btree tree, btree parent, int treeindex) {
        /*
         * apply changes: tree is a non-leaf and tree = parent-
>ptr[treeindex].
         * also tree->nval < K/2.
         */
        int parentvalindex;
        int adjtreevalindex;
        btree returntree;
        btree adjtree;
        int offset = -1;  // predecessor.
        btree adjtreeleft;

        if(parent && tree->nval >= K/2)
            return tree;
        else if(parent == NULL && tree->nval == 0) {
            free(tree);
            return tree->ptr[0];
        }
        else if(parent == NULL)
            return tree;

        // parent is NOT NULL.

        adjtreeleft = adjtree = getAdjTree(tree, offset, parent,
treeindex);       // predecessor.
        if(!adjtree || (adjtree && adjtree->nval <= K/2)) {   // no extra
val.
            offset = 1;           // successor.
            adjtree = getAdjTree(tree, offset, parent, treeindex);
// successor.
            if(!adjtree || (adjtree && adjtree->nval <= K/2)) {      // no
```

```
extra val here too.
                btree parentparent;
                int parentindex;
                type parentval;
                btree returntree;

                //printf("combine tree, parent median val, adjtree.\n");
                //printf("also check parent for having <K/2 vals.\n");
                // it is NOT possible that adjtreeleft and right both are
NULL.
                // make adjtree point to the one which is NOT NULL.
                if(!adjtree) {
                      adjtree = adjtreeleft;
                      offset = -1;
                }
                // adjtree points to the sibling: left or right.
                parentvalindex = (2*treeindex+offset)/2;
                // the parent val is indexed.

                parentval = parent->val[parentvalindex];
                deleteVal(parent, parentvalindex);
   // the return val is tree/adjtree. Hence dont worry.
                if(offset == -1) {
                      // combine adjtree, parentval and tree.
                      adjtree->val[adjtree->nval++] = parentval;
                      combineNodes(adjtree, tree);
                      free(tree);
                      returntree = adjtree;
                }
                else { // offset == 1: right sibling.
                      // combine tree, parentval and adjtree.
                      tree->val[tree->nval++] = parentval;
                      combineNodes(tree, adjtree);
                      free(adjtree);
                      returntree = tree;
                }
                return returntree;
          }
          else {
                adjtreevalindex = 0;
                returntree = adjtree;
          }
```

```
        }
        else {
            adjtreevalindex = adjtree->nval-1;
            returntree = tree;
        }
        // adjtree has an extra val.
        parentvalindex = (2*treeindex+offset)/2;
        // insert parent val in tree.
        insertVal(tree, parent->val[parentvalindex], returntree->ptr[0]);
        // now promote val in adjtree to parent.
        parent->val[parentvalindex] = adjtree->val[adjtreevalindex];
        returntree->ptr[0] = deleteVal(adjtree, adjtreevalindex);

        return tree;
    }

    btree delete(btree tree, type val, int i, btree parent, int treeindex)
{
        /*
         * delete val from tree. val == tree->val[i].
         * and tree == parent->nptr[treeindex].
         */
        int parentvalindex;
        int adjtreevalindex;
        btree adjtree;
        btree bDelete(btree tree, type val, btree parent, int treeindex);

        if(tree->nptr == 0) {    // leaf.
            deleteVal(tree, i);

        // find two adjacent nodes to tree(succ and pred) and
        // see whether any of them has >K/2 vals. if yes then bring
        // the parent value between tree and the node into tree and
        // promote the value in the adjacent node into the parent.
        // if no such adjacent node exists, combine tree, parent val and
        // the adjacent node.
        // if this leaves parent node having <K/2 vals, go into else.
        // this is done in bApplyChanges().
        }
        else {                                    // NOT a leaf.
            // find the next val in inorder traversal of this val and
replace
```

```
            // this with that val. then delete that val from tree.
            type nextval = getNextVal(tree->ptr[i+1]); // since tree->val[i]
exists

    // tree->val[i+1] exists.
            tree->val[i] = nextval;
            bDelete(tree->ptr[i+1], nextval, tree, i+1);
        }
        return bApplyChanges(tree, parent, treeindex);
    }

    btree bDelete(btree tree, type val, btree parent, int treeindex) {
        /*
         * delete val from tree if exists.
         * tree == parent[treeindex].
         */
        int i;

        for(i=0; i<tree->nval && tree->val[i]<val; ++i)       // since K is
usually < 10, dont use binsrch.
            ;
        if(tree->val[i] == val) {
            printf("val=%d found: to be deleted.\n", val);
            return delete(tree, val, i, parent, treeindex);
        }
        else if(tree->nptr > 0) {        // the val should be in a tree
pointed to by tree->ptr[i].
            //printf("val=%d may be in tree->ptr[%d].\n", val, i);
            bDelete(tree->ptr[i], val, tree, i);
            //printf("now check tree for <50% values.\n");
            return bApplyChanges(tree, parent, treeindex);
        }
        else {     // leaf reached.
            printf("val=%d does NOT exist.\n", val);
            return tree;
        }
    }
    void bPrintFormatted(btree tree) {
        btree ptr;
        int i;
        queue q = {NULL};
        qPush(&q, tree);
        qPush(&q, NULL);
```

```
    printf("————————————————————————\n");

    while(qEmpty(&q) == FALSE) {
        ptr = qPop(&q);
        if(ptr) {
                for(i=0; i<ptr->nval-1; ++i)
                        printf("%d-", ptr->val[i]);
                if(i<ptr->nval)
                        printf("%d  ", ptr->val[i]);
                for(i=0; i<ptr->nptr; ++i)
                        qPush(&q, ptr->ptr[i]);
        }
        else {
                printf("\n");
                if(qEmpty(&q) == FALSE)
                        qPush(&q, NULL);
        }
    }
    printf("————————————————————————\n");
}

int main() {
    btree tree = bNew();
    tree = bInsert(tree, 1);
    tree = bInsert(tree, 2);
    tree = bInsert(tree, 4);
    tree = bInsert(tree, 3);
    tree = bInsert(tree, 5);
    tree = bInsert(tree, 6);
    tree = bInsert(tree, 7);
    tree = bInsert(tree, 8);
    tree = bInsert(tree, 9);
    tree = bInsert(tree, 10);
    tree = bInsert(tree, 11);
    tree = bInsert(tree, 12);
    tree = bInsert(tree, 13);
    tree = bInsert(tree, 14);
    tree = bInsert(tree, 15);
    tree = bInsert(tree, 16);
```

```
    tree = bInsert(tree, 17);
    bPrintFormatted(tree);
    tree = bDelete(tree, 6, NULL, -320);
    bPrintFormatted(tree);
    tree = bDelete(tree, 9, NULL, -320);
    bPrintFormatted(tree);

    return 0;
}
```

Explanation

1. A B-tree of order K is a K-way search tree that is either empty or is of height > 0 and satisfies the following properties:
   (a) the root node has at least two children.
   (b) all nodes other than the root node and leaf nodes have at least ceil(K/2) children.
   (c) all leaf nodes are at the same level.

   It is easier to insert and delete nodes into a B-tree retaining the B-tree properties than it is to maintain the best possible K-way search tree at all times. Thus, the reasons for using B-trees rather than optimal K-way search trees for indices are the same as those for using AVL trees, as opposed to optimal binary search trees, when maintaining dynamic internal tables.
2. Each B-tree node of order K has space for K pointers and K – 1 values. We maintain an extra space for one pointer and one value which will be useful in insertion. We maintain counts of the number of values and number of pointers present in the node. Thus the node structure for a B-tree is as follows:

```
typedef struct node *btree;
struct node {
        type val[K];
        btree ptr[K+1];
        int nptr;
        int nval;
};
```

   nptr is the number of non-null pointers and nval is the number of valid values in a node. For a leaf-node, nptr == 0, while for all other nodes, nptr == nval + 1.

3. Insertion in a B-tree takes place by traversing the search tree for the value starting from root. When the search terminates in a leaf node, the value is inserted in the node. The node is then checked to see if thus more than K – 1 values. If the node has less than K values, the insertion is done. Otherwise, the node is split into two nodes of sizes floor(K/2) and the middle value is propagated into the parent node. This procedure is repeated until either the insertion stops (as there is space for the new value in a node) or nodes at all the levels in the tree get split and, finally, a new root node needs to be created. The value propagated from the original root is then put into the new root node and its pointers are adjusted accordingly.

   Examples: Original tree of order K = 3

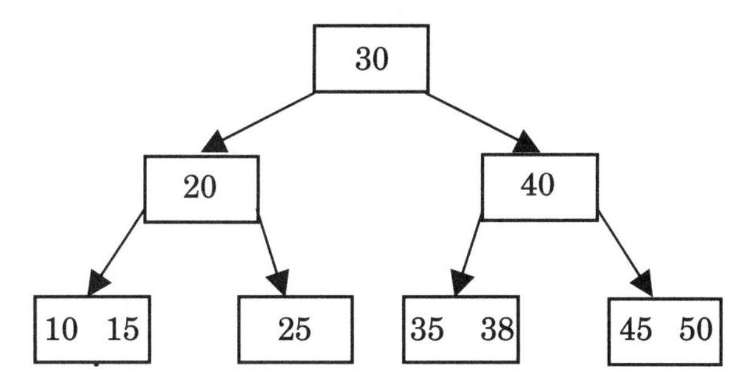

Insertion of value 28

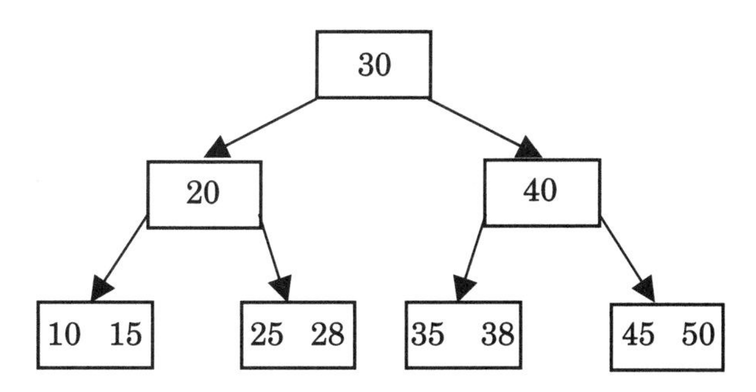

Insertion of value 55

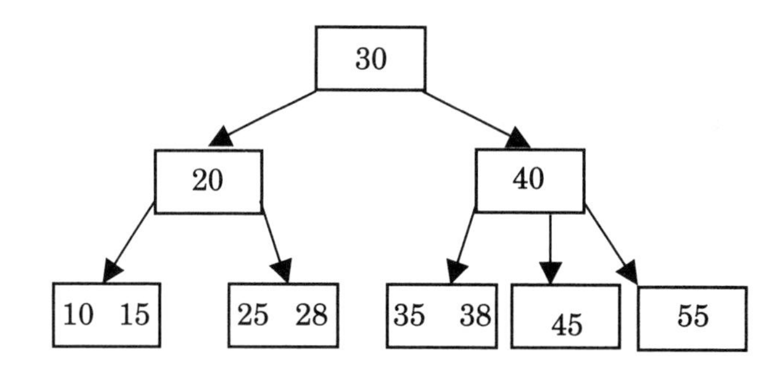

Insertion of value 37

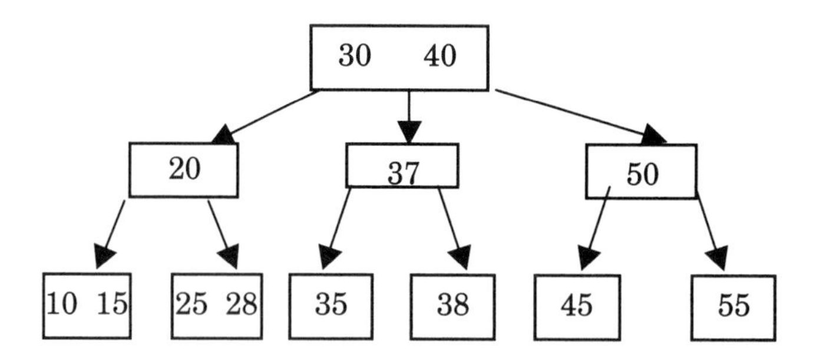

Insertion of value 5

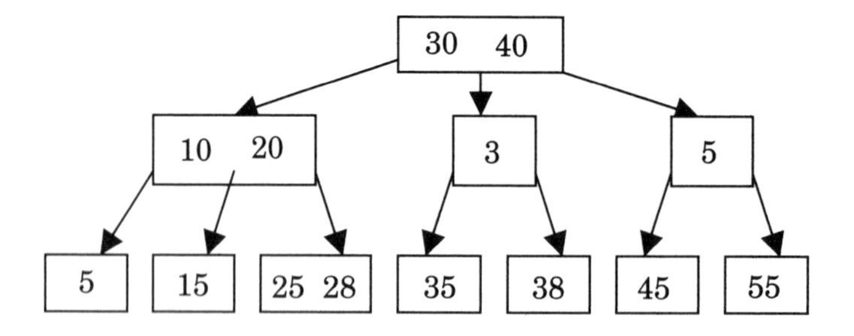

4. Deletion of a value from a B-tree can be done as simply as performing an insertion. If, even after deletion of the value from a leaf node p, the node contains >= K/2 values, then nothing needs to be done. Otherwise the sibling nodes are checked to see if they have > K/2 values. If such a sibling q exists, then the value in the parent node between the two adjacent leaf nodes is demoted to node p, and a value in node q is promoted to the parent (this is called rotation). However, if the value v to be deleted is in a non-leaf node p, then the next value (w) to v in the search order (which appears in the right subtree of v in the leaf node) is promoted at the place of v and the value w is deleted from the leaf node. If, in this promotion and demotion, the parent node r contains less than K/2 values, then its adjacent node s is promoted to its parent t and the value in t is demoted to r. Also, the adjacent pointer to the value promoted from s becomes the adjacent pointer to the parent value that has been newly inserted in r.

Examples: Original tree of order K = 3

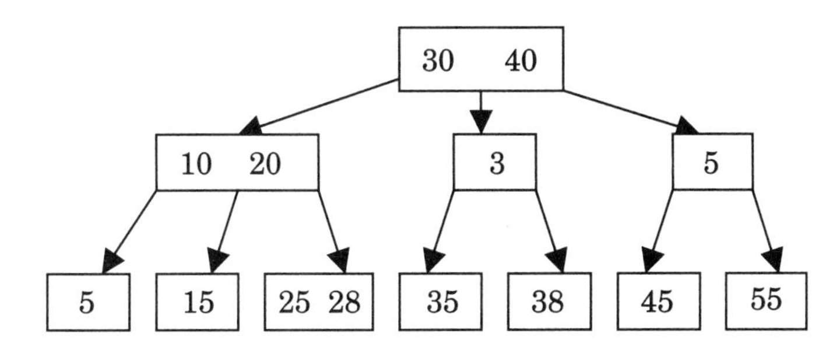

Deletion of value 25

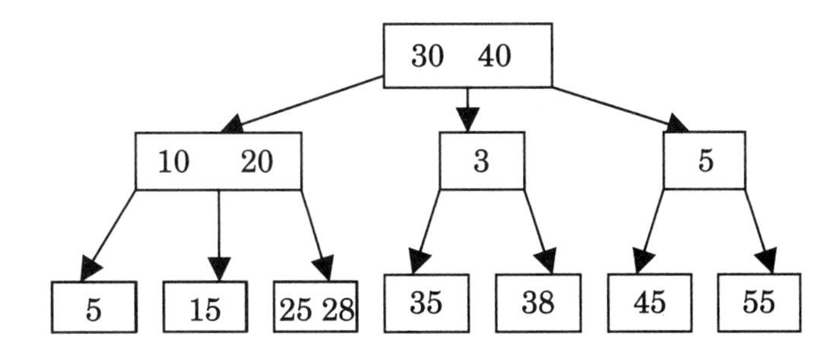

Deletion of value 35

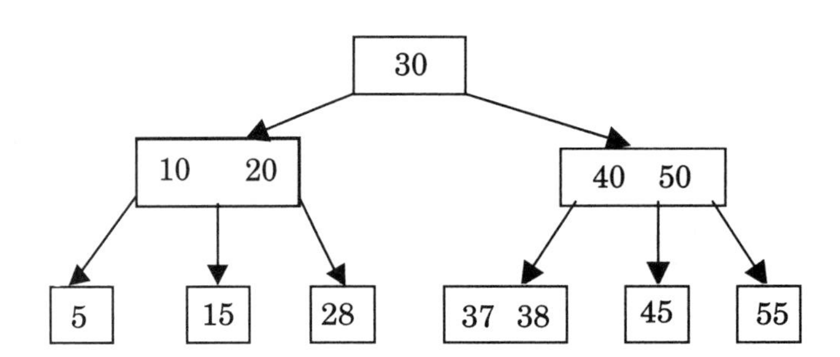

Deletion of value 30

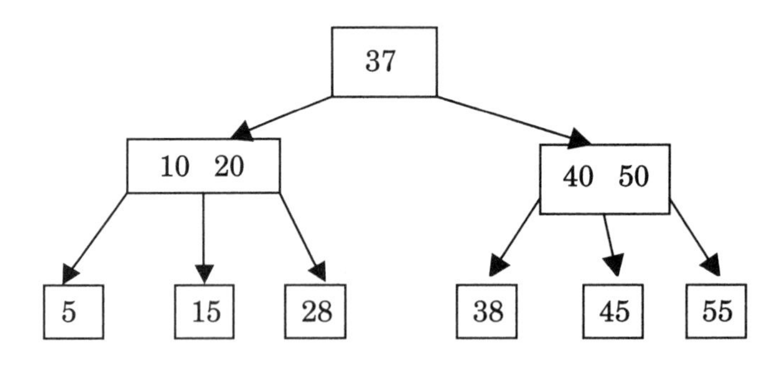

Deletion of value 37

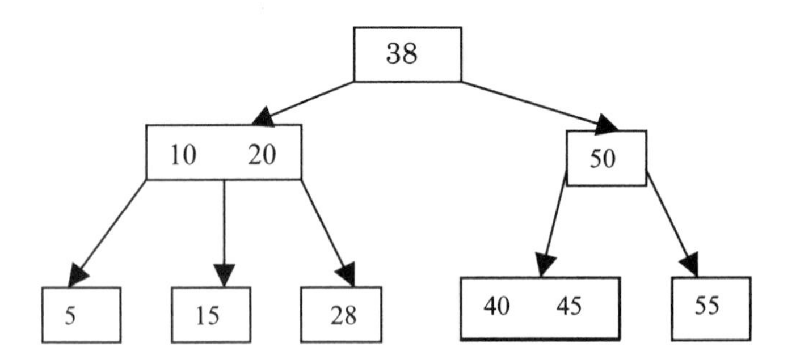

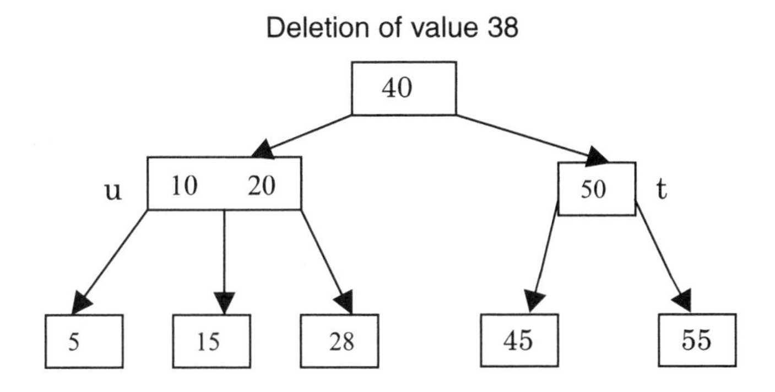

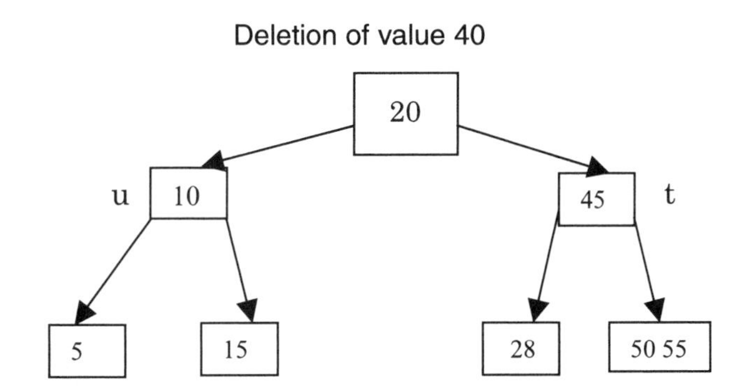

Note that in the deletion of value 40, since it is a non-leaf node, its next value in the search order, 45, is promoted to its place so that 45 appears in the root. Then 45 is deleted from the leaf node r. After deletion, r has less than K/2 values. So its sibling s is checked for any extra values. s, with a value of 55, has only K/2 values. So the values of r, s, and 50 (the value of the parent node t between the two children) are combined to get a node with the values (50, 55). Now t contains less than K/2 values. Since it is a non-leaf node, it checks its adjacent node to see if it has more than K/2 values. The node u, with the values (10, 20), actually has more than K/2 values. So the last value of u, 20, is promoted to the root and the value 45 is demoted to t. The rightmost sibling of u (the node containing a value of 28) becomes the leftmost sibling of t. 20, the rightmost sibling of u, becomes the leftmost sibling of t. A symetric transformation is done when the adjacent node appears in the right. If none of the adjacent nodes of t contain any extra values, then t, u, and the value in their parent node between these two pointers are combined to form one node. Then their parent node is checked to see if it has less than K/2 values, and the procedure continues.

5. The function hierarchy for insertion and deletion appears the same. For insertion, the main driver function is bInsert(), which calls insert() and checks its return value to see whether the level was changed in the function. If it was, then bMakeChanges(), which propagates this change upwards in the tree is called. insertVal() is the function that puts a value (along with its right pointer) in a node. The function getSplitNode() splits a node in two parts, each containing K/2 values. The $(K+1)^{th}$ value in the node being split is stored in the first node as the $(K/2)^{th}$ value, which is then used in propagation. insert() is the main function that checks whether the node is a leaf or a non-leaf, and accordingly calls the different functions that have jsut been described.

   For deletion, the main driver function is bDelete(), which calls itself recursively until it reaches a node that contains the value being deleted. It then calls delete(), which checks for the node to be a leaf or a non-leaf node and takes appropriate action as just described.

   During insertion, bApplyChanges() is called from bDelete() and delete(). bApplyChanges is similar to bMakeChanges, and does the job of demotion and propagation of values in the current node and its siblings. It uses a function called getAdjTree(), which returns a node adjacent to a node with the same parent. combineNodes(), a procedure complementary to getSplitNode(), combines the second node with the first. In order to get the next value in the search order from the tree, the function getNextVal() is called. deleteVal() is a function that is complementary to the function insertVal(), and is used to delete a value, along with its right pointer, from a node. It then returns the pointer.

6. The procedure bPrintFormatted() uses a queue to print the tree in the breadth-first manner.

Points to Remember

1. In a B-tree of order K, there can be at most K pointers in a node and K – 1 values.
2. All the leaf nodes of a B-tree are at the same level.
3. The complexity of both insertion and deletion is $O(h)$, where h is the height of the B-tree. This is because of the propagation of values upwards during insertion and downwards during deletion.

4. By writing the functions of insertion and deletion in a similar manner, the code becomes easier to write and understand.
5. A queue is required for printing a tree in the breadth-first format.
6. Searching for a value inside a node can be done using binary search as the values are sorted. However, if K is small enough, even the linear search will give a similar performance.

PROBLEM: IMPLEMENTATION OF A B+ TREE

Write procedures for insertion and deletion in a B+ tree.

Program

```
/*********************** bplus.q.c *************************/
#include <malloc.h>
#include <stdlib.h>

typedef struct qnode qnode;
typedef btree qtype;
typedef struct {
    qnode *front;
    qnode *rear;
} queue;

struct qnode {
    qtype op;
    qnode *next;
};
bool qEmpty( queue *q ) {
    return (q->front == NULL);
}

void qPush( queue *q, qtype op ) {
    /*
     * pushes op at q.rear.
     */
    qnode *ptr = (qnode *)malloc( sizeof(qnode) );
```

```
    ptr->op = op;
    ptr->next = NULL;
    if( qEmpty(q) )   // first element in q.
        q->front = ptr;
    else
        q->rear->next = ptr;
    q->rear = ptr;
}

qtype qPop( queue *q ) {
    /*
     * pops op from q->front.
     */
    qnode *ptr = q->front;
    qtype op;

    if( qEmpty(q) )
        return (qtype)-1;
    q->front = q->front->next;
    if( qEmpty(q) )
        q->rear = NULL;

    op = ptr->op;
    free(ptr);

    return op;
}

/************************** bplus.c *************************/
#include <malloc.h>
#include <stdio.h>
#define K 5

typedef struct node node;
typedef int type;
typedef node *btree;
typedef enum {FALSE, TRUE} bool;

#include "bplusq.c"

struct node {
    type val[K];    // data in the node.
```

```
// actually first K-1 vals are valid. Kth entry is used
                    // during breaking of the node.
    btree ptr[K+1]; // pointers to other nodes in the node.
 // one extra ptr to be used in some loops.
    int nptr;       // no of pointers in the node = non-null ptrs.
   // this can also be used to see whether a node is a leaf.
    int nval;       // no of values in the node.
   // this is reqd even when nptr is present: for leaf nodes.
   // for non-leaf nodes: nptr = nval+1.
   // for leaf nodes: nptr = 0.
   node *right;    // if this is a leaf node, this points to its
sibling.
};

btree bNew() {
    /*
     * returns a new initialized node.
     */
    btree tree = (btree)malloc(sizeof(node));
    tree->nval = tree->nptr = 0;
    tree->right = NULL;
    return tree;
}

void insertVal(btree tree, type val, btree ptr) {
    /*
     * insert (val, ptr) in node pointed to by tree without any checks.
     */
    int i, j;

    for(i=0; i<tree->nval && tree->val[i]<val; ++i) ;
// since K is usually < 10,dont use binsrch.

// the val should be inserted at tree->val[i].
// shift-next the next values by one position.
    for(j=tree->nval-1; j>=i; --j) {
        tree->val[j+1] = tree->val[j];
        tree->ptr[j+2] = tree->ptr[j+1];
    }
    // insert val now at i.
    tree->val[i] = val;
    tree->nval++;
```

```
        tree->ptr[i+1] = ptr;
        if(ptr != NULL) // tree is NOT a leaf.
            tree->nptr++;
         printf("\tval %d inserted at position %d, tree->nptr=%d tree-
>nval=%d.\n", val, i, tree->nptr, tree->nval);
    }

    btree getSplitNode(btree tree) {
        /*
         * returns a new node containing vals and pointers from tree.
         */
        int i, j;
         btree ptr = bNew();

        // copy vals.
        for(i=(K-1)/2+1, j=0; i<K; ++i, ++j)
            ptr->val[j] = tree->val[i];
        ptr->nval = K/2;
        if(tree->nptr > 0)  // non-leaf node.
        tree->nval = K-K/2-1;   // temporarily this node contains an extra
val.
        else                 // leaf node.
            tree->nval = K-K/2;

        // copy ptrs.
        //for(i=(K-1)/2, j=0; i<=K; ++i, ++j)
        for(i=(K-1)/2+1, j=0; i<=K; ++i, ++j)
            ptr->ptr[j] = tree->ptr[i];
        if(tree->nptr > 0) {    // non-leaf nodes.
            ptr->nptr = K/2+1;
            tree->nptr = K-K/2;
        }
        else {      // leaf nodes.
            tree->nptr = ptr->nptr = 0;
            tree->right = ptr;      // bplus tree: list of leaves.
        }

        return ptr;
    }

    btree bMakeChanges(btree tree, btree ptr, btree parent) {
        /*
```

```
         * last val of tree should be inserted in parent.
         * if parent == NULL, then tree was root.
         * tree = parent->ptr[i] if parent is NOT NULL.
         */
        // extract the last value from tree.
        type val = (tree->nptr>0 ? tree->val[tree->nval] : tree->val[tree-
>nval-1]);
        printf("in bMakeChanges().\n");

        if(parent == NULL) {
             parent = bNew();
            parent->ptr[0] = tree;
            parent->nptr = 1;
        }
        if(parent->nval < K-1) {    // parent has space.
            insertVal(parent, val, ptr);
        }
        else {                      // parent full.
            printf("parent is full.\n");
            insertVal(parent, val, ptr);
            return getSplitNode(parent);
        }
        return parent;
    }

    btree bInsert(btree tree, type val) {
        /*
         * calls insert to insert val in tree.
         * if the return node is diff from tree, that means a new node has
         * been created. This calls bMakeChanges().
         */
        btree insert(btree tree, type val);

        btree ptr = insert(tree, val);
        if(ptr != tree) // node was split.
            return bMakeChanges(tree, ptr, NULL);
        return tree;
    }

    btree insert(btree tree, type val) {
        /*
         * inserts val in tree.
```

```
        * returns tree if there is no change.
        * if there is creation of a new node then the new node is
returned.
        */
       int i;
       btree ptr;

       if(tree->nptr > 0) {    // non-leaf.
            for(i=0; i<tree->nval && tree->val[i]<val; ++i);
   // since K is usually <  10, even sequential search is fine.

   // the val should be in a tree pointed to by tree->ptr[i].
           printf("\tval should be inserted in tree->ptr[%d].\n", i);
           ptr = insert(tree->ptr[i], val);
           if(ptr != tree->ptr[i])
               return bMakeChanges(tree->ptr[i], ptr, tree);
           return tree;
       }
       else {  // tree is a leaf.
           if(tree->nval < K-1) {  // space is available in the leaf.
               insertVal(tree, val, NULL);
               return tree;
           }
           else {      // leaf full!
               printf("\tleaf is full.\n");
               insertVal(tree, val, NULL);
               // now break the leaf node.
               return getSplitNode(tree);
           }
       }
   }

       tree = bInsert(tree, 6);
       tree = bInsert(tree, 7);
       tree = bInsert(tree, 8);
       tree = bInsert(tree, 9);
       tree = bInsert(tree, 10);
       tree = bInsert(tree, 11);
       tree = bInsert(tree, 12);
       tree = bInsert(tree, 13);
       tree = bInsert(tree, 14);
       tree = bInsert(tree, 15);
```

```
    tree = bInsert(tree, 16);
    tree = bInsert(tree, 17);
    bPrintFormatted(tree);

    tree = bDelete(tree, 6, NULL, -320);
    bPrintFormatted(tree);
    tree = bDelete(tree, 5, NULL, -320);
    bPrintFormatted(tree);
    tree = bDelete(tree, 7, NULL, -320);
    bPrintFormatted(tree);
    tree = bDelete(tree, 8, NULL, -320);
    bPrintFormatted(tree);
    tree = bDelete(tree, 9, NULL, -320);
    bPrintFormatted(tree);
    tree = bDelete(tree, 10, NULL, -320);
    bPrintFormatted(tree);
    tree = bDelete(tree, 11, NULL, -320);
    bPrintFormatted(tree);

    printf("The linked list of leaf nodes:\n");
    for(; ptr; ptr=ptr->right) {
        for(i=0; i<ptr->nval; ++i)
            printf("%d-", ptr->val[i]);
        printf("  ");
    }
    printf("\n");
    return 0;
}
```

Explanation

1. A B+ tree of order K is a K-way search tree that is either empty or is of height > 0 and satisfies the following properties:
   (a) The root node has at least two children.
   (b) All nodes other than the root node and leaf nodes have at least ceil(K/2) children.
   (c) All leaf nodes are at the same level.
   (d) All values appear in the leaf nodes.
   (e) The leaf nodes are linked from left to right.

2. Each B+ tree node of order K has space for K pointers and K – 1 values. We maintain extra space for one pointer and one value, which will be useful in insertion. We maintain counts of the number of values and the number of pointers present in the node. For the linked list of leaf nodes, another pointer is needed. Thus the node structure for a B+ tree is as follows:

```
typedef struct node *btree;
struct node {
    type val[K];
    btree ptr[K+1];
    int nptr;
    int nval;
    node *right;
};
```

   `nptr` is the number of non-null pointers and `nval` is the number of valid values in a node. For a leaf node, `nptr = 0` while for all other nodes, `nptr = nval + 1`. The pointer right points to the node to the right of this node.

3. Insertion in a B+ tree takes place by traversing the search tree for the value starting from the root. The search always terminates in a leaf node. The value is inserted in the leaf node. The node is then checked to see if it has more than K–1 values. If the node has less than K values, the insertion is performed. Otherwise, the node is split into two nodes of sizes floor(K/2) and the middle value is 'copied' into the parent node. In case of a non-leaf node, the middle value is 'moved' to the parent node. This procedure is repeated until either the insertion stops (as there is space for the new value in a node) or nodes at all the levels in the tree get split, and finally a new root node needs to be created. The value propagated from the original root is then put into the new root node and its pointers are adjusted accordingly.

Examples:

Original tree of order K = 3

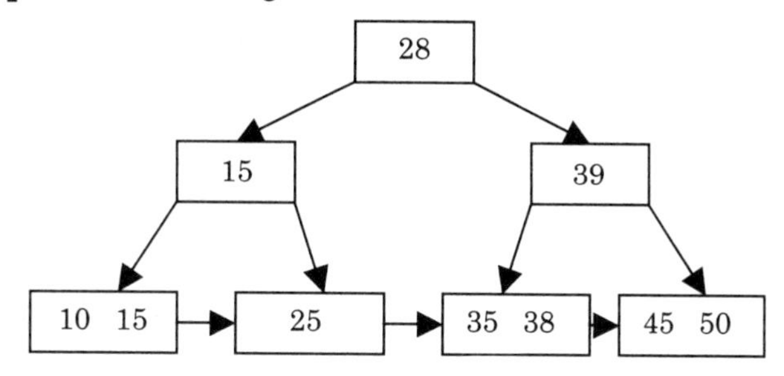

Note that for a value v in a node and its two adjacent pointers on left and right, all values on left are £ v and all values on right are > v.

Insertion of value 28

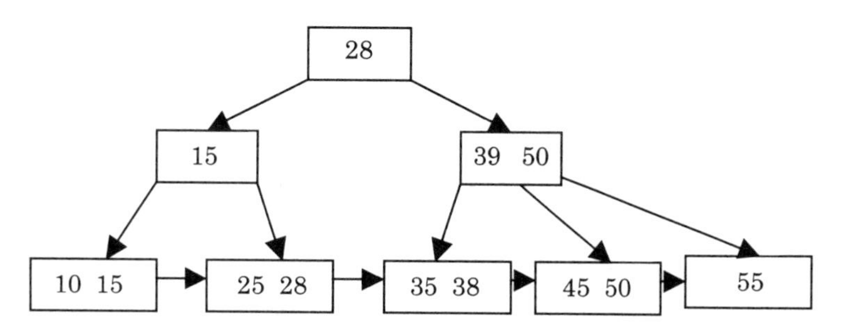

Insertion of value 55

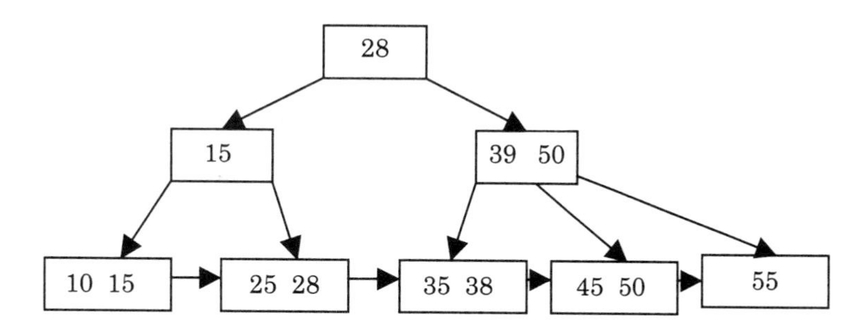

Insertion of value 37

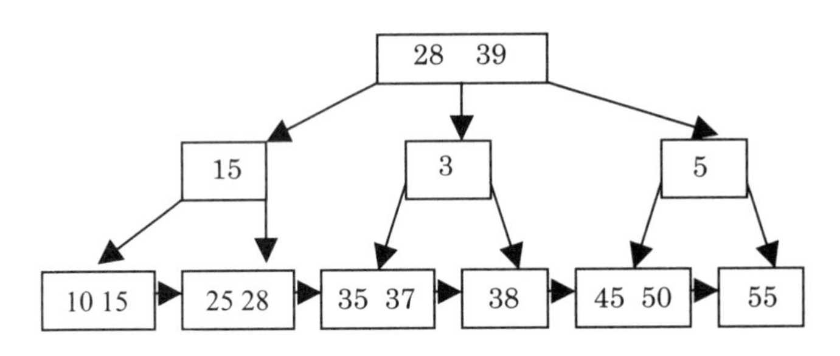

Insertion of value 5

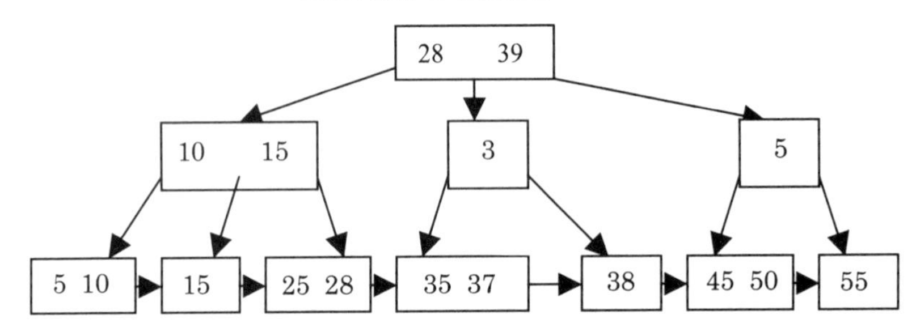

4. Deletion of a value from a B+ tree can be done as simply as insertion is done. If, even after deletion of the value from a leaf node p, the node contains >= K/2 values, then nothing needs to be done. Otherwise the sibling nodes are checked for > K/2 values. If such a sibling q exists, then the adjacent value in q is moved to the node p and the value in the parent node between the two leaf nodes is replaced by the last value in its left child. If such a sibling q does not exist, then nodes p and q are combined, and the value in the parent node between these two leaf nodes is deleted. If in this process, the parent node r contains less than K/2 values, then its adjacent node s is promoted to its parent t and the value in t is demoted to r. Also, the adjacent pointer to the value promoted from s becomes the adjacent pointer to the parent value newly inserted in r.

Examples

Original tree of order K = 3

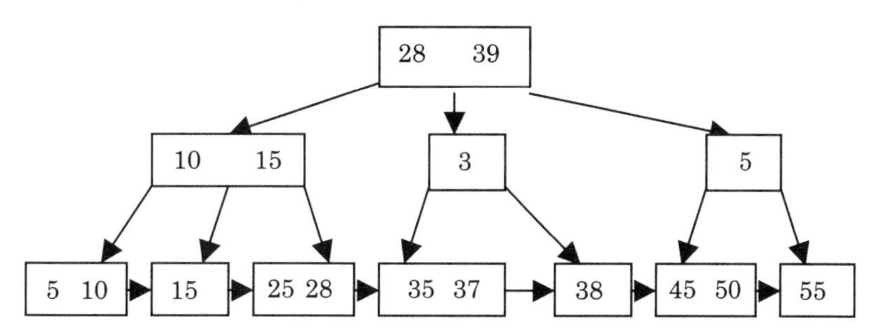

Deletion of value 25

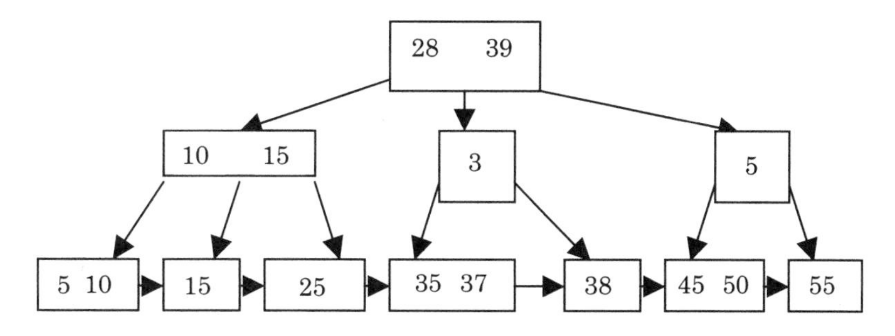

Deletion of value 35

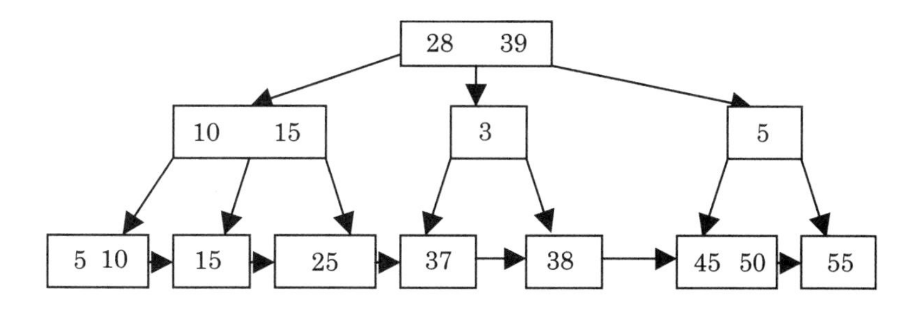

Deletion of value 37

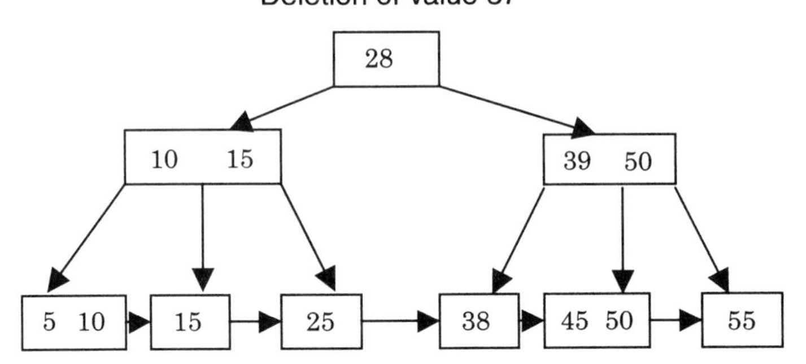

Deletion of value 38

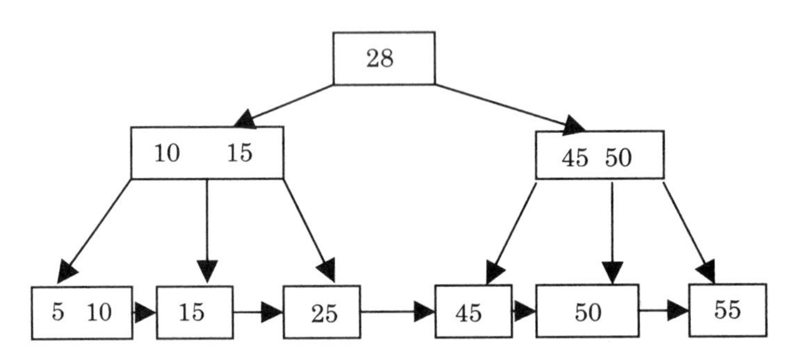

Deletion of value 50

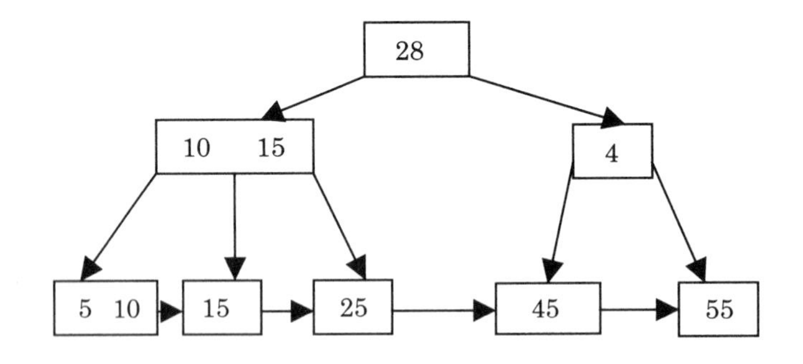

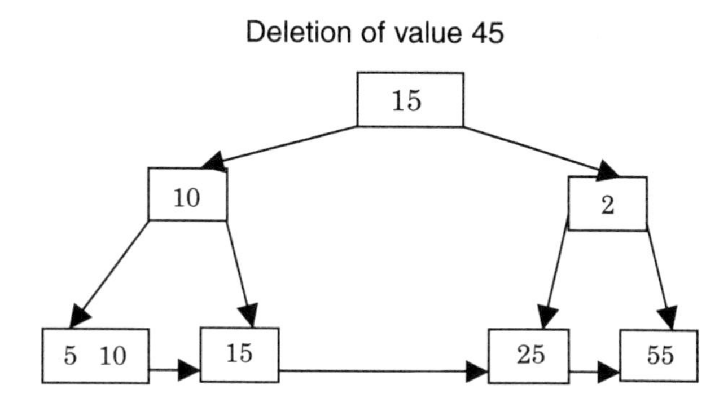

Note that the value 45 is deleted from the leaf node r. After deletion, r has less than K/2 values. So its sibling s is checked for any extra values. s, with a value of 55, has only K/2 values. So the values of r, s, and 50 (the value of the parent node t between the two children), are combined to get a node with the value 55, and the value 45 is deleted from t, along with its right pointer. r's right pointer is appropriately updated. Now t contains less than K/2 values. Since it is a non-leaf node, it checks its adjacent node to see if it has more than K/2 values. The node u, with the values (10, 15) actually has more than K/2 values. So the last value of u, 15, is promoted to the root and the value 28 is demoted to t. The rightmost sibling of u (the node containing a value of 25) becomes the leftmost sibling of t. A symmetric transformation is done when the adjacent node appears in the right. If none of the adjacent nodes of t contain any extra values, then the values of r, u, and the value in their parent node between these two pointers are combined to form one node. Then their parent node is checked to see if it has less than K/2 values, and the procedure continues.

5. The function hierarchy for insertion and deletion appears the same. For insertion, the main driver function is `bInsert()`, which calls `insert()` and checks its return value to see whether the level was changed in the function. If it was, `bMakeChanges()` is called, which propagates this change upwards in the tree. `insertVal()` is the function that puts a value (along with its right pointer) in a node. The function `getSplitNode()` splits a node in two parts, each containing K/2 values. The $(K+1)^{th}$ value in the node being split is stored in the first node as the $(K/2)^{th}$ value, which is then used in propagation. `insert()` is the main function that checks whether the node is a leaf or a non-leaf and accordingly calls the different functions that have just been described.

For deletion, the main driver function is bDelete(), which calls itself recursively until it reaches the leaf node that contains the value being deleted. It then calls delete(), which takes appropriate action as just described. During insertion, bApplyChanges() is called from bDelete() and delete(). bApplyChanges is similar to bMakeChanges, and does the job of demotion and propagation of values in the current node and its siblings. It uses a function called getAdjTree(), which returns a node adjacent to a node with the same parent. combineNodes(), a procedure complementary to getSplitNode(), combines the second node with the first. In order to get the next value in the search order from the tree, the function getNextVal() is called. deleteval() is a function that is complementary to the function insertVal(), and is used to delete a value, along with its right pointer, from a node. It then returns the pointer.

6. The procedure bPrintFormatted() uses a queue to print the tree in the breadth-first manner.

Points to Remember

1. In a B+ tree of order K, there can be at most K pointers in a node and K–1 values.
2. All the leaf nodes of a B+ tree are at the same level.
3. All the inserted values in a B+ tree are present in leaf nodes. The non-leaf nodes may contain values that are not present in leaf nodes.
4. The complexity of both insertion and deletion is $O(h)$, where h is the height of the B-tree. This is because of the propagation of values upwards during insertion and downwards during deletion.
5. A queue is required for printing a tree in the breadth-first format.
6. Searching for a value inside a node can be done using binary search as the values are sorted. However, if K is small enough, even the linear search will give a similar performance.
7. The updation of a value can be simulated by deletion followed by insertion.

28 Problems in Graphs

PROBLEM: THE DFS METHOD FOR GRAPH TRAVERSAL

Write a function dfs(v) to traverse a graph in a depth-first manner starting from vertex v. Use this function to find connected components in the graph. Modify dfs() to produce a list of newly visited vertices. The graph is represented as an adjacency matrix.

Program

```
#include <stdio.h>

#define MAXVERTICES 20
#define MAXEDGES    20

typedef enum {FALSE, TRUE, TRISTATE} bool;

struct graph {
    int matrix[MAXVERTICES][MAXEDGES];
    int vertices, edges;
}graph;

void buildINC( int edges[][MAXEDGES], int nedges ) {
    /*
     * fills graph.matrix with information from edges.
     * graph.edges = nedges.
     * graph.vertices is maxEntry in edges.
     */
    int i, j;

    graph.vertices = -1;
    graph.edges    = nedges;
```

```
    // init matrix to FALSE.
    for( i=0; i<MAXVERTICES; ++i )
        for( j=0; j<MAXEDGES; ++j )
            graph.matrix[i][j] = FALSE;

    // now enter values into it.
    for( i=0; i<2; ++i )
        for( j=0; j<nedges; ++j ) {
            graph.matrix[ edges[i][j] ][j] = TRUE;
            if( edges[i][j] > graph.vertices )
                graph.vertices = edges[i][j];
        }
    graph.vertices++;           // no of vertices = maxvertes + 1;
}

void printINC() {
    /*
     * prints graph.
     */
    int i, j;
    for( i=0; i<graph.vertices; ++i ) {
        for( j=0; j<graph.edges; ++j )
            printf( "%d ", graph.matrix[i][j] );
        printf( "\n" );
    }
    printf( "\n" );
}

int getOtherVertex( int edge, int v ) {
    /*
     * returns vertex at the other end of edge whose one vertex is v.
     */
    int i;

    for( i=0; i<graph.vertices; ++i )
        if( i != v && graph.matrix[i][edge] == TRUE )
            return i;
    printf( "getOtherVertex(): This should not be printed.\n" );
    return -1;
}
```

```
int getAdj( int v, int *adjv ) {
    /*
     * using graph, finds adj nodes of v and stores them in adjv.
     * returns no of such adj vertices.
     */
    int j;
    int *adjvptr = adjv;

    for( j=0; j<graph.edges; ++j )
        if( graph.matrix[v][j] == TRUE )
                *adjvptr++ = getOtherVertex( j, v );
    return adjvptr-adjv;
}

void dfs( int v, int *visited ) {
    /*
     * recursively traverse graph from v using visited.
     * and mark all the vertices that come in dfs path to TRISTATE.
     */
    int adjv[ MAXVERTICES ];
    int i;

    visited[v] = TRISTATE;

    for( i=getAdj(v,adjv)-1; i>=0; --i )
        if( visited[ adjv[i] ] == FALSE )
                dfs( adjv[i], visited );
}

void printSetTristate( int *visited ) {
    /*
     * prints all vertices of visited which are TRISTATE.
     * and set them to TRUE.
     */
    int i;

    for( i=0; i<graph.vertices; ++i )
        if( visited[i] == TRISTATE ) {
                printf( "%d ", i );
                visited[i] = TRUE;
```

```
        }
    printf( "\n\n" );
}

void compINC() {
    /*
     * prints all connected components of graph represented using INC.
     */
    int *visited;
    int i;

    visited = (int *)malloc( graph.vertices );
    for( i=0; i<graph.vertices; ++i )
        visited[i] = FALSE;

    for( i=0; i<graph.vertices; ++i )
        if( visited[i] == FALSE ) {
                dfs( i, visited );
                // print all vertices which are TRISTATE.
                // and mark them to TRUE.
                printSetTristate( visited );
        }
    free( visited );
}

int main() {
    int edges[][MAXEDGES] = {        {0,2,4,5,5,4},
                            {1,1,3,4,6,6}
                        };
    buildINC( edges, 6 );
    printINC();
    compINC();

    return 0;
}
```

Explanation

1. The graph is represented as an incidence matrix. The rows correspond to each vertex and the columns correspond to each edge in the graph. An entry

matrix[i][j] is TRUE if edge j contains vertex i, otherwise it is FALSE. For example, if the graph is as follows;

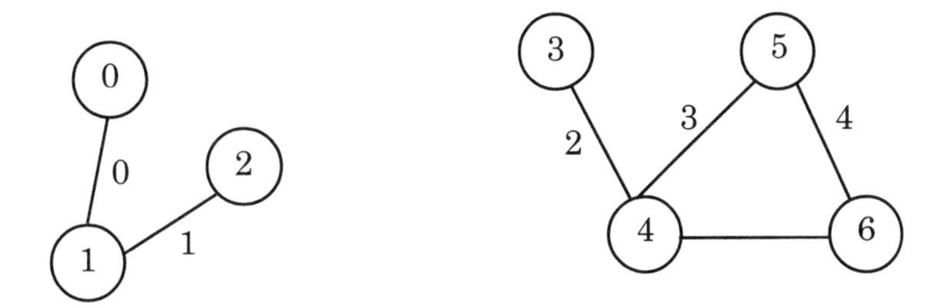

and the edges are numbered as shown, then the incidence matrix for the graph is as shown here:

	0	1	2	3	4	5
0	1	0	0	0	0	0
1	1	1	0	0	0	0
2	0	1	0	0	0	0
3	0	0	1	0	0	0
4	0	0	1	1	0	1
5	0	0	0	1	1	0
6	0	0	0	0	1	1

The rows of the incidence matrix are vertices and the columns are edges.

2. dfs(v) is implemented recursively. A Boolean vector visited[] is maintained, whose all entries are initially FALSE. dfs(v) marks v as visited by making visited[v] = TRUE. It then finds all the adjacent nodes of v and starts dfs() from those nodes which have not yet been visited.
3. compINC() is a function that finds all the connected components of a graph. It maintains a local copy of the vector visited[] and passes it as a parameter to dfs(v). compINC() passes that vertex as a parameter to dfs(), which is not yet visited. Thus each invoking of dfs() finds one connected component of the graph.
4. In order to modify dfs() to produce a list of newly visited vertices, we tag the vertices visited using dfs() as TRISTATE. In compINC(), all these TRISTATE vertices will form one connected component. This status is then converted to TRUE. The next invocation of dfs() returns another set of vertices tagged as TRISTATE, which forms another connected component, and so on.

For example, in this graph, first all vertices are tagged as FALSE. After invoking dfs(0), the vertices tagged as TRISTATE are {0, 1, 2}. These are output and their tags are changed from TRISTATE to TRUE. The next invoking of dfs(3) tags vertices {3, 4, 5, 6} as TRISTATE. These are then output and their tags are changed from TRISTATE to TRUE. Since there is no vertex remaining whose tag is FALSE, the algorithm stops.

Points to Remember

1. All the reachable vertices can be traversed from a source vertex by using the depth-first search.
2. The data representation (graph in this case) should be such that it should make algorithms operating on the data efficient.
3. Note how a simple recursive procedure solves the problem of finding all the reachable vertices from a vertex.
4. Note the use of descriptive words such as FALSE, TRUE, and TRISTATE, rather than integers 0, 1, and 2. It makes the program easily understandable.

PROBLEM: CONNECTED COMPONENTS IN A GRAPH

Write a function dfs(v) to traverse a graph in a depth-first manner starting from vertex v. Use this function to find connected components in the graph. Modify dfs() to produce a list of the newly visited vertices. The graph is represented as adjacency lists.

Program

```
#include <stdio.h>

#define MAXVERTICES 20
#define MAXEDGES    20

typedef enum {FALSE, TRUE, TRISTATE} bool;
typedef struct node node;

struct node {
    int dst;
```

```
    node *next;
};

void printGraph( node *graph[], int nvert ) {
    /*
     * prints the graph.
     */
    int i, j;

    for( i=0; i<nvert; ++i ) {
        node *ptr;
        for( ptr=graph[i]; ptr; ptr=ptr->next )
            printf( "[%d] ", ptr->dst );
        printf( "\n" );
    }
}

void insertEdge( node **ptr, int dst ) {
    /*
     * insert a new node at the start.
     */
    node *newnode = (node *)malloc( sizeof(node) );
    newnode->dst = dst;
    newnode->next = *ptr;
    *ptr = newnode;
}

void buildGraph( node *graph[], int edges[2][MAXEDGES], int nedges ) {
    /*
     * fills graph as adjacency list from array edges.
     */
    int i;
    for( i=0; i<nedges; ++i ) {
        insertEdge( graph+edges[0][i], edges[1][i] );
        insertEdge( graph+edges[1][i], edges[0][i] );   // undirected
graph.
    }
}

void dfs( int v, int *visited, node *graph[] ) {
    /*
     * recursively traverse graph from v using visited.
```

```
         * and mark all the vertices that come in dfs path to TRISTATE.
         */
        node *ptr;

        visited[v] = TRISTATE;
        //printf( "%d \n", v );

        for( ptr=graph[v]; ptr; ptr=ptr->next )
            if( visited[ ptr->dst ] == FALSE )
                dfs( ptr->dst, visited, graph );
    }

    void printSetTristate( int *visited, int nvert ) {
        /*
         * prints all vertices of visited which are TRISTATE.
         * and set them to TRUE.
         */
        int i;

        for( i=0; i<nvert; ++i )
            if( visited[i] == TRISTATE ) {
                printf( "%d ", i );
                visited[i] = TRUE;
            }
        printf( "\n\n" );
    }

    void compINC(node *graph[], int nvert) {
        /*
         * prints all connected components of graph represented using INC
  lists.
         */
        int *visited;
        int i;

        visited = (int *)malloc( nvert*sizeof(int) );
        for( i=0; i<nvert; ++i )
            visited[i] = FALSE;

        for( i=0; i<nvert; ++i )
            if( visited[i] == FALSE ) {
                dfs( i, visited, graph );
```

```
            // print all vertices which are TRISTATE.
            // and mark them to TRUE.
            printSetTristate( visited, nvert );
        }
    free( visited );
}

int main() {
    int edges[][MAXEDGES] = { {0,2,4,5,5,4},
                              {1,1,3,4,6,6}
                            };
    int nvert = 7;      // no of vertices.
    int nedges = 6; // no of edges in the graph.
    node **graph = (node **)calloc( nvert, sizeof(node *) );

    buildGraph( graph, edges, nedges );
    printGraph( graph, nvert );
    compINC( graph, nvert );

    return 0;
}
```

Explanation

1. The graph is represented as adjacency lists. The graph contains an array of n pointers where n is the number of vertices in the graph. Each entry i in the array contains a list of vertices to which i is connected. For example, if the graph is as follows:

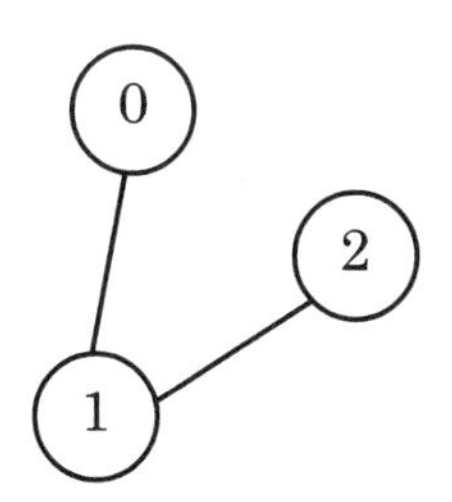

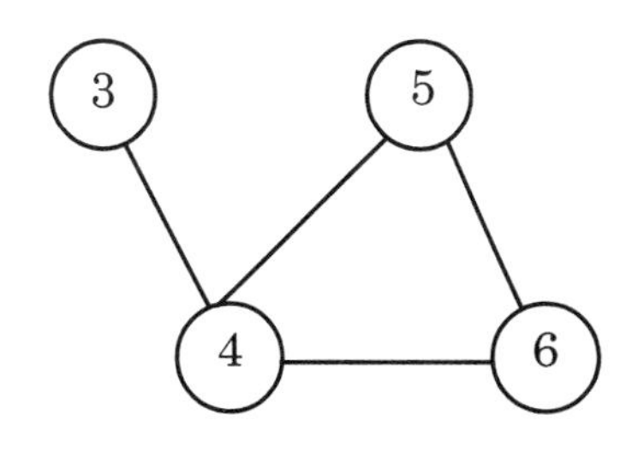

then the adjacency lists for the graph are as shown here:

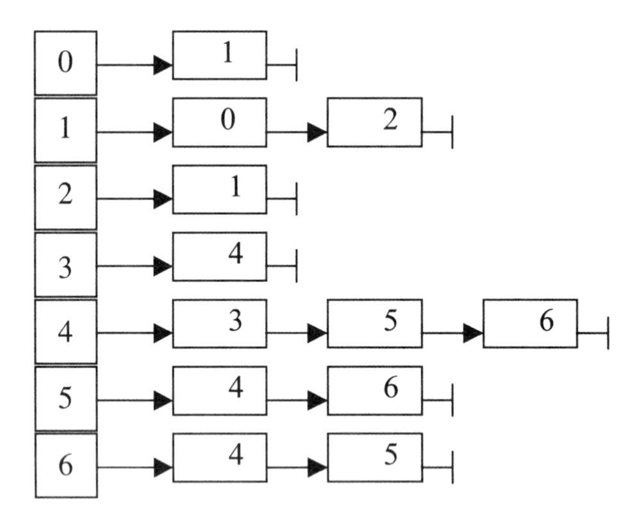

Each node in list i contains a vertex to which i is connected.

2. dfs(v) is implemented recursively. A Boolean vector visited[] is maintained, whose all entries are initially FALSE. dfs(v) marks v as visited by making visited[v] = TRUE. It then finds all the adjacent nodes of v and starts dfs() from those nodes which have not yet been visited.

   For example, if dfs(v) is called with v = 0, it marks 0 and then it traverses the adjacency list graph[0] and calls dfs(1). This marks 1 and traverses the adjacency list graph[1]. But since 0 is already marked, dfs(2) is called. It marks 2 and starts traversal of graph[2]. But since 1 is marked, it returns. All the previous invocations return as there are no nodes being considered in the lists. Thus, the marked vertices are {0, 1, 2}.

3. compINC() is a function that finds all the connected components of a graph. It maintains a local copy of the vector visited[] and passes it as a parameter to dfs(v). compINC() passes that vertex as a parameter to dfs() which has not yet been visited. Thus each invoking of dfs() finds one connected component of the graph.

4. In order to modify dfs() to produce a list of newly visited vertices, we tag the vertices visited by using dfs() as TRISTATE. In compINC(), all these TRISTATE vertices will form one connected component. This status is then converted to TRUE. The next invocation of dfs() returns another set of vertices tagged as TRISTATE, which forms another connected component, and so on.

For example, in this graph, first all vertices are tagged as FALSE. After invoking dfs(0), the vertices tagged as TRISTATE are {0, 1, 2}. These are output and their tags are changed from TRISTATE to TRUE. The next invocation of dfs(3) tags vertices {3, 4, 5, 6} as TRISTATE. These are then output and their tags are changed from TRISTATE to TRUE. Since there is no vertex remaining whose tag is FALSE, the algorithm stops.

Points to Remember

1. All the reachable vertices can be traversed from a source vertex by using the depth-first search.
2. The data representation (a graph, in this case) should be such that it makes algorithms operating on the data efficient. Being represented as adjacency lists, we could easily traverse the list to get the vertices adjacent to a particular vertex.
3. Note how a simple recursive procedure solves the problem of finding all the reachable vertices from a vertex.
4. Note the use of descriptive words such as FALSE, TRUE, and TRISTATE, rather than integers 0, 1, and 2. It makes the program easily understandable.

PROBLEM: MINIMUM SPANNING TREE

Write a program to find a minimum spanning tree in a graph.

Program

```
#include <stdio.h>

#define MAXVERTICES 10
#define MAXEDGES 20

typedef enum {FALSE, TRUE} bool;

int getNVert(int edges[][3], int nedges) {
    /*
     * returns no of vertices = maxvertex + 1;
     */
```

```
    int nvert = -1;
    int j;
    for( j=0; j<nedges; ++j ) {
        if( edges[j][0] > nvert )
                nvert = edges[j][0];

        if( edges[j][1] > nvert )
                nvert = edges[j][1];
    }
    return ++nvert;          // no of vertices = maxvertex + 1;
}

bool isPresent(int edges[][3], int nedges, int v) {
    /*
     * checks whether v has been included in the spanning tree.
     * thus we see whether there is an edge incident on v which has
     * a negative cost. negative cost signifies that the edge has been
     * included in the spanning tree.
     */

    int j;

    for(j=0; j<nedges; ++j)
        if(edges[j][2] < 0 && (edges[j][0] == v || edges[j][1] == v))
                return TRUE;

    return FALSE;
}

void spanning(int edges[][3], int nedges) {
    /*
     * finds a spanning tree of the graph having edges.
     * uses kruskal's method.
     * assumes all costs to be positive.
     */
    int i, j;
    int tv1, tv2, tcost;
    int nspanedges = 0;
    int nvert = getNVert(edges, nedges);
```

```
        // sort edges on cost.
        for(i=0; i<nedges-1; ++i)
            for(j=i; j<nedges; ++j)
                    if(edges[i][2] > edges[j][2]) {
                            tv1 = edges[i][0]; tv2 = edges[i][1]; tcost =
edges[i][2];
                            edges[i][0] = edges[j][0]; edges[i][1] =
edges[j][1]; edges[i][2] = edges[j][2];
                            edges[j][0] = tv1; edges[j][1] = tv2; edges[j][2]
= tcost;
                    }

        for(j=0; j<nedges-1; ++j) {
            // consider edge j connecting vertices v1 and v2.
            int v1 = edges[j][0];
            int v2 = edges[j][1];

            // check whether it forms a cycle in the uptil now formed
spanning tree.
            // checking can be done easily by checking whether both v1 and
v2 are in
            // the current spanning tree!
            if(isPresent(edges, nedges, v1) && isPresent(edges, nedges, v2))
// cycle.
                    printf("rejecting: %d %d %d...\n", edges[j][0],
edges[j][1], edges[j][2]);
            else {
                    edges[j][2] = -edges[j][2];
                    printf("%d %d %d.\n", edges[j][0], edges[j][1], -
edges[j][2]);
                    if(++nspanedges == nvert-1)
                            return;
            }
        }

        printf("No spanning tree exists for the graph.\n");
    }

    main() {
        int edges[][3] = {
                                        {0,1,16},
                                        {0,4,19},
```

```
                                        {0,5,21},
                                        {1,2,5},
                                        {1,3,6},
                                        {1,5,11},
                                        {2,3,10},
                                        {3,4,18},
                                        {3,5,14},
                                        {4,5,33}
                                 };
    int nedges = sizeof(edges)/3/sizeof(int);
    spanning(edges, nedges);

    return 0;
}
```

Program Description

1. A tree consisting solely of edges in a graph G and including all vertices in G is called a spanning tree. A minimum spanning tree of a weighted graph is the spanning tree with the minimum total cost of its edges.

   Example:

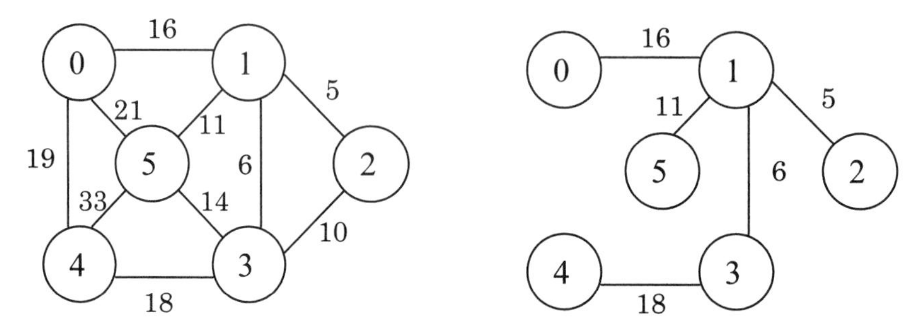

An example graph and its minimum spanning tree.

2. The graph is represented as an array of edges. Each entry in the array is a triplet representing an edge consisting of source vertex, destination vertex, and the cost associated with the edge. The method used in finding a minimum spanning tree is that given by Kruskal. In this approach, a minimum spanning tree T is built edge by edge. Edges are considered for inclusion in T in a non-decreasing order of their costs. An edge is included if it does not form a cycle with the edges already in T. Since graph G is connected, and has $n > 0$ vertices, exactly $n-1$ edges will be selected for inclusion in T.

3. Kruskal's algorithm is as follows:

```
T={};      // empty set.
while T contains less than n-1 edges and E not empty do
   choose an edge (v, w) from E of lowest cost.
   delete (v, w) from E.
   if (v, w) does not create a cycle in T
         add (v, w) to T.
   else
         discard (v, w).
endwhile.
if T contains less than n-1 edges
   print("no spanning tree exists for this graph.");
```

4. In order for the choice of the lowest cost edge from E to become efficient, we sort the edge array over the cost of the edge. To check whether an edge (v,w) forms a cycle, we simply need to check whether both v and w appear in any of the previously added edges in T. We assume that all the costs are positive and we make them negative to signify that the edge has been included in T.

5. **Example:**

   For the example graph, the run of the algorithm is as follows:

step	edge	cost	action	spanning-tree
0	—	—	—	{}
1	(1, 2)	5	accept	{(1, 2)}
2	(1, 3)	6	accept	{(1, 2), (1, 3)}
3	(2, 3)	10	reject	{(1, 2), (1, 3)}
4	(1, 5)	11	accept	{(1, 2), (1, 3), (1, 5)}
5	(3, 5)	14	reject	{(1, 2), (1, 3), (1, 5)}
6	(0, 1)	16	accept	{(1, 2), (1, 3), (1, 5), (0, 1)}
7	(3, 4)	18	accept	{(1, 2), (1, 3), (1, 5), (0, 1), (3, 4)}

Points to Remember

1. A minimum spanning tree of a weighted graph G is a tree that consists of edges solely from the edges of G, which covers all the vertices in G, and which has the minimum combined cost of its edges.

2. The complexity of Kruskal's method used for finding the minimum spanning tree of a graph G is O(e loge), where e is the number of edges in G.
3. Note that the `union` and `find` algorithms for set representation can be used to check for cycle and inclusion of an edge in a set.
4. There can be multiple minimum spanning trees in a graph.

PROBLEM: TOPOLOGICAL SORT

Write a program to find the topological order of a digraph G represented as adjacency lists.

Program

```
#include <stdio.h>

#define N 11          // no of total vertices in the graph.

typedef enum {FALSE, TRUE} bool;
typedef struct node node;

struct node {
    int count;        // for arraynodes : in-degree.
                      // for listnodes  : vertex no this vertex is
connected to.
                      // if this node is out of graph : -1.
                      // if this has 0 indegree then it occurs in
zerolist.
    node *next;
};
node graph[N];
node *zerolist;

void addToZerolist( int v ) {
    /*
     * adds v to zerolist as v has 0 predecessors.
     */
    node *ptr = (node *)malloc( sizeof(node) );
    ptr->count = v;
```

```
    ptr->next  = zerolist;
    zerolist   = ptr;
}

void buildGraph( int a[][2], int edges ) {
    /*
     * fills global graph with input given in a.
     * a[i][0] is src vertex and a[i][1] is dst vertex.
     */
    int i;

    // init graph.
    for( i=0; i<N; ++i ) {
        graph[i].count = 0;
        graph[i].next  = NULL;
    }

    // now add the list entries.
    for( i=0; i<edges; ++i ) {
        // add new node to src list.
        node *ptr = (node *)malloc( sizeof(node) );
        ptr->count = a[i][1];
        ptr->next = graph[ a[i][0] ].next;
        graph[ a[i][0] ].next = ptr;
        // increase indegree of dst.
        graph[ a[i][1] ].count++;
    }

    // now create list of zero predecessors.
    zerolist = NULL;  // list of vertices having 0 predecessors.
    for( i=0; i<N; ++i )
        if( graph[i].count == 0 ) {
                addToZerolist(i);
        }
}

void printGraph() {
    int i;
    node *ptr;

    for( i=0; i<N; ++i ) {
        node *ptr;
```

```
            printf( "%d: pred=%d: ", i, graph[i].count );
            for( ptr=graph[i].next; ptr; ptr=ptr->next )
                    printf( "%d ", ptr->count );
            printf( "\n" );
        }
        printf( "zerolist: " );
        for( ptr=zerolist; ptr; ptr=ptr->next )
            printf( "%d ", ptr->count );
        printf( "\n" );
    }

    int getZeroVertex() {
        /*
         * returns the vertex with zero predecessors.
         * if no such vertex then returns -1.
         */
        int v;
        node *ptr;

        if( zerolist == NULL )
            return -1;
        ptr = zerolist;
        v = ptr->count;
        zerolist = zerolist->next;
        free(ptr);

        return v;
    }

    void removeVertex( int v ) {
        /*
         * deletes vertex v and its outgoing edges from global graph.
         */
        node *ptr;
        graph[v].count = -1;
        // free the list graph[v].next.
        for( ptr=graph[v].next; ptr; ptr=ptr->next ) {
            if( graph[ ptr->count ].count > 0 ) // normal nodes.
                    graph[ ptr->count ].count--;
    if( graph[ ptr->count ].count == 0 )        // this is NOT else of above
if.
```

```
                addToZerolist( ptr->count );
        }
}

void topsort( int nvert ) {
    /*
     * finds recursively topological order of global graph.
     * nvert vertices of graph are needed to be ordered.
     */
    int v;

    if( nvert > 0 ) {
        v = getZeroVertex();
        if( v == -1 ) {         // no such vertex.
                fprintf( stderr, "graph contains a cycle.\n" );
                return;
        }
        printf( "%d.\n", v );
        removeVertex(v);
        topsort( nvert-1 );
    }
}

int main() {
    int a[][2] = {
                              {0,1},
                              {0,3},
                              {0,2},
                              {1,4},
                              {2,4},
                              {2,5},
                              {3,4},
                              {3,5}
                          };
    buildGraph( a, 8 );
    printGraph();
    topsort(N);
}
```

Explanation

1. A linear ordering of vertices of a digraph G, with the property that if `i` is a predecessor of `j`, then `i` precedes `j` in the linear ordering, is called a topological order of G.

2. The digraph G is maintained as adjacency lists. In this representation, G is an array graph[0...n-1], where each element graph[i] is a linked list of vertices to which vertex i is connected, and n is the number of vertices in G.
3. We also maintain a zerolist, which is a list of vertices that have zero predecessors. The necessity for this list will be clear from an algorithm for a topological sort, as follows:

```
    topsort(n) {
        if( n > 0 ) {
    if every vertex has a predecessor then
        error( "graph contains a cycle." ).
    pick a vertex v that has no predecessors. //
getZeroVertex()
    output v.
    delete v and all the edges leading out of v in the graph.
// removeVertex()
    topsort(n-1).
    }
}
```

5. The algorithm topsort() is tail-recursive. From zerolist, it removes a vertex v containing zero predecessors, and outputs it. This vertex v either has no predecessors in G, or all its predecessors have already been output. Thus all the vertices in zerolist are candidates for the next output. After v is output, all the vertices to which v points may become the candidates for the next output. Thus we remove all the edges starting from v and rerun topsort() over the remaining vertices.

Example: Let the digraph be as shown here.

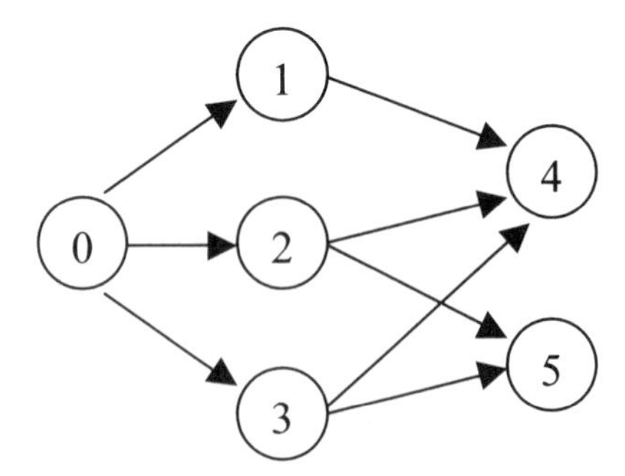

Step	Zerolist	Output
0	{0}	nil
1	{1, 2, 3}	0
2	{2, 3}	1
3	{3}	2
4	{4,5}	3
5	{5}	4
6	{}	5

Points to Remember

1. A linear ordering of vertices of a digraph G, with the property that if `i` is a predecessor of `j`, then `i` precedes `j` in the linear ordering, is called the topological order of G.
2. The complexity of topological order is $O(n+e)$ where n is the number of vertices and e is the number of edges in the digraph.
3. Removal of an edge results in a decrease in the predecessor count of the destination vertex. If this count reaches 0, the vertex should be inserted in the `zerolist`.
4. By maintaining a list of vertices with zero predecessors, the computing time of the algorithm decreases.

PROBLEM: FINDING THE SHORTEST PATH BY USING AN ADJACENCY MATRIX

Given a digraph represented as an adjacency matrix, find the shortest path from a vertex `v` to all the other vertices in the graph.

Program

```
#include <stdio.h>

#define MAXINT 99999
#define MAXVERTICES 10
typedef enum {FALSE, TRUE} bool;
```

```
void print( int cost[][MAXVERTICES], int nvert ) {
    /*
     * prints cost matrix.
     */
    int i, j;

    for( i=0; i<nvert; ++i ) {
        for( j=0; j<nvert; ++j )
                printf( "%6d", cost[i][j] );
        printf( "\n" );
    }
}

int choose( int dist[], bool s[], int nvert ) {
    /*
     * returns vertex u such that:
     * dist[u] = min{ dist[w] } where s[w] == FALSE.
     */
    int i;
    int u=-1;
    int mindist = MAXINT;

    for( i=0; i<nvert; ++i )
        if( s[i] == FALSE && dist[i] <= mindist )
                u=i, mindist=dist[i];
    return u;
}

void sssp( int v, int cost[][MAXVERTICES], int dist[], int nvert ) {
    /*
     * finds shortest path from v to all other vertices.
     * cost is the cost matrix.
     * dist is the vector in which output will be written.
     * nvert is no of vertices in the graph.
     */
    bool s[MAXVERTICES];
    int i, u, num, w;

    for( i=0; i<nvert; ++i )
        s[i] = FALSE, dist[i] = cost[v][i];
    s[v] = TRUE;
    dist[v] = 0;
```

```
        num = 1;

        while( num < nvert-1 ) {
            u = choose( dist, s, nvert );
            s[u] = TRUE;
            num++;
            for( w=0; w<nvert; ++w )
                    if( s[w] == FALSE && dist[u]+cost[u][w] < dist[w] )
                            dist[w] = dist[u] + cost[u][w];
        }
    }

    void printDist( int v, int dist[], int nvert ) {
        /*
         * prints distance matrix which shows min distance of each vertex
         * from v.
         */
        int i;

        printf( "min dist from vertex %d...\n", v );
        for( i=0; i<nvert; ++i )
            printf( "dist[%d]=%d.\n", i, dist[i] );
    }

    int main() {
        int cost[][MAXVERTICES] =
{{0,2,2,1},{3,0,4,1},{5,16,0,9},{1,1,2,0}};
        int dist[MAXVERTICES];
        int nvert = 4;             // no of vertices.

        sssp( 2, cost, dist, nvert );
        printDist( 2, dist, nvert );

        return 0;
    }
```

Explanation

1. The digraph is represented as an adjacency matrix. The size of the matrix is N×N where N is the number of vertices in the digraph. An entry `matrix[i][j]` specifies the cost of the directed edge from vertex `i` to vertex `j`. If such an edge does not exist, then the cost is set to `MAXINT`. All diagonal entries of the matrix `matrix[i][i]` are set to 0.
2. The function `sssp(v, cost, dist, nvert )` implements the single-source-shortest-path algorithm to find the shortest path from a vertex `v` in a digraph that is represented as an adjacency matrix cost. `nvert` is the number of vertices in the digraph and `dist` is the vector that will finally contain the output. Thus an entry `dist[i]` will be the minimum distance of vertex `i` from vertex `v`.
3. The function uses a vector `s[]` of Boolean to represent the status of each vertex: whether processed or not. It is initialized to `FALSE`. `s[v]` is marked, meaning it is processed. `dist[v]` is set to 0. Then the function chooses a vertex `u` that is yet to be processed and which is at a minimum distance from `v`. Vertex `u` is marked as processed. It then processes vector `dist[]`, which may get updated because of the addition of `u`. The function checks whether `dist[u]+cost[u][w] < dist[w]` for each yet-to-be-processed node `w` in the digraph. If it is, then `dist[w]` is updated as `dist[u]+cost[u][w]`.

   When all the vertices are processed, `dist[]` contains the required result.

   Example:

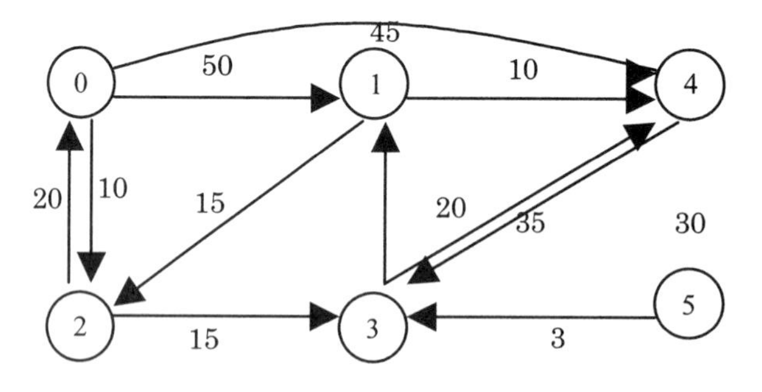

Let `v=0`. Then, different steps in the run of the algorithm are as follows:

M signifies MAXINT, which means that the vertex cannot be reached from v.

step	u	dist
0	nil	0 50 10 M 45 M
1	2	0 50 10 35 45 M
2	3	0 45 10 35 45 M
3	1	0 45 10 35 45 M
4	4	0 45 10 35 45 M
5	5	0 45 10 35 45 M

Points to Remember

1. The complexity of the sssp() is O(N^2) where N is the number of vertices in the digraph.
2. The constant of proportionality of the complexity of choose() may be improved by using a list of yet-to-be-processed vertices.
3. Representing the digraph as an adjacency matrix makes the retrieval of the cost of an edge O(1).

PROBLEM: FINDING THE SHORTEST PATH BY USING AN ADJACENCY LIST

Rewrite the function sssp() for finding the shortest path, such that the digraph is represented by its adjacency lists instead of the adjacency matrix. Also, instead of maintaining set S of vertices to which the shortest paths have already been found, the set V(G)–S is represented by using a linked list where V(G) is the set of vertices in digraph G.

Program

```
#include <stdio.h>

#define MAXINT 99999
#define MAXVERTICES 10
```

```
typedef struct node node;
typedef struct setnode setnode;
typedef enum {FALSE, TRUE} bool;

struct node {
    int dst;
    int cost;
    node *next;
};

struct setnode {
    int v;
    setnode *next;
};

void printGraph( node *cost[], int nvert ) {
    /*
     * prints the graph.
     */
    int i, j;

    for( i=0; i<nvert; ++i ) {
        node *ptr;
        for( ptr=cost[i]; ptr; ptr=ptr->next )
                printf( "[%d,%d] ", ptr->dst, ptr->cost );
        printf( "\n" );
    }
}

void insertEdge( node **ptr, int dst, int cost ) {
    /*
     * insert a new node at the start.
     */
    node *newnode = (node *)malloc( sizeof(node) );
    newnode->dst = dst;
    newnode->cost = cost;
    newnode->next = *ptr;
    *ptr = newnode;
}

void buildGraph( node *cost[], int costnew[][3], int nedges ) {
    /*
```

```
     * fills cost as adjacency list from array costnew.
     */
    int i;
    for( i=0; i<nedges; ++i )
        insertEdge( cost+costnew[i][0], costnew[i][1], costnew[i][2] );
}

int choose( int dist[], setnode *s ) {
    /*
     * returns vertex u such that:
     * dist[u] = min{ dist[w] } where w is in set s.
     */
    int u=-1;
    int mindist = MAXINT;
    setnode *ptr;

    for( ptr=s->next; ptr; ptr=ptr->next )
        if( dist[ptr->v] <= mindist )
                u=ptr->v, mindist=dist[ptr->v];
    return u;
}

int getCost( node **cost, int src, int dst ) {
    /*
     * return cost of edge from src to dst.
     */
    node *ptr;
    for( ptr=cost[src]; ptr; ptr=ptr->next )
        if( ptr->dst == dst )
                return ptr->cost;
    return MAXINT;
}

void removeFromSet( setnode *s, int v ) {
    /*
     * remove vertex v from set s.
     */
    setnode *prev, *ptr;
    for( prev=s, ptr=prev->next; ptr; prev=ptr, ptr=ptr->next )
        if( ptr->v == v ) {
                prev->next = ptr->next;
```

```
                free(ptr);
                return;
        }
    // v does NOT exist in the set.
}

void insertIntoSet( setnode *s, int v ) {
    /*
     * add vertex v to the set s.
     */
    setnode *ptr = (setnode *)malloc( sizeof(setnode) );
    ptr->v = v;
    ptr->next = s->next;
    s->next = ptr;
}

bool isInSet( setnode *s, int v ) {
    /*
     * returns TRUE if vertex v is in set s.
     */
    setnode *ptr;
    for( ptr=s->next; ptr; ptr=ptr->next )
        if( ptr->v == v )
                return TRUE;
    return FALSE;
}

void sssp( int v, node **cost, int dist[], int nvert ) {
    /*
     * finds shortest path from v to all other vertices.
     * cost is the cost adjacency list.
     * dist is the vector in which output will be written.
     * nvert is no of vertices in the graph.
     */
    setnode s; // list of vertices yet to be considered.
    int i, u, num, w;
    node *ptr;

    for( i=0; i<nvert; ++i ) {
        insertIntoSet( &s, i );
        dist[i] = MAXINT;
    }
```

```
    for( ptr=cost[v]; ptr; ptr=ptr->next )
        dist[ptr->dst] = ptr->cost;

    removeFromSet( &s, v );
    dist[v] = 0;
    num = 1;

    while( num < nvert-1 ) {
        u = choose( dist, &s );
        removeFromSet( &s, u );
        num++;
        for( w=0; w<nvert; ++w ) {
                int c = getCost( cost, u, w );
                if( isInSet( &s, w ) && dist[u]+c < dist[w] )
                        dist[w] = dist[u] + c;
        }
    }
}

void printDist( int v, int dist[], int nvert ) {
    /*
     * prints min dist vector.
     */
    int i;

    printf( "min dist from vertex %d...\n", v );
    for( i=0; i<nvert; ++i )
        printf( "dist[%d]=%d.\n", i, dist[i] );
}

int main() {
    int costnew[][3] =      {                              {0,1,50},
                                        {0,2,10},
                                        {0,4,45},
                                        {1,2,15},
                                        {1,4,10},
                                        {2,0,20},
                                        {2,3,15},
                                        {3,1,20},
                                        {3,4,35},
                                        {4,3,30},
```

```
                                    {5,3,3}
                                };
    int dist[MAXVERTICES];
    int nvert = 6;            // no of vertices.
    int nedges = 11;  // no of edges in costnew.
    node **cost = (node **)calloc( nvert, sizeof(node *) );

    buildGraph( cost, costnew, nedges );
    printGraph( cost, nvert );
    sssp( 4, cost, dist, nvert );
    printDist( 4, dist, nvert );

    return 0;
}
```

Explanation

1. Cost is maintained as an array containing adjacency lists for each vertex in the digraph. The function buildGraph() fills this array. An adjacency list for a vertex is a linked list where each node contains the vertex, the cost of the corresponding edge, and a pointer to the next node.

   For example, consider the following graph:

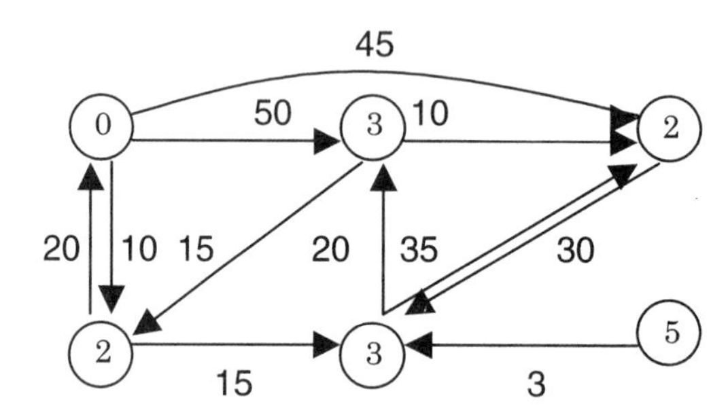

The adjacency lists are maintained as follows:

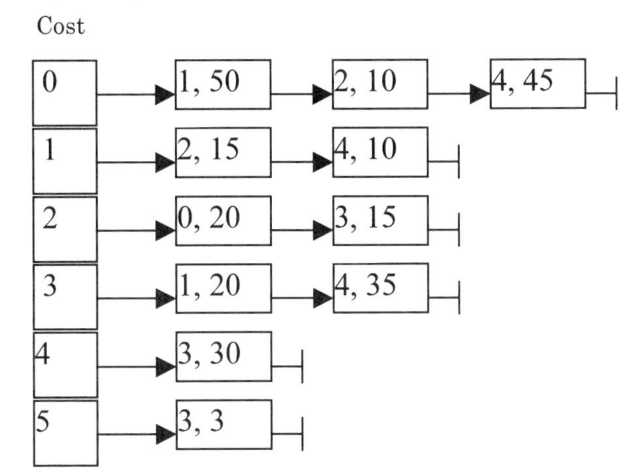

Note that each node contains a pair of integers, (vertex, cost).

2. The function sssp() remains nearly the same except getting the cost of an edge and accesses the set S. The cost of an edge can be retrieved using the function getCost(). It needs O(n) traversal of the adjacency list corresponding to the source vertex of the edge, where n is the number of nodes in the list. In the cost matrix-based algorithm, this was O(1). Removal of a vertex (removeFromSet()) from the list requires O(n) time. This time was also constant earlier as we flagged the vertex as FALSE. Furthermore, earlier we could check the status of a vertex to be TRUE or FALSE using s[i], but now the function isInSet() needs to traverse the list s to see whether the node has been processed or NOT. This makes it O(n). However, the function choose(), which was earlier O(n), now becomes O(1), as we need to remove only the first element from the list s.
3. The overall complexity of sssp() increases from O(n^2) to O(n^3) as getCost() and isInSet() become O(n), where n is the number of vertices in the digraph.

Points to Remember

1. Note how the change in the data structures affected the algorithm efficiency.
2. In order to make choose() O(1), we had to make removeFromSet() and isInSet() O(n). This trade-off should be studied carefully before implementation.

3. The change from an adjacency matrix to adjacency lists may reduce some space, but it increases complexity of the function getCost().

PROBLEM: THE M SHORTEST PATH

Write a function mshortest(cost, src, dst, m) that finds m shortest paths 7(if any exist) from src to dst in a digraph represented by its cost adjacency matrix cost.

Program

```
#include <stdio.h>

#define M 8
#define MAXVERTICES 10
#define MAXINT 99999

typedef struct node node;
typedef struct ansnode ansnode;
typedef enum {FALSE, TRUE} bool;

struct node {
    int src;   // for head node: this is cost of the edges in the list.
    int dst;
    int dummy; // used for temporarily saving the cost of this edge.
    node *next;
};

struct ansnode {
    node path;
    node inedges, exedges;
};

ansnode answers[M];
int indexans;         // no of paths generated so far: >=0 & <= M.

void init() {
```

```
        /*
         * some initialization.
         */
        int i;
        for( i=0; i<M; ++i ) {
            ansnode *a = answers+i;
            a->path.src = 0;     // this is combined cost of the edges in
this list.
            a->path.next = NULL;
            a->inedges.next = a->exedges.next = NULL;
        }
        indexans = 0;
    }

    void pushFront( node *edges, int src, int dst ) {
        /*
         * adds a new node containing (src,dst) at start of edges.
         */
        node *ptr = (node *)malloc( sizeof(node) );
        ptr->src = src;
        ptr->dst = dst;
        ptr->next = edges->next;
        edges->next = ptr;
    }

    void pushBack( node *edges, int src, int dst ) {
        /*
         * adds a new node containing (src,dst) at end of edges.
         */
        node *ptr;
        for( ptr=edges; ptr->next; ptr=ptr->next )
            ;
        // (src,dst) should be inserted after ptr;
        pushFront( ptr, src, dst );             // another hack!

    }

    void popFront( node *edges, int *src, int *dst ) {
        /*
         * remove a node from start of edges.
         */
        node *ptr = edges->next;
```

```
    if( ptr == NULL ) {
        *src = *dst = -1;
        return;
    }
    *src = ptr->src;
    *dst = ptr->dst;
    edges->next = ptr->next;
    free(ptr);
}

int choose( int dist[], bool s[], int nvert ) {
    /*
     * returns vertex u such that:
     * dist[u] = min{ dist[w] } where s[w] == FALSE.
     */
    int i;
    int u=-1;
    int mindist = MAXINT;

    for( i=0; i<nvert; ++i )
        if( s[i] == FALSE && dist[i] <= mindist )
                u=i, mindist=dist[i];
    return u;
}

void printList( char *str, node *edges ) {
    /*
     * prints a node list.
     */
    node *ptr;
    printf( "%s: ", str );
    for( ptr=edges->next; ptr; ptr=ptr->next )
        printf( "(%d,%d) ", ptr->src, ptr->dst );
    printf( "\n" );
}

node *revList( node *list ) {
    /*
     * returns reverse of list: modifies list.
     */
    node *ptr=list->next, *temp, *prev=NULL;
    for( ; ptr; prev=ptr, ptr=temp ) {
        temp = ptr->next;
```

```
        ptr->next = prev;
    }
    list->next = prev;
    return list;
}

int getInsertIndex( node *answer, int start ) {
    /*
     * returns index in answers where answer can be inserted.
     * uses binsrch.
     */
    int mid, end=indexans-1;
    int midcost=-1, anscost=answer->src;

    while( start <= end ) {
        mid = (start+end)/2;
        midcost = answers[mid].path.src;
        if( midcost == anscost )
                return mid;
        else if( midcost < anscost )
                start = mid+1;
        else
                end = mid-1;
    }
    if( midcost == -1 )
        return start;
    if( midcost < anscost )
        return mid+1;
    else
        return mid;
}

void shiftInsert( int index ) {
    /*
     * shifts answers[index..indexans-1] by one down.
     */
    int i;
    for( i=index; i<indexans; ++i )
        answers[i+1] = answers[i];
}

void insertAnswer( node *answer, node *inedges, node *exedges, int
start ) {
```

```
    /*
     * inserts answer in answers sorted on cost of the answer.
     * uses binsrch and then linearly shifts answers down.
     * start helps in reducing no of iterations of binsrch.
     * it is possible that the cost of the answer is >= MAXINT.
     */
    int index;

    if( answer->src >= MAXINT )     // path length is infinite.
        return;
    index = getInsertIndex( answer, start );
    shiftInsert( index );
    answers[index].path = *answer;
    answers[index].inedges = *inedges;
    answers[index].exedges = *exedges;
    printf( "—%d cost=%d.\n", index, answers[index].path.src );
    printList( "——inedges: ", inedges );
    printList( "——exedges: ", exedges );
    indexans++;
}

node *copied( node *edges ) {
    /*
     * return a copy of the list of edges.
     */
    node *ret = (node *)malloc( sizeof(node) );
    node *retptr = ret;
    node *ptr;

    ret->src = edges->src;
    ret->next = NULL;

    for( ptr=edges->next; ptr; ptr=ptr->next ) {
        retptr->next = (node *)malloc( sizeof(node) );
        retptr = retptr->next;
        retptr->src = ptr->src;
        retptr->dst = ptr->dst;
        retptr->next = NULL;
    }
    return ret;
}
```

```
void findDstSrc( node *edges, int *dst1, int *src2, int dst ) {
    /*
     * returns the first src and last dst of the path ptr.
     */
    node *ptr = edges->next;
    if( ptr == NULL ) {
        *dst1 = *src2 = dst;
        return;
    }
    *dst1 = ptr->src;
    for( ; ptr->next; ptr=ptr->next )
        ;
    *src2 = ptr->dst;
}

void attachPaths( node *path1, node *inedges, node *path2 ) {
    /*
     * path1 = path2+inedges+path1;
     */
    node *ptr;

    for( ptr=inedges; ptr->next; ptr=ptr->next )
        ;
    ptr->next = path1->next;
    if( path2 != NULL ) {
        for( ptr=path2; ptr->next; ptr=ptr->next )
                ;
        ptr->next = inedges->next;
        path1->next = path2->next;
        path1->src += path2->src;
    }
    else
        path1->next = inedges->next;
    path1->src += inedges->src;
}

node *advanceByInc( node *path, node *inedges ) {
    /*
     * return a ptr to the first edge in path which is not in inedges.
     * simply traverse no of nodes in path equal to that in inedges and
return
```

```
         * the next pointer.
         */
        node *ptrpath, *ptrin;

        for( ptrpath=path->next, ptrin=inedges->next; ptrin;
ptrpath=ptrpath->next, ptrin=ptrin->next )
            ;
        return ptrpath;
    }

    node *sssp( int cost[][MAXVERTICES], int v, int finalv, int nvert ) {
        /*
         * finds shortest path from v to all other vertices and thus to
finalv.
         * cost is the cost matrix.
         * nvert is no of vertices in the graph.
         * returns the shortest path.
         */
        bool s[MAXVERTICES];
        int i, u, num, w;
        int spath[MAXVERTICES];
        int dist[MAXVERTICES];
        node *newpath = (node *)malloc( sizeof(node) );
        int src, dst;

        printf( "solving sssp(%d,%d).\n", v, finalv );
        newpath->next = NULL;
        for( i=0; i<nvert; ++i )
            spath[i] = -1;
        for( i=0; i<nvert; ++i )
            s[i] = FALSE, dist[i] = cost[v][i];

        s[v] = TRUE;
        dist[v] = 0;
        num = 1;

        while( num < nvert-1 ) {
            u = choose( dist, s, nvert );
            s[u] = TRUE;
            num++;
            if( spath[u] == -1 )
```

```
                spath[u] = v;
        for( w=0; w<nvert; ++w )
                if( s[w] == FALSE && dist[u]+cost[u][w] < dist[w] ) {
                        dist[w] = dist[u] + cost[u][w];
                        spath[w] = u;
                }
    }
    printf( "path=%d ", finalv );
    dst = finalv;
    for( w=spath[finalv]; w!=-1; w=spath[w] ) {
        printf( "%d ", w );
        src = w;
        pushBack( newpath, src, dst );
        dst = src;
    }
    newpath->src = dist[finalv];    // cost of this path.
    printf( ".\n" );

    return newpath;
}

void solveshortest( int index, int cost[][MAXVERTICES], int src, int
dst, int nvert ) {
    /*
     * driver for sssp().
 * sets constraints for the new path. The constraints are
 * inclusion and exclusion of some edges.
 * exclusion is done by temporarily making entry cost[i][j] equal to 0
     * and then restoring it after sssp().
 * inclusion of some path is carried out by calling sssp() twice with
the included
 * path removed. The two paths and the inclusion list together contain
the new
     * shortest path.
     * the global array of answers is updated as new paths are being
calculated.
     */
    node *path = &answers[index].path;
    node *inedges = &answers[index].inedges;
    node *exedges = &answers[index].exedges;
```

```
        node *ptr;

        for( ptr=exedges->next; ptr; ptr=ptr->next ) {// for each exclusion
    edge.
            ptr->dummy = cost[ptr->src][ptr->dst];     // saving.
            cost[ptr->src][ptr->dst] = MAXINT;
        }
        ptr = advanceByInc( path, inedges );
        for( ; ptr; ptr=ptr->next ) {   // for each edge.
            int dummy1, dummy2;
            int dst1, src2;
            node *path1, *path2=NULL;

            printList( "path: ", path );
            printList( "inedges: ", inedges );
            printList( "exedges: ", exedges );

        // exclusion edges.
            printf( "exedge=(%d,%d).\n", ptr->src, ptr->dst );
            pushFront( exedges, ptr->src, ptr->dst );
            exedges->next->dummy = cost[ptr->src][ptr->dst];  // saving.
            cost[ptr->src][ptr->dst] = MAXINT;  // temporarily removed.

            // inclusion edges.
            findDstSrc( inedges, &dst1, &src2, dst );
            printf( "inedges: dst1=%d, src2=%d.\n", dst1, src2 );

            path1 = sssp( cost, src, dst1, nvert );
            if( src2 != dst )
                   path2 = sssp( cost, src2, dst, nvert );
            attachPaths( path1, revList(copied(inedges)), path2 );
            printList( "attachpath: ", path1 );
            insertAnswer( path1, copied(inedges), copied(exedges), index+1
    );

            // now restore the cost matrix.
            cost[ptr->src][ptr->dst] = exedges->next->dummy;

            // remove the exclusion edge.
            popFront( exedges, &dummy1, &dummy2 );
```

```
            // update inclusion list with that edge.
            pushFront( inedges, dummy1, dummy2 );
            inedges->src += cost[dummy1][dummy2];
        }
        for( ptr=exedges->next; ptr; ptr=ptr->next )  // for each exclusion
edge.
            cost[ptr->src][ptr->dst] = ptr->dummy;
    }

    void mshortest( int cost[][MAXVERTICES], int src, int dst, int nvert )
{
        /*
         * finds m shortest paths between src and dst.
         */
        int i=0;
        node *inedges = (node *)malloc( sizeof(node) );
        node *exedges = (node *)malloc( sizeof(node) );
        node *newpath = sssp( cost, src, dst, nvert );

        inedges->src = exedges->src = 0;
        inedges->next = exedges->next = NULL;
        insertAnswer( newpath, inedges, exedges, i );

        while( i<indexans && indexans <= M ) {
            printf( "——————————————\nsolving i=%d.\n", i );
            solveshortest( i++, cost, src, dst, nvert );
        }
    }

    int main() {
        int cost[][MAXVERTICES] = {
{0,3,2,MAXINT,MAXINT,MAXINT},
                                            {MAXINT,0,5,7,4,MAXINT},
                                            {MAXINT,MAXINT,0,4,1,MAXINT},
                                            {MAXINT,MAXINT,MAXINT,0,5,3},
{MAXINT,MAXINT,MAXINT,MAXINT,0,5},

{MAXINT,MAXINT,MAXINT,MAXINT,MAXINT,0
                            };
```

```
    mshortest( cost, 0, 5, 6 );
    return 0;
}
```

Explanation

1. The digraph G is maintained as its cost adjacency matrix. `cost[i][i] = 0` and `cost[i][j] = MAXINT` for `i != j`, and if there is no edge from `i` to `j`.
2. We assume that every edge has a positive cost. Let `p1 = v0, v1, ..., vk` be the shortest path from `src` to `dst`. If `P` is the set of all simple `src`-to-`dst` paths in G, then it is easy to see that every path in `P-{p1}` differs from `p1` in exactly one of the following `k` ways.

   (1) It contains the edges `(v1, v2), ..., (vk-1, vk)` but not `(v0, v1)`.

   (2) It contains the edges `(v2, v3), ..., (vk-1, vk)` but not `(v1, v2)`.

   ... ...

   (`k`) It does not contain the edge `(vk-1, vk)`.

   Thus we see that if we put constraints on a shortest path as for the aforementioned `k` conditions, then we get more next-shortest paths satisfying different constraints. These constraints either require certain edges to be included or excluded from the original shortest path as just given.
3. The algorithm `mshortest()` is as follows:

   `Q` = {(shortest src-to-dst-path, phi)}. // `phi` denotes an empty set.

   for `i = 1` to *m* do // generate *m* shortest paths.

   Let `(p,C)` be the tuple in `Q` such that path `p` is of minimal length with set of constraints `c`.

   Output path `p`.

   Delete path `p` from `Q`.

   Determine the shortest paths in `G` under the constraints `C` and the additional constraints imposed for the new path being generated as described previously.

   Add these shortest paths together with their constraints to `Q`.

   `Q` is a set of `(p, C)` pairs where `p` is the shortest path generated after imposing constraints `C`. Initially, we find the shortest path from `src` to `dst` under no constraints. We then generate *m* paths in a loop. Inside the loop, we select from `Q` that `(p, C)` pair that has the minimum path length. After outputting, we remove the pair from `Q`. We then find the next shortest paths after imposing the constraints on C and additional constraints by removing an edge from the exclusion-list of edges, and adding that edge to the inclusion list. We also add the next edge of the path to the exclusion list. Thus we form a new set of constraints. We add the shortest paths generated along with the new constraints to the set Q and reiterate.

4. Use the function sssp() to find a single-source shortest path. Make minor modifications to the function to follow the constraints and exclusion of certain edges. Exclusion is carried out by temporarily removing that edge's cost from the cost matrix and restoring the cost after running sssp() over cost. Inclusion of consecutive edges is carried out by calling sssp() twice, once for the path between the first vertex of the shortest path and before the first vertex in the inclusion list, and then for the path between the last vertex of the included path and the last vertex of the shortest path.

   For example, if the shortest path is v0 v1 v2 v3 v5, and the inclusion list is (v2 v3), then sssp() is called for the path v0 v1 v2, and then for the path v3 v5. The new shortest path generated is returned by sssp(). This new path, along with its constraints and cost, is added to a global array[M] of answers sorted on the cost of the path. A binary search is used to get the index (getInsertIndex()) where the new path should be added. Insertion into the array at position i requires shifting of the elements from i forward by one position (shiftInsert()).

5. **Example:**

   Consider the following digraph.

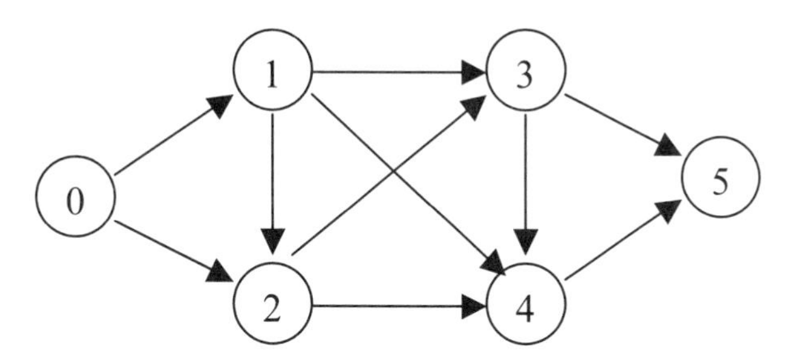

The shortest paths generated for the above graph for M = 8 are shown here:

shortest path	cost	included edges	excluded edges	new path
v0 v2 v4 v5	8	none	none.	
		none	(v4 v5)	v0 v2 v3 v5 = 9
		(v4 v5)	(v2 v4)	v0 v1 v4 v5 = 12
		(v2 v4) (v4 v5)	(v0 v2)	v0 v1 v2 v4 v5 = 14
v0 v2 v3 v5	9	none	(v4 v5).	
		none	(v3 v5) (v4 v5)	infinity
		(v3 v5)	(v2 v3) (v4 v5)	v0 v1 v3 v5 = 13
		(v2 v3) (v3 v5)	(v0 v2) (v4 v5)	v0 v1 v2 v3 v5 = 15
v0 v1 v4 v5	12	(v4 v5)	(v2 v4).	
		(v4 v5)	(v1 v4) (v2 v4)	v0 v2 v3 v4 v5 = 16
		(v1 v4) (v4 v5)	(v0 v1) (v2 v4)	infinity
v0 v1 v3 v5	13	(v3 v5)	(v2 v3) (v4 v5)	
		(v3 v5)	(v1 v3) (v2 v3) (v4 v5)	infinity
		(v1 v3) (v3 v5)	(v0 v1) (v2 v3) (v4 v5)	infinity
v0 v1 v2 v4 v5	14	(v2 v4) (v4 v5)	(v0 v2)	
		(v2 v4) (v4 v5)	(v1 v2) (v0 v2)	infinity
		(v1 v2) (v2 v4) (v4 v5)	(v0 v1) (v0 v2)	nfinity
v0 v1 v2 v3 v5	15	(v2 v3) (v3 v5)		(v0 v2) (v4 v5)
		(v2 v3) (v3 v5)	(v1 v2) (v0 v2) (v4 v5)	infinity
		(v1 v2) (v2 v3) (v3 v5)	(v0 v1) (v0 v2) (v4 v5)	infinity
v0 v2 v3 v4 v5	16	(v4 v5)	(v1 v4) (v2 v4)	
		(v4 v5)	(v3 v4) (v1 v4) (v2 v4)	infinity
		(v3 v4) (v4 v5)	(v2 v3) (v1 v4) (v2 v4)	v0 v1 v3 v4 v5 = 20
		(v2 v3) (v3 v4) (v4 v5)	(v0 v2) (v1 v4) (v2 v4)	v0 v1 v2 v3 v4 v5 = 22

Points to Remember

1. The complexity of `mshortest()` is O(mn^3), where m is the number of shortest paths generated, and n is the number of vertices in the digraph.

2. Note how we reused the single-source-shortest-path algorithm with some modifications in mshortest(). This not only eased programming but also simplified the analysis.
3. Maintaining a reverse list of edges helps in updating the exclusion and inclusion lists easily.

PROBLEM: THE ALL-COST SHORTEST PATH

Write a function allCosts() to find the shortest path between any two vertices in a digraph.

Program

```
#include <stdio.h>

#define MAXINT 99999
#define MAXVERTICES 10

typedef enum {FALSE, TRUE} bool;

void print( int cost[][MAXVERTICES], int nvert ) {
    /*
     * prints the cost matrix.
     */
    int i, j;

    for( i=0; i<nvert; ++i ) {
       for( j=0; j<nvert; ++j )
              printf( "%6d", cost[i][j] );
       printf( "\n" );
    }
}

void printCosts( int a[][MAXVERTICES], int nvert ) {
    /*
     * prints min cost matrix a.
     */
    int i, j;
```

```
    for( i=0; i<nvert; ++i )
        for( j=0; j<nvert; ++j )
            printf( "cost[%d][%d]=%d.\n", i, j, a[i][j] );
}

void allCosts( int cost[][MAXVERTICES], int a[][MAXVERTICES], int nvert
) {
    /*
     * finds all pairs shortest paths and store in a[][].
     */
    int i, j, k;

    for( i=0; i<nvert; ++i )
        for( j=0; j<nvert; ++j )
            a[i][j] = cost[i][j];
    for( k=0; k<nvert; ++k )
        for( i=0; i<nvert; ++i )
            for( j=0; j<nvert; ++j )
                if( a[i][k]+a[k][j] < a[i][j] )
                    a[i][j] = a[i][k] + a[k][j];
}

int main() {
    int cost[][MAXVERTICES] =     { {0,50,10,MAXINT,45,MAXINT},
                                   {MAXINT,0,15,MAXINT,10,MAXINT},
                                   {20,MAXINT,0,15,MAXINT,MAXINT},
                                   {MAXINT,20,MAXINT,0,35,MAXINT},
                                   {MAXINT,MAXINT,MAXINT,30,0,MAXINT},
                                   {MAXINT,MAXINT,MAXINT,3,MAXINT,0}
                                 };
    int a[MAXVERTICES][MAXVERTICES];
    int nvert = 6;          // no of vertices.

    allCosts( cost, a, nvert );
    printCosts( a, nvert );

    return 0;
}
```

Explanation

1. One way to solve the all-costs shortest path problem is to apply the algorithm of shortest path sssp() n times for each vertex in the digraph G. The complexity would be O(n^3). For the all-costs problem, we can obtain a conceptually simpler algorithm that will work even if some edges in G have negative weights, as long as G has no cycles with negative lengths. The computing time will still be O(n^3), although the constant factor will be smaller.
2. The digraph G is represented as a cost adjacency matrix with cost[i][i] = 0 and cost[i][j] = MAXINT, in case edge <i,j>, i != j is not in G.
3. We define Ak[i][j] to be the shortest path from i to j going through no intermediate vertex of index greater than k. Then, An[i][j] will be the cost of the shortest i to j path in G, since G contains no vertex with an index greater than n. A0[i][j] is cost[i][j].
4. The basic idea in the algorithm is to successively generate matrices A0, A1, ..., An. If we have already generated A(k-1), then we may generate Ak by realizing that for any pair of vertices i, j either:

- the shortest path from i to j going through no vertex with index greater than k does not go through the vertex with index k, and so its cost is A(k-1)[i][j]; or
- the shortest such path does go through vertex k. Such a path consists of a path from i to k and another one from k to j. These paths must be the shortest paths from i to k and from k to j going through no vertex with index greater than k-1, and so their costs are A(k-1)[i][k] and A(k-1)[k][j].

Note that this is true only if G has no cycle with negative length containing vertex k. If this is not true, then the shortest i to j path going through no vertices of index greater than k may make several cycles from k to k, and thus have a length substantially less than A(k-1)[i][k]+A(k-1)[k][j]. So, we have the following formulas:

```
Ak[i][j] = min{ A(k-1)[i][j], A(k-1)[i][k]+A(k-1)[k][j] },
k > 0
```

and

```
A0[i][j] = cost[i][j]
```

5. **Example:** Let the graph be as shown here.

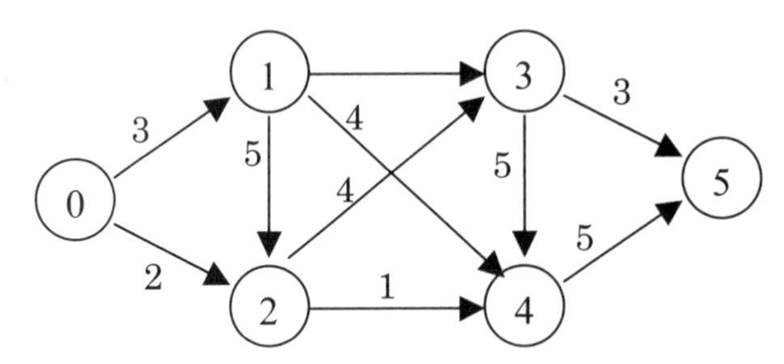

Here, n = 6. The values of `Ak`, `0<=k<6` are shown next. `M` signifies `MAXINT`.

$$A0 = \begin{matrix} 0 & 50 & 10 & M & 45 & M \\ M & 0 & 15 & M & 10 & M \\ 20 & M & 0 & 15 & M & M \\ M & 20 & M & 0 & 35 & M \\ M & M & M & 30 & 0 & M \\ M & M & M & 3 & M & 0 \end{matrix}$$

$$A1 = \begin{matrix} 0 & 50 & 10 & M & 45 & M \\ M & 0 & 15 & M & 10 & M \\ 20 & 70 & 0 & 15 & 65 & M \\ M & 20 & M & 0 & 35 & M \\ M & M & M & 30 & 0 & M \\ M & M & M & 3 & M & 0 \end{matrix}$$

$$A2 = \begin{matrix} 0 & 50 & 10 & M & 45 & M \\ M & 0 & 15 & M & 10 & M \\ 20 & 70 & 0 & 15 & 65 & M \\ M & 20 & 35 & 0 & 30 & M \\ M & M & M & 30 & 0 & M \\ M & M & M & 3 & M & 0 \end{matrix}$$

	0	50	10	25	45	M
	35	0	15	30	10	M
A3 =	20	70	0	15	65	M
	55	20	35	0	30	M
	M	M	M	30	0	M
	M	M	M	3	M	0

	0	45	10	25	45	M
	35	0	15	30	10	M
A4 =	20	35	0	15	45	M
	55	20	35	0	30	M
	85	50	65	30	0	M
	58	23	38	3	33	0

	0	45	10	25	45	M
	35	0	15	30	10	M
A5 =	20	35	0	15	45	M
	55	20	35	0	30	M
	85	50	65	30	0	M
	58	23	38	3	33	0

An entry in `A5[i][j]` gives the minimum path length from vertex `i` to `j`.

Points to Remember

1. The complexity of the algorithm `allCosts()` is O(n^3).
2. Different options should be considered before implementation, as we did for this problem. This can result not only in getting an efficient algorithm but also in effectively handling special cases.
3. See how the problem of size `k` is recursively defined in terms of a problem of size `k-1`.

29 Miscellaneous Problems

PROBLEM: THE TWO-CLASS CLASSIFICATION PROBLEM

Given a training data set having two classes and the functional form of the decision boundary between them, find the decision boundary of that form that best separates the two classes. Assume that a linear decision boundary will be used to classify samples into two classes, and that each sample has M features.

Program

```
#include <stdio.h>

typedef enum {FALSE, TRUE} bool;

int computeD(int *x, int n, int *w) {
    /*
     * compute the value of the discriminant function D using the xis
and wis.
     * D = w0 + w1x1 + w2x2 + ... + wnxn.
     */
    int i;
    int d = w[0];

    for(i=1; i<=n; ++i)
        d += w[i]*x[i];

    return d;
}

int signum(int d) {
    /*
     * signum(d) = 1 if d>=0
     *           = -1 otherwise.
```

```
        */
        if(d >= 0)
            return 1;
        return -1;
    }

    void updateW(int *x, int n, int *w, const int c, const int k, int d) {
        /*
         * update the ws as the discriminant function was different than
the class
         * value.
         * wi += cdxi for 1<=i<=n.
         * w0 += cdk.
         */
        int i;

        w[0] += c*d*k;

        for(i=1; i<=n; ++i)
            w[i] += c*d*x[i];
    }

    void printHeader(int n) {
        /*
         * print header of the output.
         */
        int i;

        for(i=1; i<=n; ++i)
            printf("x%d\t", i);
        printf("d\t");
        for(i=0; i<=n; ++i)
            printf("old w%d\t", i);
        printf("D\tError?\t");
        for(i=0; i<=n; ++i)
            printf("new w%d\t", i);
        printf("\n");
    }

    int findK(FILE *fp, int n) {
        int tempk, k=0, count=0;
```

```
    int i;

    while(fscanf(fp, "%d", &tempk) == 1) { // x1.
        k += tempk;
        for(i=2; i<=n; ++i) {// x2 to xn.
                fscanf(fp, "%d", &tempk);
                k += tempk;
        }
        fscanf(fp, "%d", &tempk);    // d.
        count++;
    }
    k /= count*n;
    k = abs(k);

    return k;
}

void printArray(int *xorw, int n) {
    /*
     * print xorw[0..n-1].
     */
    int i;

    for(i=0; i<n; ++i)
       printf("%d\t", xorw[i]);
}

int main() {
    int *x, *w;
    int n, i, D;
    const int c=1;
    int k = 0, tempk, count = 0;
    bool wChanged = TRUE;
    FILE *fp = fopen("adb.dat", "r");

    fscanf(fp, "%d", &n);
    x = (int *)malloc((n+1)*sizeof(int));  // 1 d + n xs.
    w = (int *)calloc(n+1, sizeof(int));   // n(w) = n(x) + 1.

    // make first pass in the file to calculate value of k.
    k = findK(fp, n);
    printf("n=%d, k=%d.\n", n, k);
```

```
    // now print the header.
    printHeader(n);
    // now the loop until there is no change in ws in any iteration.

    while(wChanged) {
        wChanged = FALSE;
        fseek(fp, 0L, SEEK_SET);
        fscanf(fp, "%d", &n);// again read it.

        // the xs are stored as d followed by n xs.
        // the ws are stored as w0, w1, ..., wn.
        // thus xi and wi have the same indices.

        // now read each row.
        while(fscanf(fp, "%d", x+1) == 1) { // x1.
                for(i=2; i<=n; ++i)  // x2 to xn.
                        fscanf(fp, "%d", x+i);
                printArray(x+1, n);  // print xis.
                fscanf(fp, "%d", x); // the d value.
                printf("%d\t", x[0]);
                printArray(w, n+1);  // print old wis.

                D = computeD(x, n, w);
                printf("%d\t", D);
                if(signum(D) != x[0]) {
                        wChanged = TRUE;
                        updateW(x, n, w, c, k, x[0]);
                        printf("yes\t");
                }
                else
                        printf("no\t");
                printArray(w, n+1);  // print new wis.
                printf("\n");
        }
    }
    fclose(fp);

    return 0;
}
```

Explanation

1. If the discriminant function is of the form D = w0 + w1x1 + … + wMxM, then D = 0 is the equation of the decision boundary between the two classes. The weights w0, w1,…, wM are to be chosen to provide good performance on the training set. A sample with feature vector x = (x1,…, x2) is classified into one class, say class 1 if D >= 0, and into the other class, say class –1 if D < 0. Geometrically, D = 0 is the equation of a hyperplane decision boundary that divides the M-dimensional feature space into two regions. Two classes are said to be linearly separable if there exists a hyperplane decision boundary such that D > 0 for all the samples in class 1, and D < 0 for all the samples in class –1.
2. The adaptive decision boundary algorithm consists of the following steps:
   (a) Initialize the weights w0,…, wM to zero or to small random values or to some initial guesses. Choosing good initial guesses for the weights will speed convergence to a perfect solution if one exists.
   (b) Choose the next sample x = (x1,…, xM) from the training set. Let the true class of desired value of D be d, so that d = 1 or -1 represents the true class of x.
   (c) Compute D = w0 + w1x1 + … + wMxM.
   (d) If signum(D) != d, replace wi by wi + cdxi, for i = 1,…, M, where c is a positive constant that controls the step size for weight adjustment. signum(D) == 1 if D >= 0 and signum(D) == -1 if D < 0. Also replace w0 by w0 + cdk where k is a positive constant. If signum(D) == d, then no change in the weights should be made. We take k to be the average absolute value of all the features for fast convergence.
   (e) Repeat steps b through d with each of the samples in the training set. When finished, run through the entire training data set again. Stop and report perfect classification when all the samples are correctly classified during one complete pass of the entire training set through the training procedure. An additional stopping rule is also needed since this process would never terminate if the two classes were not linearly separable. A fixed maximum number of iterations could be tried, or the algorithm could be terminated when the running average error rate ceases to decrease significantly. We assume that the classes are linearly separable.
3. The data is read from a file. The first line of the file should contain a single number n signifying the number of features. From line 2 onwards, each line contains n+1 numbers representing one sample. It contains n feature values

followed by the class value (1 or –1). One pass of the file is done in `findK()` to find the average absolute value of the features to be assigned to the constant `k` for fast convergence to the solution, if it exists. `main()` then contains a loop that runs as long as there is at least one misclassification in the whole data set. The inner `while` loop reads every sample and stores the vector `x` in an array `x[]` from indices 1 to n. `x[0]` stores the class value d. `w[0..n]` is another array that stores the values of weights `w0, …, wn`. Initially, all `ws` are set to zero. The function `computeD()` computes the value of the discriminant function `D` by using `x[]` and `w[]`. We then check whether `signum(D) == class value d`. If they are equal, nothing is done; otherwise, `w[]` vector is changed as given in the preceding algorithm. The function `printArray()` prints a vector.

4. Example: Let the training file be

   1
   –4 –1
   –1 1.

   This means there is only one feature and there are two samples. In the first, the value of `x1` is –4 and its class is –1. The second contains the value of `x1` as –1, and its class is 1.

   Let `c = 1.`

   `k` = average absolute value of features = `abs((-4-1)/2) = abs(-2) = 2.`

   `w0 = w1 = 0.`

   Using the first sample x1 = –4 results in D = w0 + w1x1 = 0 + 0(–4) = 0, where we have arbitrarily assigned the sign of 0 to be 1. However, d = –1, so the sample is misclassified and we adapt the weights as follows:
   w0 = w0 + cdk = 0 + 1(–1)2 = –2.
   w1 = w1 + cdx1 = 0 + 1(– 1)(–4) = 4.
   The algorithm is repeated until both samples are correctly classified.
   Different steps of the algorithm are as follows:

step	x1	d	old w0	old w1	D	Error?	new w0	new w1
1	–4	–1	0	0	0	yes	–2	4
2	–1	1	–2	4	–6	yes	0	3
3	–4	–1	0	3	–12	no	0	3
4	–1	1	0	3	–3	yes	2	2
5	–4	–1	2	2	–6	no	2	2
6	–1	1	2	2	0	no	2	2

Points to Remember

1. One way to assure a sort of convergence, even when the classes are not linearly separable, is to replace the constant c by a variable step size that decreases with the number of iterations. However, convergence to a decision boundary that best separates the classes is not guaranteed when the classes are not linearly separable.
2. Nonlinear decision boundaries that are more complex than hyperplanes can also be found by this adaptive technique.
3. If there are more than two classes, but each pair of classes is linearly separable, then the same algorithm can be used to find the boundary between each two pairs of classes.

PROBLEM: THE *N*-COINS PROBLEM

Given a string of 0s and 1s as tails and heads obtained from tossing n coins of arbitrary biases, find the (approximate) points where the change of coin could have taken place.

Program

```
#include <stdio.h>
#include <math.h>

#define MINGROUPSIZE 10
#define EPSILON 0.08

int adjustPartition(int *a, int n, double *p, int m, int i) {
    /*
     * slide the partition boundary from i forward or backward to get
     * better partition and return the index of the last element in the
     * partition.
     * i >= 1.
     */
    int aindex = i*MINGROUPSIZE;
    int j, startj = aindex-MINGROUPSIZE/4, finalj =
aindex+MINGROUPSIZE/4;
```

```
    int k;
    int ones, nelem;

    for(k=(i-1)*MINGROUPSIZE, nelem=0; k<startj; ++k, ++nelem)
        ones += a[k];
    // the bias of the coin uptil now here is ones/nelem.

    for(j=startj; j<finalj; ++j) {
        ones += a[j]; ++nelem;
        if(fabs(ones/(double)nelem-p[i]) >= EPSILON)     // found the
partition.
            break;
    }
    return j;
}

int findChange(int *a, int n, double *p, int m) {
    /*
     * the running probabilities are given in p[m].
     * find the points of coin change from them.
     */
    int i;

    for(i=1; i<m; ++i) {
        if(fabs(p[i]-p[i-1]) >= EPSILON)     // found the change.
            return adjustPartition(a, n, p, m, i);
    }
    return -1;
}

int findProb(int *a, int n, double *p) {
    /*
     * find the average probabilities of groups in a[n] and store in p.
     */
    int i, j=-1, k, ones, nelem;

    for(i=0; i<n; ++i, ++nelem) {
        if(i%MINGROUPSIZE == 0)
            ones=0, nelem=0, ++j;
        p[j] = (a[i]==1 ? ++ones, ones/(double)(nelem+1) : ones/
(double)(nelem+1
));
```

```
        }
        return j+1;
    }

    void printAverage(int *a, int partindex) {
        /*
         * print average of this new partition a[0..partindex-1];
         */
        int ones = 0, i;

        for(i=0; i<partindex; ++i)
            ones += a[i];
        printf("average = %.2lf.\n", ones/(double)partindex);
    }

    void findPartition(int *a, int n, double *p, int startoff) {

        int m = findProb(a, n, p);
        int partindex = findChange(a, n, p, m);

        if(partindex != -1) {
            printf("partition at %d.\n", startoff+partindex);
            printAverage(a, partindex);
            findPartition(a+partindex, n-partindex, p, startoff+partindex);
        }
        else
            printAverage(a, n);
    }

    int main() {
        int a[] = {0,0,0,1,0,0,0,1,1,1,1,1,0,1,1,1,1,0,0,
0,0,1,0,1,0,0,0,1, 0,0, 1,0,0,1,0,0,0,1,0};
        double *p;
        int n = sizeof(a)/sizeof(int);
        int m;

        p = (double *)malloc((n+MINGROUPSIZE-1)/
MINGROUPSIZE*sizeof(double));
        findPartition(a, n, p, 0);

        return 0;
    }
```

Explanation

1. Consider the following input string:

   0,0,0,1,0,0,0,1,1,1,1,1,0,1,1,1,1,0,0,0,0,1,0,1,0,0,0,1,0,0,1,0,0,1,0,0,0,1,0.

   Visually, we may partition it as follows:

   0,0,0,1,0,0,0 1,1,1,1,1,0,1,1,1,1 0,0,0,0,1,0,1,0,0,0,1,0,0,1,0,0,1,0,0,0,1,0.

   We define bias of a coin as the probability of tossing a head (1).

   Thus, bias of coin 1 = 1/7 = 0.14.

   bias of coin 2 = 9/10 = 0.9.

   bias of coin 3 = 6/22 = 0.27.

   Thus, we have to find such partitions in the input string.

   Note two extreme cases. First, there was a single coin whose bias is 16/39 = 0.41. Second, there were 39 coins with biases equal to their outcomes (0 or 1). These conclusions are valid but of no practical importance. What we are interested in are patterns of 0s and 1s that appear uniformly over a length of the string.

2. Since this is a problem with an approximate solution, there can be various ways to get an answer. We have chosen the following strategy. We partition the elements (0s and 1s) into various groups of a fixed size `MINGROUPSIZE` (perhaps except the last group) and find the average biases of each group. We then compare adjacent group biases, and if the difference in the biases is >= an error term `EPSILON`, we know that these two partitions have different patterns, and so a possible change of coin took place somewhere near the boundary between the two partitions. We then try to shift the boundary by some fixed amount and find a suitable partition point. Thus, we found out one possible change of coin. We run this algorithm recursively to find the next partitions in the input.

3. The function `findPartition()` is the driver function. It calls `findProb()` to group the elements of the input string into various partitions and find their average biases. It then calls `findChange()` to check whether there are consecutive biases having a difference of `>= EPSILON`. The function `findChange()` returns –1 if there is no such change. Otherwise, it calls `adjustPartition()` to make a smooth change to the partition index for better accuracy. The function `adjustPartition()` slides the end of the partition from `-MINGROUPSIZE/3` to `0` to `+MINGROUPSIZE/3` across the original boundary and stops as soon as the difference in the biases of the original partition and the sliding partition is `>= EPSILON`. The function

printAverage() is a dummy function used to print values required for understanding and debugging.

4. Example: Let MINGROUPSIZE=10 and EPSILON=0.08.

   For this input string, the function findProb() groups the elements into the following partitions having the given biases:

   group 0 = 0,0,0,1,0,0,0,1,1,1 bias = 4/10 = 0.4.

   group 1 = 1,1,0,1,1,1,1,0,0,0 bias = 6/10 = 0.6.

   group 2 = 0,1,0,1,0,0,0,1,0,0 bias = 3/10 = 0.3.

   group 3 = 1,0,0,1,0,0,0,1,0 bias = 3/9 = 0.33.

   The first call to findChange() tells that there is a change of >= EPSILON between group 0 and group 1. Thus, the partition wall is assumed between the two groups. This wall is shifted from its position from negative to positive direction in the function adjustPartition(). The offset within which this is done is fixed at MINGROUPSIZE/4. If the difference between the average bias of the partition before this wall and the average bias of the partition before the original partition wall becomes >= EPSILON, then the partition wall is fixed at that point.

   Thus the first partition is fixed at index 8, and is 0,0,0,1,0,0,0,1 with average bias of 2/8 = 0.25.

   The remaining array is then partitioned recursively in a similar manner. Thus we get the remaining partitions as 1,1,1,1,0,1,1,1 with average bias of 7/8 = 0.88 and 1,0,0,0,0,1,0,1,0,0,0,1,0,0,1,0,0,1,0,0,0,1,0 with average bias of 7/23 = 0.3.

Points to Remember

1. One can find an input string that can fool this algorithm for a fixed group size MINGROUPSIZE and error term EPSILON.
2. There will be greater accuracy if some adaptive algorithm is used that will refine the partitions.
3. This algorithm can be used in noise detection and correction of the input signal.

PROBLEM: ALL COMBINATIONS OF STRINGS

Write a program to print all the combinations of a string.

Program

```
#include <stdio.h>

#define MAXLEN 80

void init(char *answer, int slen) {
    /*
     * initialize first slen entries in answer[] to 0.
     */
    int i;

    for(i=0; i<slen; ++i)
        *answer++ = 0;
    *answer = 0;    // eos.
}

void printComb(char *s, int slen, char *answer) {
    /*
     * fixes a character of s and then calls printComb() recursively
     * to get all combinations of the remaining chars.
     */
    int i;
    static int count = 0;

    if(*s == 0) {
        count++;
        printf("%2d: %s.\n", count, answer);
        return;
    }
    for(i=0; i<slen; ++i)
        if(answer[i] == 0) {
            answer[i] = *s;
            printComb(s+1, slen, answer);
            answer[i] = 0;
        }
}
```

```
int main() {
    char s[MAXLEN];
    char answer[MAXLEN];

    printf("Enter characters for combination: ");
    gets(s);

    while(*s) {
        init(answer, strlen(s));
        printComb(s, strlen(s), answer);
        printf("Enter characters for combination(press enter to end):
");

       gets(s);
    }
    return 0;
}
```

Explanation

1. Consider an input string abc. Various combinations of this string are abc, acb, bac, cab, bca, and cba. To generate these combinations, we once again make use of recursion.
2. Note that combinations of the string "abc" are obtained by fixing character 'a' at different positions and finding combinations of string 'bc'. Fixing 'a' at a position means, in the answer string, the character 'a' will occur at a fixed position. Similarly, combinations of string 'bc' can be found by fixing character 'b' at different positions and finding combinations of string 'c'. If string 'c' has only one combination, we get the following:

   ```
   comb('c') = c.
   comb('bc') = bc   /* character 'b' fixed at position 1. */
      and cb /* character 'b' fixed at position 2. */
   comb('abc')       = a fixed at different places in comb('bc').
      = abc and acb /* character 'a' fixed at position 1. */
   and bac and cab   /* character 'a' fixed at position 2. */
   and bca and cba   /* character 'a' fixed at position 3. */
   ```

3. The program repeatedly asks for a string until the entered string is empty and calls `printComb()`. `printComb()` fixes the position of a character and calls `printComb()` recursively for the remaining string.

4. The complexity of printComb() can be easily derived from the following recurrence relation:

   T(n) = n*T(n-1).

   This is because, for finding combinations of a string of length n, we fix the position of a character at a position out of n possible positions and then find combinations of the remaining (n-1) characters. After solving the equation, we get the complexity of printComb() as O(n!).

Points to Remember

1. The complexity of printComb() is O(n!) where n is the length of the input string.
2. A simple recursive procedure, such as printComb(), can solve a larger problem in an elegant manner.
3. Care should be taken to end a string with '\0', as done in function init().

PROBLEM: THE 8-KNIGHTS PROBLEM

Solve the problem of a knight's tour without backtracking. Assume that it starts from a corner of the chessboard.

Program

```
#include <stdio.h>

#define M 8
#define NDIR 8// no of directions.

typedef enum {FALSE, TRUE} bool;

int rowchange[] = {-2, -1,  1,  2, 2,  1, -1, -2};
int colchange[] = {-1, -2, -2, -1, 1,  2,  2,  1};

void printMatrix(int a[][M], int n) {
    /*
     * print the final solution.
     */
```

```
        int i, j;

        for(i=0; i<n; ++i) {
           for(j=0; j<n; ++j)
                 printf("%2d ", a[i][j]);
           printf("\n");
        }
    }

    int getCost(int a[][M], int row, int col, int n) {
        /*
         * find the number of positions which can be visited from
a[row][col].
         */
        int i;
        int count = 0;

        if(!(row>=0 && row<n && col>=0 && col<n && a[row][col]==FALSE))
           return NDIR+1;

        for(i=0; i<NDIR; ++i) {
           int newrow = row+rowchange[i];
           int newcol = col+colchange[i];
           if(newrow>=0 && newrow<n && newcol>=0 && newcol<n &&
a[newrow][newcol]==FALSE)
                 count++;
        }
        return (count ? count : NDIR+1);
    }

    void knight(int a[][M], int n, int row, int col, int num) {
        /*
         * find the next position in a of size n*n.
         * next position is the position from where min number of positions
         * can be reached.
         * current position is (row, col): unmarked.
         */
        int mincost = NDIR+2;    // infinity.
        int mindir = -1;
        int i;
```

```
    a[row][col] = num;
    if(num == n*n)
        return;
    for(i=0; i<NDIR; ++i) {
        int newrow = row+rowchange[i];
        int newcol = col+colchange[i];
        int cost   = getCost(a, newrow, newcol, n);

        if(cost < mincost)
                mincost=cost, mindir=i;
    }
    knight(a, n, row+rowchange[mindir], col+colchange[mindir], num+1);
}

int main() {
    int a[M][M] ; //here you have to initialize matrix
    int i=0,j=0;
    for(i=0;i<M;i++)
     for(j=0;j<M;j++)
     a[i][j]=0;

    knight(a, M, 0, 0, 1);
    printMatrix(a, M);

    return 0;
}
```

Explanation

1. On an 8 × 8 chessboard, a knight can travel two-and-a-half positions at a time. The problem is to start from a corner of the chessboard and visit all 64 positions without visiting any position more than once. This travel is called a knight's tour.
2. There is more than one way to solve this problem. We implement here the strategy suggested by J.C. Warnsdorff in 1823. His rule is that the knight must always be moved to one of the positions from which the minimum number of not-yet-visited positions can be traversed.
3. The knight can visit one of the eight (at most) positions from a fixed position. Those are shown in Figure 29.1. It is possible that the number of positions that can be traversed is less than eight, depending on the chessboard boundaries and visit of the knight to other squares.

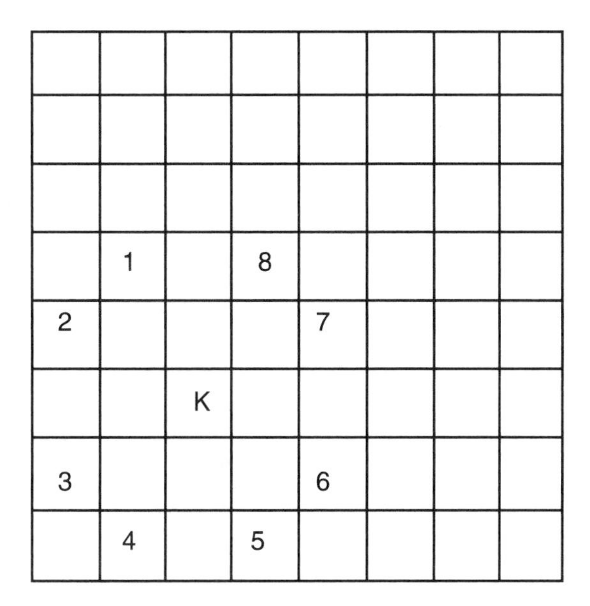

FIGURE 29.1 A knight can move to a maximum of eight positions.

If the board is represented (naturally) by 8×8 matrix of integers `a[8][8]`, and if the current position of the knight is `a[i][j]`, then the next position could be one the eight: `a[i-2][j-1]`, `a[i-1][j-2]`, `a[i+1][j-2]`, `a[i+2][j-1]`, `a[i+2][j+1]`, `a[i+1][j+2]`, `a[i-1][j+2]`, `a[i-2][j+1]`. We maintain two arrays of these changes in row and column in `rowchange[8]` and `colchange[8]`. Thus, if the current position is `a[i][j]`, then the next position in direction `d`, `1<=d<=8`, as shown in the figure, is given by `a[i+rowchange[d-1]][j+colchange[d-1]]`.

4. The program initializes the array by zeros and calls function `knight()` with the values of (row. column) as (0, 0), which forms the first position of the knight. Then, for each of the eight directions we find the new position `a[newrow][newcol]` using the rule in Number 3. From each of these entries, we find the number of positions to which the knight can travel (`getCost()`). We select the minimum of such values and call `knight()` recursively for that position and so on until we finish all the squares. As we visit a position, we mark it to print the path at the end of the traversal.

Points to Remember

1. The knight's tour can be solved with backtracking by the trial-and-error method in which we select the next move randomly, and, if we find it to be

non-feasible, we backtrack and try another direction. Its complexity is exponential.

2. The problem of the knight's tour can also be solved without backtracking in O($N \times N$) time where N is the size of the chessboard.
3. There are other heuristics available that can solve the problem of the knight's tour with only a little backtracking.
4. Note how maintaining two vectors `rowchange[]` and `colchange[]` avoids the use of a long `switch` statement containing 8 cases, and keeps the code concise.

PROBLEM: *N*-QUEENS PROBLEM

Solve the N-queens problem in which in an array of $N \times N$ positions, N queens are placed such that no two queens are in attacking positions.

Program

```
#include <stdio.h>
#define N 10

typedef enum {FALSE, TRUE} bool;

void printMatrix(int a[][N], int n) {
    /*
     * print one solution.
     */
    int i, j;

    for(i=0; i<n; ++i) {
        for(j=0; j<n; ++j)
                printf("%d ", a[i][j]);
        printf("\n");
    }
    //getchar();
    printf("\n");
}

int getMarkedCol(int a[][N], int n, int row) {
```

```
    /*
     * returns the column marked in row.
     */
    int j;

    for(j=0; j<n; ++j)
        if(a[row][j] == TRUE)
            return j;
    printf("ERROR: No col marked in the row %d.\n", row);
    return -1;
}
bool feasible(int a[][N], int n, int row, int col) {
    /*
     * checks whether next queen can be kept at a[row][col].
     */
    int i;
    int markedCol;

    for(i=0; i<row; ++i) {   // for all rows before this row.
        markedCol = getMarkedCol(a, n, i);
        if(markedCol == col || abs(row-i) == abs(col-markedCol))
            return FALSE;
    }
    return TRUE;
}

void NQueens(int a[][N], int n, int row) {
    /*
     * solve n-queens problem. the solution is obtained using matrix a.
     * the procedure is recursive. current row being considered is row.
     * that means all the rows from 0 to row-1 are considered.
     */
    int j;

    if(row < n) {
        for(j=0; j<n; ++j)   // for each col.
            if(feasible(a, n, row, j)) {
                a[row][j] = TRUE;
                NQueens(a, n, row+1);
                a[row][j] = FALSE;
            }
    }
    else
```

```
        printMatrix(a, n);
}

int main() {
    int a[N][N] ;
    int i=0,j=0;
    for(i=0;i<8;i++)
     for(j=0;j<8;j++)
     a[i][j]=0;
    NQueens(a, 8, 0);

    return 0;
}
```

Program Description

1. The $N \times N$ positions are naturally represented by using an $N \times N$ array `a[N][N]` of integers (even an array of Boolean is sufficient). A value of `0` (`FALSE`) of `a[i][j]` indicates absence of a queen in row `i` and column `j`, while a value of `1` (`TRUE`) indicates her presence.
2. Two queens are in attacking position if they are in the same row, column, or diagonal. Thus, queens placed at `a[1][2]` and `a[7][2]` are attacking as they are in the same column (2). Similarly, `a[1][2]` and `a[5][6]` are attacking positions as they are present on the same diagonal. Thus, the positions `a[i][j]` and `a[k][l]` are attacking if:
   - `i == k`, or
   - `j == l`, or
   - `abs(i-k) == abs(j-l)` where `abs()` returns the absolute value of a number.
3. The program starts by initializing the array `a[][]` by 0. It then recursively calls function `NQueens()` to place each queen in the next rows. Thus in each recursive call to `NQueens()`, one queen is placed at the non-attacking position. Our aim is not to find a single solution but to find all the combinations of positions in which the N queens can be placed. So we repeat it for each column of the array. Thus, the algorithm is as follows:

```
NQueens(row) {
    if(row < N) {
        for j=0 to N-1
```

```
            if(feasiblePosition(row, j)) {
                a[row][j] = TRUE; // queen placed.
                NQueens(row+1);
    a[row][j] = FALSE;    // queen removed from this row.
            }
    }
    else {
                    printMatrix();
        }
    }
```

The function feasibleSolution(row, col) simply checks whether the next queen can be put at a[row][col]. This is done by checking whether none of the previously placed queens are attacking this queen.

4. The complexity of NQueens() is $O(N!)$ if it finds all solutions. This can be derived easily from the following recurrence relations:

   $T(N) = N*(N\wedge 2+T(N-1))$

   and

   $T(1) = 1.$

   The complexity of feasibleSolution() is $O(N\wedge 2)$. It can be made $O(N)$ by maintaining a vector of columns in which the queens are placed.

Points to Remember

1. Once again a recursive procedure has played the trick of solving an apparently larger problem easily, by dividing a problem of size N into a problem of size N–1.
2. By separating the procedure feasibleSolution(), the function NQueens() itself looks like the algorithm. This suggests the usefulness of writing a pseudo-code before the implementation details.
3. The complexity of NQueens() is O(N!).

PROBLEM: MAPPING OF *N*-QUEUES IN AN ARRAY

Design a data representation sequentially mapping N-queues into an array nqueue[0...N-1]. Represent each queue as a circular queue within the array. Write functions qAdd(), qDelete(), qIsFull() for this representation.

Program

```
#include <stdio.h>

#define N 50                  // combined size of all queues.
#define NQ 5                  // number of queues.
/* ASSUMPTION : NQ is a divisor of N. */

#define ILLEGALINDEX -1            // illegal index -- for special
cases.
#define EINDEXOUTOFBOUND -1 // error code on overflow in the queue.
#define SUCCESS 0                  // success code.

typedef int type;                  // type of each data item.

type nqueue[N];               // queue implemented using array.
int front[NQ]; // points to first element in the queue.
int rear[NQ]; // points to last element in the queue.

void qInit() {
    /*
     * initialize front[] to contain ILLEGALINDEX.
     * ILLEGALINDEX specifies empty queue.
     */
    int i;
    for( i=0; i<NQ; ++i ) {
        front[i] = ILLEGALINDEX;
    }
}

int qAdd( int queue, type data ) {
    /*
     * adds 'data' at the end of queue.
     */
    int maxelem, nelem;

    if( queue < 0 || queue >= NQ )  // invalid queue number.
        return EINDEXOUTOFBOUND;
    maxelem = N/NQ;
    nelem = qGetNElements( queue );
    if( nelem == 0 ) {        // empty queue.
        front[queue] = rear[queue] = maxelem*queue;
    }
```

```
    else if( nelem == maxelem )      // queue full.
        return EINDEXOUTOFBOUND;
    else
        rear[queue] = (rear[queue]+1)%maxelem + maxelem*queue;

    printf( "inserting at %d\n", rear[queue] );
    nqueue[rear[queue]] = data;
    return SUCCESS;
}

int qGetNElements( int queue ) {
    /*
     * returns no of elements in queue.
     */
    int start, end;

    if( front[queue] == ILLEGALINDEX )      // queue empty.
        return 0;
    if( front[queue] <= rear[queue] )
        return rear[queue]-front[queue]+1;

    start = N/NQ*queue;
    end   = N/NQ*(queue+1);

    return (end-front[queue]) + (rear[queue]-start+1);
}

int qDelete( int queue ) {
    /*
     * removes front element of queue.
     */
    int nelem = qGetNElements(queue);
    printf( "deleting from queue %d...\n", queue );
    if( nelem == 0 )  // empty queue.
        return EINDEXOUTOFBOUND;
    else if( nelem == 1 )    // last element getting deleted.
        front[queue] = ILLEGALINDEX;
    else
        front[queue] = (front[queue]+1)%(N/NQ) + N/NQ*queue;
    return SUCCESS;
}

int qIsFull( int queue ) {
```

```
    /*
     * returns 1 if queue is full, otherwise 0.
     */
    return ( qGetNElements(queue) == N/NQ );
}

void qPrint( int queue ) {
    /*
     * prints the queue.
     */
    int nelem = qGetNElements(queue);
    int maxelem = N/NQ;
    int start = maxelem*queue;
    int i;

    for( i=0; i<nelem; ++i )
       printf( "%d ", nqueue[(front[queue]+i)%maxelem+start] );
    printf( "\n" );
}
int main() {
    qInit();
    printf( "nelem of 3=%d\n", qGetNElements(3) );
    qAdd(3,0);
    qPrint(3);
    qDelete(3);
    qPrint(3);
    printf( "nelem of 3=%d\n", qGetNElements(3) );
    qAdd(3,1);
    qAdd(3,2);
    qAdd(3,3);
    qPrint(3);
    qDelete(3);
    qPrint(3);
    qAdd(3,4);
    qAdd(3,5);
    qAdd(3,6);
    qAdd(3,7);
    qAdd(3,8);
    qPrint(3);
    qDelete(3);
    qPrint(3);
    qAdd(3,9);
    qPrint(3);
}
```

Explanation

1. Along with the array nqueue[], front and rear indices for each queue are maintained in arrays front[0...NQ-1] and rear[0...NQ-1]. front[i] signifies the first element in queue i. The maximum number of elements in each queue is *N*/*N*Q. It is assumed that *N*Q is a divisor of *N*. The index of the first element in queue i is given by the formula *N*/*NQ**i.
2. The program starts by initializing the front and rear of all the queues to a sentinel value (–1). The number of elements in queue i is calculated as follows:

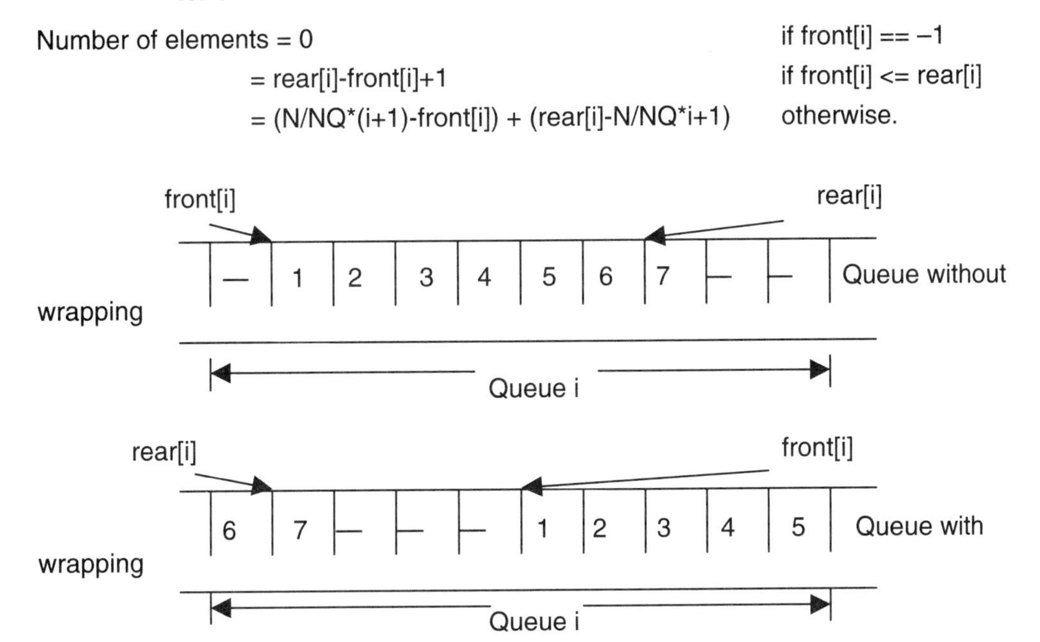

3. The function qAdd(queue, data) adds data to the end of the queue. It checks for a valid queue number (0...*N*/*NQ*–1) and queue-overflow condition. If space is available for the new element in the queue, then rear[queue] is incremented and the new element is inserted at the position. The end condition that the queue was initially empty is also checked as it needs updation of front[queue]. It is possible for rear[queue] to get wrapped to *N*/*NQ**queue.
4. The function qDelete(queue) removes the front element of the queue. It checks for underflow by using the number of elements in the queue. If the element being deleted is the last element of the queue, rear[queue] is also

updated. Otherwise, only front[queue] points to the next element in the queue. It is possible for front[queue] to get wrapped to *N/NQ**queue.

5. The function qIsFull(queue) checks whether there is any space for a new element. It is implemented by calculating the number of elements in the queue.

Points to Remember

1. The index of the first element in queue i is calculated as (i*size of each queue).
2. The complexity of insertion and deletion as described is O(1). Instead, if we do not allow wrapping of elements, then every deletion will cost shifting of elements. That will keep insertion O(1) but deletion would become O(n) where n is the number of elements in the queue.
3. Since all the queues operate on distinct memory areas, the logic of functions remains the same as in a single queue implementation, except that the calculations are w.r.t. the index of the first element in the queue, compared to an assumed zero for the single queue case.
4. Note how a single procedure qGetNElements(i), which returns the number of elements in queue i, simplifies corner cases of insertion and deletion, and also simplifies the function qIsFull().

PROBLEM: IMPLEMENTATION OF A* ALGORITHM

Given n integers and a sum m, write a program to find the set of integers summing to m by using the A* algorithm.

Program

```
#include <stdio.h>

typedef enum {FALSE, TRUE} bool;

int subsetsum(int *a, int n, int sum, bool *selected, int startoff) {
```

```
    /*
     * for those elements a[i] which have selected[i] == FALSE,
     * solve subsetsum problem for sum.
     * the elements before startoff are of no use.
     */
    int i;

    if(sum == 0)
        return TRUE;
    for(i=startoff; i<n; ++i)
        if(selected[i] == FALSE && a[i] <= sum) {
            selected[i] = TRUE;
            if(subsetsum(a, n, sum-a[i], selected, i+1))
                return TRUE;
            selected[i] = FALSE;
        }
    return FALSE;
}

int subsetsumori(int *a, int n, int sum) {
    /*
     * check whether there is any subset in a[n] having sum sum.
     */
    int i;

    for(i=0; i<n; ++i)
        if(a[i] == sum || (a[i] < sum && !subsetsumori(a+i+1, n-i-1,
sum-a[i])))
  {
            printf("%d ", a[i]);
            return 0;
        }
    return 1;
}
int compare(void *e1, void *e2) {
    /*
     * function used in qsort().
     * returns values <0, ==0, >0 if e2 is <e1, ==e1, >e1.
     * thus we need elements in non-ascending order.
     */
    return *(int *)e2-*(int *)e1;
}
```

```
void printAnswer(int *a, int n, bool *selected) {
    /*
     * print those a[i] which have selected[i] == TRUE.
     */
    int i;

    for(i=0; i<n; ++i)
        if(selected[i])
            printf("%d ", a[i]);
    printf("\n");
}

int main() {
    int a[] = {5, 3, 4, 8, 9, 6};
    int sum = 15;
    int size = sizeof(a)/sizeof(int);
    bool *selected = (bool *)calloc(size, sizeof(bool));    // init
with FALSE.

    qsort(a, size, sizeof(int), compare);
    if(subsetsum(a, size, sum, selected, 0))
        printAnswer(a, size, selected);

    return 0;
}
```

Explanation

1. Inputs are an array `a[]` of n integers and the sum as another integer. We need to find out a subset of `a[]` that has the sum of its elements equal to sum. This problem is called a 'subset sum problem.'
2. The A* algorithm is an AI technique which chooses the best of the available options to proceed. It uses heuristics that apply to most real-world situations but do not guarantee the best solution. We use the heuristic that courses the element not nearest to the required sum to be added to the current sum.
3. The list of elements in array `a[]` is sorted in non-ascending order using the `qsort()` library function. `main()` then calls `subsetsum()`. The function `subsetsum()` uses the Boolean array `selected[n]`, where `selected[i]` indicates whether `a[i]` was included in the sum. The recursive algorithm of `subsetsum()` follows.

```
boolean subsetsum(a[], n, sum, selected[], startoffset) {
    if(sum == 0)
   return TRUE;
    for i=startoffset to n-1
        if(selected[i] == FALSE && a[i] <= sum) {
   selected[i] = TRUE;
   if(subsetsum(a, n, sum-a[i], selected, i+1) == TRUE)
       return TRUE;
   selected[i] = FALSE;
        }
    return FALSE;
}
```

Thus, if an element has not yet been selected and it is less than the sum, then we choose it and call subsetsum() recursively for the remaining sum by using the elements after this element. The variable startoffset is used to indicate the start of elements which may be of interest to add to the current sum. Since a[i] is selected, no element before a[i+1] will be of interest because all of them are more than a[i], and so are either already selected or greater than the sum. Remember that the elements are in non-increasing order and the sum in different invocations of subsetsum() is also non-increasing. When the sum reaches 0, it indicates the termination condition.

4. But where in the algorithm have we used the A* algorithm? It is used by first sorting the list in non-increasing order and then traversing it from the highest to the lowest elements. This way, at every step, we choose the option which is nearest to the required sum. If it fails, we deselect the option and choose another option.
5. Example:

   Let a[] = {5, 3, 4, 8, 9, 6} and sum = 10.

   selected[] = {FALSE, FALSE, FALSE, FALSE, FALSE, FALSE}.

   a[] is sorted to {9, 8, 6, 5, 4, 3} and subsetsum(a, 6, 10, selected, 0) is called.

   subsetsum() loops over elements 0 to 5 of a.

   for i=0, selected[0] = TRUE i.e. a[0] = 9 is selected and subsetsum(a, 6, 10-9, selected, 0+1) is called.

   subsetsum() loops over elements 1 to 5 of a.

   for i=1, a[1] = 8 > sum 1.

   for i=2, a[2] = 6 > sum 1.

   ...

for i=5, a[5] = 3 > sum 1.

So subsetsum() returns FALSE.

Since subsetsum() returned FALSE, selected[0] = FALSE.

for i=1, selected[1] = TRUE that is, a[1] = 8 is selected and subsetsum(a, 6, 10-8, selected, 1+1) is called.

subsetsum() returns FALSE in the similar manner as above.

for i=2, selected[2] = TRUE, i.e. a[2] = 6 is selected and subsetsum(a, 6, 10-6, selected, 2+1) is called.

subsetsum() loops over elements 3 to 5 of a.

for i=3, a[3] = 5 > sum 4.

for i=4, a[4] = 4 <= sum 4.

So selected[4] = TRUE that is a[4] = 4 is selected and subsetsum(a, 6, 4-4, selected, 4+1) is called.

Since sum is 0, subsetsum() returns TRUE.

Since subsetsum() returned TRUE, this subsetsum() returns TRUE.

Since subsetsum() returned TRUE, this subsetsum() returns TRUE to main().

The function printAnswer() then prints the elements a[i] for which selected[i] == TRUE.

Points to Remember

1. The A* algorithm chooses the best possible of the currently available paths for the next exploration of options.
2. A* algorithm uses a heuristic. So it does not guarantee the best outcome in all cases. However, in most of the instances of the problem, it is able to reach the solution faster than the algorithm generating all the combinations of the elements.
3. The complexity of subsetsum() is exponential.
4. By using the variable startoff, the search space is reduced.

Index

Symbols

A

B

C

D

M

N

O

P

Q

R

S

T

U

V

W